The Sources of Invention

Revised and Enlarged, Second Edition

John Jewkes, David Sawers, and Richard Stillerman

The first edition of *The Sources of Invention,* published in 1958, has been described as "a classic in science policy which has had a very considerable influence on both economists and scientists in Europe and in the United States." The authors set out to study the causes and consequences of industrial innovation – one, if not the main, spring of economic progress. They examined the important inventions of the nineteenth and twentieth centuries in order to discover just how far recent inventions have emerged from conditions different from those of the past. The evidence collected threw light on many questions, such as the influence of large research institutions and the concept of teamwork, the arguments for monopoly in industry, and the possibility of predicting inventions.

The second edition is a considerable enlargement of the first. To the original group of fifty-one case histories – which included Automatic Transmissions, Fluorescent Lighting, the Helicopter, Kodachrome, Polyethylene, Synthetic Detergents, the Transistor, and Xerography – have now been added ten other recent important cases, each of which has its own fascinating peculiarity: Air Cushion Vehicles; Chlordane, Aldrin, and Dieldrin; Electronic Digital Computers; Float Glass; the Moulton Bicycle; Oxygen Steelmaking; Photo-Typesetting; the Cure for Rhesus Haemolytic Disease; Semi-Synthetic Penicillins; and the Wankel Engine. A new chapter evaluates the relevant literature of the last ten years.

About the Authors

John Jewkes is Fellow of Merton College and Professor of Economic Organization, University of Oxford. He was graduated from Manchester University, where he subsequently held the post of Professor of Social Economics from 1936 to 1946. Professor Jewkes was a Rockefeller Foundation Fellow and a Visiting Professor at the University of Chicago and Princeton University. His other publications include *The New Ordeal by Planning, Juvenile Unemployment* (with W. A. Winterbottom), and *Value for Money in Medicine* (with Sylvia Jewkes).

David Sawers joined Professor Jewkes in the preparation of *The Sources of Invention* in 1954, after reading Philosophy, Politics, and Economics at Oxford. He spent the next four years working with Professor Jewkes, until he joined the editorial staff of *The Economist* in 1959 to write about industry. He joined the Ministry of Aviation as an economic adviser in 1966, and is now a senior adviser to the Ministry of Technology.

Richard Stillerman was born in Chicago. He is a graduate of the University of Chicago and the University of Chicago Law School. Following graduation, he held the position of Research Associate at the University of Chicago Law School. Since 1956 he has been engaged in the private practice of law in Chicago.

THE SOURCES OF INVENTION

THE SOURCES OF INVENTION

BY

JOHN JEWKES

DAVID SAWERS

RICHARD STILLERMAN

SECOND EDITION

W · W · NORTON & COMPANY · INC ·

New York

SBN 393 05408 x CLOTHBOUND EDITION
SBN 393 00502 x PAPERBOUND EDITION

Library of Congress Catalog Card No. 79-90986. Printed in the United States of America.

1 2 3 4 5 6 7 8 9 0

To S. J.

Whose advice and assistance was no less with the second edition than with the first

'The human mind is often so awkward and ill-regulated in the career of invention that it is at first diffident, and then despises itself. For it appears at first incredible that any such discovery should be made, and when it has been made, it appears incredible that it should so long have escaped men's research. All which affords good reason for the hope that a vast mass of inventions yet remains, which may be deduced not only from the investigation of new modes of operation, but also from transferring, comparing and applying those already known, by the methods of what we have termed literate experience.'

FRANCIS BACON, *Novum Organum*, Book I.

CONTENTS

Preface to the Second Edition 11
Preface to the First Edition 13
Acknowledgements 15

PART I

I Introduction 19
II Modern Views on Invention 34
III Inventors and Invention in the Nineteenth Century 40
IV Some Recent Important Inventions 65
V The Individual Inventor 79
VI Research in the Industrial Corporation: I 104
VII Research in the Industrial Corporation: II 117
VIII The Development of Inventions 152
IX Conclusions and Speculations 169
X The Past Ten Years: A Retrospect 194

PART II

SUMMARIES OF ORIGINAL CASE HISTORIES

Automatic Transmissions, p. 231; Bakelite, p. 233; Ball-point Pen, p. 234; Catalytic Cracking of Petroleum, p. 235; 'Cellophane', p. 237; Continuous Casting of Steel, p. 239; Continuous Hot Strip Rolling, p. 241; Cotton Picker, p. 243; Crease-resisting Fabrics, p. 245; Cyclotron, p. 248; DDT, p. 249; Diesel-Electric Railway Traction, p. 250; Fluorescent Lighting, p. 252; Gyro-Compass, p. 254; Hardening of Liquid Fats, p. 256; Helicopter, p. 257; Insulin, p. 260; Jet Engine, p. 262; Kodachrome, p. 266; Long-playing Record, p. 268; Magnetic Recording, p. 269; Methyl Methacrylate Polymers: Perspex, etc., p. 272; Neoprene, p. 274; Nylon and Perlon, p. 275; Penicillin, p. 278; Polyethylene, p. 279; Power Steering, p. 281; Radar, p. 283; Radio, p. 286; Rockets, p. 289; Safety Razor, p. 292; Self-winding

Wrist-Watch, p. 293; Shell Moulding, p. 295; Silicones, p. 296; Stainless Steels, p. 299; Streptomycin, p. 301; Sulzer Loom, p. 303; Synthetic Detergents, p. 304; Television, p. 307; 'Terylene' Polyester Fibre, p. 310; Tetraethyl Lead, p. 312; Titanium, p. 314; Transistor, p. 317; Tungsten Carbide, p. 319; Xerography, p. 321; Zip Fastener, p. 324

PART III

SUMMARIES OF NEW CASE HISTORIES

Air Cushion Vehicles, p. 329; Chlordane, Aldrin and Dieldrin, p. 332; Float Glass, p. 334; Moulton Bicycle, p. 337; Oxygen Steel-making, p. 338; Electronic Digital Computers, p. 341; Photo-Typesetting, p. 345; Rhesus Haemolytic Disease Treatment, p. 348; Semi-Synthetic Penicillins, p. 351; Wankel Engine, p. 354

INDEX 357

PREFACE TO THE SECOND EDITION

IN this second edition we have taken the opportunity of introducing some important changes in the book as originally published.

1. In Chapter X, which has been added, the relevant literature of the past ten years is examined and our earlier conclusions tested in the light of it.
2. Ten new case histories of recent date have been inserted. Each has its own fascinating peculiarities and all satisfy the criteria adopted in the selection of the earlier fifty cases. Float Glass and Semi-Synthetic Penicillins are instances of successful invention and development by firms of considerable size. Air-Cushioned Vehicles, Computers, Photo-Typesetting and the Wankel engine are cases where independent inventors made important advances at the earlier stages and both large and small companies made contributions in the later development. With Oxygen Steel-making and Chlordane, Aldrin and Dieldrin the discovery and development can be largely attributed to comparatively small firms. The histories of the treatment for Rhesus Haemolytic Disease and the Moulton Bicycle are largely records of the achievements of independent workers.
3. In Chapter III, a section has been added summarising the results of much important recent work on the history of inventors in aeronautics in the nineteenth century.

With the original fifty case histories much has happened since we drew up our stories. Some of the inventions have been more and others less successful than was expected. It is interesting and significant to recall that in 1958 we could speak of the Transistor as 'still in its infancy'; that we could speak of Rockets as no more than 'a deadly weapon of war and a useful instrument for the study of conditions in the upper air'; that Xerography (Xerox) was still being developed by comparatively small firms; that the Electronic Digital Computer seemed to have so uncertain a commercial future that we decided to exclude it from our case histories then. But in nearly every case improvements and developments have occurred in the last ten years. We had, therefore, to choose between leaving the original case histories substantially as they were first published or embarking upon a general revision of the whole list. We have chosen the first of these alternatives partly because our own resources for research were limited, partly because the later improvements have often been small, numerous and of no great novelty, but mainly because it is usually the earlier parts of the histories of inventions which, as time passes, tend to accumulate distortions. 'Historical revisionism is common in the field of innovation'.[1] In order to economise in space, however, four of the original case histories – Cinerama, Electric Precipitation, Freon Refrigerants and the Synthetic Light Polariser have been omitted.

In the preparation of the new edition we have once again depended greatly upon the work and writings of others. Most of the acknowledgements are made in their due place but some of the help given to us calls for special mention. Mr. Alistair Pilkington was kind enough to provide us with written material regarding the discovery of the Float Glass process and to answer our subsequent questions. In the preparation of the case history for Semi-Synthetic Penicillins, Mr. C. D. W. Stafford, Professor E. B. Chain and Dr. H. Raistrick provided valuable information. Mr. R. Stanton-Jones and Mr. D. Hennessy gave us much help with the history of the Hovercraft as did Mr. F. Wankel and Mr. R. Ansdale with that of the Wankel Engine and Mr. P. Purdy, Mr. R. Mackintosh, Mr. L. Moyroud and Mr. H. W. Larkem with Photo-Typesetting. Mrs. A. G. Clarke spent much time in reading the literature and helping to prepare the story of the discovery of the treatment for Rhesus Haemolytic Disease and in this case we also had generous help from Professor P. M. Sheppard and Dr. J. G. Gorman, who were intimately connected with this discovery. Mr. Moulton opened up to us his records relating to the invention of the bicycle which bears his name. We wish also to offer our thanks for the assistance given to us so willingly and over a long period by the staff of the John Crerar Library in Chicago. But, of course, we take full and final responsibility for the form in which the material is set down and the conclusions stated.

July 1968

J. J., *Oxford*
D. S., *London*
R. S., *Chicago*

[1] Donald A. Schon, *Technology and Change* (1967), p. 40.

PREFACE TO THE FIRST EDITION

IN compiling the information upon which this essay is based it has been highly encouraging to have received such a ready response to our enquiries from scientists, technologists, patent lawyers, historians, inventors, administrators and businessmen. They have often given much time and gone to great trouble to indicate to us the more reliable printed sources, to detail their own experiences, to explain in non-technical language highly complex subjects, to examine hypotheses set before them and to try to steer us clear of errors in fact. They will, however, not be named here. Many of them have spoken to us in confidence and would prefer anonymity. Others would probably not agree with the conclusions to which we think the evidence points. There are some case histories where the experts are in conflict and where, in consequence, we have been compelled to try and strike a balance. But, although it has seemed wisest not to record our obligations in detail, our gratitude is none the less.

One special debt must, however, be acknowledged. This work was begun in 1954 in the Law School of the University of Chicago and has, throughout, been financed by that University with funds provided by the Rockefeller Foundation. Professor Edward Levi,* the Dean of the Law School, gave us encouragement at a time when it appeared that little or nothing could be done with this subject. To him and to other members of that stimulating institution we owe a great deal, although it must be understood that they are in no way responsible for the views we have expressed. Finally we would like to express to R. & R. Clark, the printers of this book, our appreciation of their high standards in care, accuracy and efficiency, standards which may well seem to them simply a part of the day's work but which to a long line of authors continue to appear remarkable.

1958

J. J.
D. S.
R. S.

* [Now President of the University of Chicago.]

ACKNOWLEDGEMENTS

We wish to acknowledge with thanks the permission of the following publishers and authors to quote extracts from the books listed below:

Science and Freedom: Martin Secker & Warburg Ltd.
American Capitalism, by J. K. Galbraith: Hamish Hamilton Ltd.
Invention and Society, by W. B. Kaempffert: The American Library Association.
Science and Industry in the 19th Century, by J. D. Bernal: Routledge & Kegan Paul Ltd.
Transistors: Theory and Applications, by A. Coblenz & H. L. Owens: McGraw-Hill Book Co. Inc.
Science and the Modern World, by A. N. Whitehead: Cambridge University Press.
The Return of Arthur, by Martyn Skinner: Chapman & Hall Ltd.
Oliver Evans, by G. and D. Bathe: The Historical Society of Pennsylvania.
Samuel Crompton, by H. C. Cameron: the Author.
My Friend Mr. Edison, by H. Ford: Ernest Benn Ltd.
Emile Berliner, by F. W. Wile: The Bobbs-Merrill Co. Inc.
Life and Work of Sir Hiram Maxim, by P. F. Mottelay: John Lane, The Bodley Head Ltd.
The Inventor and His World, by H. S. Hatfield: Routledge & Kegan Paul Ltd.
Atomic Quest, by A. H. Compton: Oxford University Press.
The Reith Lectures, by Sir Edward Appleton: the Author.
Autobiography, by Lee de Forest: the Author.
The World of Eli Whitney, by J. Mirsky and A. Nevins: The Macmillan Company of New York.
Jet, by Sir Frank Whittle: the Author.
My Life and Work, by Henry Ford: Doubleday & Co. Inc.
The Art of Scientific Investigation, by W. I. B. Beveridge: William Heinemann Ltd.
Science Advances, by J. B. S. Haldane: George Allen & Unwin Ltd.
Fleming, Discoverer of Penicillin, by L. J. Ludovici: the Author.
Miracle Drug, by D. Masters: Eyre & Spottiswoode Ltd.
Antoine Lavoisier, by D. McKie: Constable & Co. Ltd.
Atoms in the Family, by L. Fermi: George Allen & Unwin Ltd.
Development of Aircraft Engines and Fuels, by R. Schlaifer and S. D. Heron: Harvard Graduate School of Business Administration.
Dialogues of Alfred North Whitehead, by L. Price: Max Reinhardt Ltd.
Thoughts and Adventures, by W. S. Churchill: Odhams Press Ltd.
Rutherford, by A. S. Eve: Cambridge University Press.
I am a Mathematician, by N. Wiener: Victor Gollancz Ltd.
Sir Frederick Banting, by Lloyd Stevenson: The Ryerson Press.
Magnetic Recording, by S. J. Begun: Rinehart & Co. Inc.
Invention and Innovation in the Radio Industry, by W. R. Maclaurin: The Macmillan Company of New York.

Into Space, by P. E. Cleator: George Allen & Unwin Ltd.
V2, by W. R. Dornberger: Hurst & Blackett Ltd. and The Viking Press, Inc.
The Technical Development of Modern Aviation, by Ronald Miller and David Sawers: Routledge & Kegan Paul.
Petroleum Progress and Profits, by John L. Enos: The M.I.T. Press.
Invention and Economic Growth, by Jacob Schmookler: Harvard University Press.

PICTURE CREDITS

J. Bardeen, W. Shockley, and W. H. Brattain
Bell Telephone Laboratories

C. F. Carlson
Fabian Bachrach

W. H. Carothers
Du Pont Photo Library

Sir Christopher Cockerell
Hovercraft Development

G. W. Jessup and F. W. Davis
Popular Science Monthly

T. Midgley Jr.
Frederick Bradley

A. E. Moulton
Rex Coleman

L. M. Moyroud
Dwight Davis

L. A. B. Pilkington
Pilkington Brothers

I. Schonberg
E.M.I. Research Laboratories

Felix Wankel
London Express News and Feature Service

J. R. Whinfield
I.C.I.

Sir Frank Whittle
Keystone

K. Ziegler
Tita Binz

PART I

CHAPTER I

INTRODUCTION

> Looking back on this present congress I must confess in all modesty that I may not perhaps have understood precisely and in every detail what is actually meant by science. I have learnt and observed that there exists a mass of different methods and different conceptions and opinions, and this whole rich diversity has once more been clearly set before us. – ERNST REUTER, Bürgermeister of Berlin.[1]

I

FUTURE historians of economic thought will doubtless find it remarkable that so little systematic attention was given in the first half of this century to the causes and the consequences of industrial innovation. Material progress, it had long been taken for granted, was bound up with technical advance and technical advance in turn, with change, variety and novelty; but whence this novelty, how closely it was related to rising standards of living, whether and how it might be stimulated or stifled: all this ground remained largely untrodden by the economic historian or the economic theorist. The comparative disregard of one, if not the main, spring of economic progress is not altogether mysterious. The subject is not one to which economic analysis is easily applied; it may yet prove impossible to apply it so. And the descriptive economist finds his way blocked by the complexity of the subject: the growing specialisation in science and technology presents to the outside observer a barrier even to the simplest understanding of what is occurring there.

A more important reason is simply that economists have been occupied in other ways where it seemed easier to reach results or where the immediate and practical value of their ideas seemed to be greater. The way in which the distribution of a given output of goods is, or should be, determined; how the factors of production, especially labour, can be kept fully employed; how internal domestic policies can be made consistent with external solvency; how industry should best be organised; what is the right balance between centralised and diffused economic decisions; these are the controversies through which the middle-aged have passed, the subjects which have been thrashed out to the exhaustion of the protagonists and, it is to be hoped, to the enhancement of social knowledge. But hardly ever have the studies and disputes centred upon the question: how will this or that policy influence the flow of innovation?

[1] *Science and Freedom*, p. 281.

All this now seems to have changed or to be in process of changing and, with that intriguing capacity which groups of scholars and scientists seem to possess of spontaneously wheeling and pursuing a new object of greater intrinsic interest, the study of economic growth, and with it that of change and innovation, has become a major preoccupation. A new slant is being given to every economic topic. Thirty years ago, when it was asked whether the appearance of large corporations and the concentration of industry should be tolerated, the answer seemed to turn on whether the big units would manufacture more cheaply than the small: now it is much more frequently discussed in terms of whether the big unit will be more effective as a source of innovations. In recent discussions of how far monopoly can be assumed to be in the public interest, one crucial question is how far the fewness of producers will influence the rate of innovation and the speed of development. The anxious search for the proper methods and institutions through which poorer countries can improve their lot has led to speculation of how best they can reproduce the outpouring of new technical ideas which older industrial countries have already enjoyed. Even in the study of balance-of-payments problems, emphasis is frequently placed upon the possibilities that technical advance in one country will continue to outstrip that in another and thereby create chronic difficulties of adjustment.

If these portents are in one sense encouraging, as revealing that economics has an internal vitality enabling it to slough off one skin and take on another, they also carry with them dangers. For, with the enthusiasm of converts, economists are now crowding into a subject where very little is known precisely because the relevant data is not that which they have been concerned to accumulate in the past half-century. It is not, of course, a subject in which they must start entirely from scratch; economists have not always been so indifferent to the dynamics of their subject as in the present century. But the raw fact from which their hypotheses may emerge and by reference to which these hypotheses can be tested is pretty scanty. If, therefore, economists are ever to be in a position to reach conclusions or prescribe policies with any kind of authority, the first need is to recognise the almost complete state of ignorance in which they find themselves. It is more than likely that their working hypotheses, without which the systematic collection of facts cannot be entertained, will be elementary, provisional and over-simplified or incorrect.

One simple proof of the present rudimentary state of knowledge is that practically any statement can be made without fear of decisive contradiction. Technical progress, it is commonly assumed, is now going on more rapidly than heretofore. But there is no statistical proof of it. It would, in any case, be necessary to distinguish between net and gross. Much new technical improvement is called for, to alleviate the evils of earlier technical progress; we must go faster to stay where we are: if, in the future, for example, ways are invented of making the jet engine quiet, then the skies will merely be as silent as they were before the invention of the jet engine. But, putting that on one side, why should it be assumed that the scientific advances or the inventions of the nineteenth century were any less revolutionary, possessed less powers to change ways and improve standards of living, than those of recent days? To every generation the events new to it naturally loom larger in the mind than those of the past to which it has become accustomed: but there is no objective measurement by which it can be established that the jet engine was more significant than the steam engine, the discoveries in atomic energy than the discoveries of the molecular constitution of matter, the development of pre-stressed concrete than that of the gothic arch.

There is no evidence which establishes definitely that technical or economic progress receives greater contributions from the few and rare large advances in knowledge than from the many and frequent smaller improvements. Economically, it might for a period well pay a community to starve its scientific and major technical work and to devote resources to the most thorough and systematic gathering together and exploitation of all the immediate and tiny practical improvements in ways of manufacture and design.

It is not known whether there is any necessary connection between the growth of scientific knowledge and the growth of technology and invention or, if there is a connection, what are its laws. If science and technology have different motives and criteria of success, it is a rash assumption that the one immediately and proportionately stimulates the other. It is not inconceivable that for long periods scientific advance may lie wholly in fields which have no immediate, or even ultimate, utility in the narrow sense. It is not inevitable that the country with the outstanding scientific successes will be the richest country. Indeed it is often suggested that, even where scientific advance has ultimately contributed to technology, the lag has been so great that it automatically rules out the possibility either of prediction or of calculated investment to produce results. If, for instance, the growth of technology in the nineteenth century ultimately arose out of the great progress in mathematics in the seventeenth century, the gap in time is so great as to destroy completely any value that that fact might otherwise possess as a basis for foresight; no one would invest now in any branch of scientific research on the assumption that it might be useful in A.D. 2300. Any community, therefore, which deliberately invests in pure science solely as a way of producing returns in technology and invention is not merely setting out on a course which threatens the ultimate values of science itself but is also engaging in a blind gamble.

We do not know whether there is an optimum rate of invention and technical advance or, if such an optimum is accepted as a conceptual device, how it would be defined or determined. Technical changes frequently cause social upset, which may conceivably be great enough to frustrate the powers for economic advance inherent in improved techniques. That, apparently, is the argument sometimes made for monopoly: that by rendering more deliberate the absorption of new ideas into industrial methods, the real economic advantages can, over the longer period, be made the greater.

But even if there were no doubts on this score, if the community sets out to encourage innovation it is by no means clear how best it could be brought about, what institutional framework will most effectively stimulate and encourage the men with powers of originality, most swiftly distinguish between the channels open to progress and the blind alleys, and most thoroughly glean the economic harvest from innovation.

The study of these weighty and complex affairs is, therefore, at the elementary stage in which no one need be ashamed of making a mistake or being forced to go back along his tracks; the only inexcusable error is that of rushing to premature final conclusions and hanging on to them too possessively. What is needed is, first, more actual knowledge about technical and economic change: the blanks are so great and so obvious that simultaneous efforts on the historical and the theoretical sides are to be welcomed, whether or not at this stage there appears to be any connection between the two. Second, an effort should be made to clarify and test the assumptions underlying the current working hypotheses. And third, an attempt should be made to determine what it is we are actually doing, and whether the immediate observable results are those being aimed at. After all, the world will not wait while social scientists are striving to understand it and offer their guidance; it will go on pushing and struggling, snatching out of ignorance and the exigencies of the day the solutions and policies forced upon it by the pressure it most immediately feels. And if, at that juncture, the social scientist has any function, it can only be that of trying to prevent the errors of the moment from becoming deeply embedded as ineradicable mistakes. And this by asking questions: whether men of affairs are in fact doing what they say and think they are doing; whether their immediate results are what they suppose them to be; whether their interim solutions appear to be consistent with their professed long-period aims.

II

Every community has to decide how best to maintain continuity and a reasonable measure of stability while leaving open channels for new ideas and room for change. It cannot afford to be tossed about defencelessly by the demonic impulses of the innovator, yet it can ignore him or suppress him only at the risk of stagnation. It can make itself completely secure against the charlatan, but only by damping the procreant urge of the exceptional few. In every aspect of society this struggle goes on to reconcile the two, to get the best of the two worlds of authority and questioning, of tradition and novelty, conservatism and radicalism, stability and progress, continuity and change. It is a search for a compromise never finally reached because the weights in the balance must always be shifting. This is the dilemma of society: neither to fling away the confirmed knowledge of the past nor reject precipitately the opportunities of adding to it; and every community which understands its tasks will strive to embody the general doctrine in practices and institutions. To find ways of preserving the rights of minorities, of tolerating the rebel, of giving ear to the doubter, of encouraging the true innovator: these are all different ways of extracting what is good from what is new and avoiding both the explosive force of repressed opinion and the loss occasioned by suppressing originality.

There is, indeed, a deeply rooted belief that sufficient is never made of invention and inventive genius, that in one way or another the community is casual and wasteful of ideas. The grounds for these suspicions vary from time to time. Not so long ago it was common talk, although with very little evidence, that private enterprise would lead to the deliberate suppression of important inventions. This myth has now largely fallen out of fashion, and it is now more commonly asserted that there are some much less deliberate and sinister influences at work, usually described as 'resistance to change', which stand between the community and the bounties which the innovator might well confer upon it. Whatever the truth here, it is a fair comment that industrial societies have shown little originality or ingenuity in creating institutions to ensure that all new ideas will be swept into the net and that nothing will be lost. They have not been very prolific in the invention of institutions to encourage invention. In particular, it is extraordinary how the Patent System goes on largely unchallenged, much as if it were some august political institution and not an economic device directed to a specific economic end.

In one way or another, every society makes decisions about the proportion of its resources which are to be devoted to research and invention. In some countries in these days, notably in Great Britain, many would hold that the present proportion is wrong, that not enough scientists and technologists are trained, that firms do not devote sufficient funds to industrial development, that public funds for these same purposes are inadequate, that not enough is invested in the men and the institutions to bring the maximum economic returns in the long run or to guarantee national defence. Conversely, all such arguments carry with them the implication that too ample resources are being devoted to other activities in the community; fundamental questions are thereby raised about the ends for which society should strive.

Finally, among the major speculations in which existing knowledge, true or false, must be brought to bear is how far invention is a calculable, predictable element which can be embodied into the conscious planning of economic and social activity for the purpose of making that planning more feasible. After war itself, it is often said, the major predicaments of society probably arise more frequently out of technical progress than from any other cause. Social and economic disturbances which, dealt with at the start, might have been manageable, become with time progressively more intractable or call for other major shifts and changes which pile up one crisis, one emergency, upon another. This has led to the hope that inventions might be foreseen, for then at least some of their social consequences could be anticipated and at least their more undesirable consequences side-stepped.

III

The aim of this essay is to try to throw some light upon at least one fragment of such vast and intricate matters by asking what seems at first sight a simple and direct question: where and under what conditions have industrial inventions arisen in modern times? Have important changes taken place in their sources as between the present and the last century and, if so, why? It may, at first blush, seem strange that this task should have been undertaken by investigators without training in science or technology and thereby largely dependent upon second-hand knowledge in fields intricate enough to deter the novice. The apology must be that answers now being proffered to a question of great intellectual interest, and perhaps public importance, reveal most perplexing differences of opinion even among the informed and experienced; that current views are not infrequently unsystematic and inadequate, and that this study, impinging as it does upon history, science, technology, economics and law, is a kind of everyman's land, and therefore no-man's land, in which perhaps no one can hope to be fully equipped at all points for the exploration.

It was with hopes of accumulating facts and of making some contribution to a better understanding of the dynamics of the economic system and not primarily of reaching policy conclusions, that this enquiry was embarked upon. But forbidding obstacles, which of themselves fully account for the often vague and unsystematic character of many other studies of invention, immediately presented themselves, enforced a more modest target even than that originally set, and reinforced the sentiment that the present is the time for hypotheses for testing, and not *ex cathedra* theories for putting into practice. The subject seemed to have no beginning and no end; any effort to break it into parts in order to make it more manageable raised the question of where and how to draw dividing lines. Indeed the paramount difficulty throughout has lain in definition. No words are more commonly used than pure science, applied science, discovery, invention, technology, basic research, applied research, objective research and so on. Yet to determine what each connotes, or to distinguish between them, is a most slippery task, but one which cannot be evaded.

What is an invention? Technical progress is an indivisible moving stream from which it seems impossible, except in an arbitrary fashion, to isolate one fragment for independent examination. Every item seems in the last analysis to be linked with every other item, so that nothing can be thought about or explained unless everything is taken into account. The windscreen wiper, the zip fastener, the jet engine, the cyclotron, nylon: all these have been described as inventions. Yet any definition that includes them all would seem to include also every technical or product variation that has ever occurred. Is the invention the idea; or the first conception of a way of using the idea; or the actual working utilisation of the idea; or the compounding together of two existing ideas; or the effective fusion of two ideas for a useful purpose? And if an invention cannot be defined, what becomes of the attempts to classify inventions? Thus the distinction between a 'cost reducing' invention and an invention which consists of a 'new product' seems theoretically valid. But in practice, every device for reducing cost is a new product. Every new product is a method of reducing cost in one form or another. A jet-engine aircraft for crossing the Atlantic in place of a sailing-boat is a new product but it is a new product only because it reduces the cost of the journey either in money or time or personal hazard or exertion. It is, indeed, not surprising that in the long history of patent litigation the efforts of the courts to define 'invention' have produced such contradictions and confusions.

What is an *important* invention? Anyone seeking to generalise about inventions from case histories must confine himself to the salient novelties. There is no way of measuring the totality of invention; only its sporadic dominating features can be picked out. Yet there are no economic principles to which to appeal to determine whether one invention is more 'important' than another. On what grounds could it be asserted that the ball-point pen is a more important invention than the cotton picker, unsplinterable glass than colour photography, the safety razor than the steam turbine?

Who is *the* inventor in any particular case? Which of the long line of thinkers and manipulators, each of whom has added something to the final appearance of a useful thing, should take the palm? When Carothers heated chemicals together in a test tube and discovered nylon, was it the original inventor of fire or the inventor of the bunsen burner or Carothers himself who gave this particular ball the biggest push? Where, as often happens, men working independently of each other appear to have reached the same ideas at about the same time, what tests of priority can be applied? And, an even more difficult question which the persons actually concerned might often be at trouble to answer, how can the real originator of an idea be picked out from among a group of men who have been working closely together for a period?

There can be no doubt that an over-rigid insistence upon definition would immediately bring all discussion of invention, and of the part it plays in changes in ways of living, to a dead stop. The choice must be between discussing these matters with concepts that are necessarily somewhat vague and not discussing them at all. The fact that this essay has been written means, of course, it has been considered that the difficulties outlined above are not a decisive bar to useful investigation, provided only that, in interpreting the evidence, an eye is kept upon the insecure standpoint from which the start is made. Put in other words, this essay embodies preconceptions which will now be set down.

There is, first, a distinction to be made in purposes, methods and results between pure science and technology.[1] Science is directed towards understanding, and technology is directed towards use. The criteria of achievement in science can only be applied by scientists, as a professional group, who must be the judges of the significance of any claim to the extension of pure knowledge. The criteria of success in technology is that of the market, whether the new idea is a commercial asset or not: the final judgment here is exercised by the consumer. Scientists sometimes take on the role of technologists; technologists in the course of their work may make scientific discoveries; scientists and technologists may at some time be in closer *rapport* than others; some scientific discoveries may carry with them greater value than others for technologists. None of these things blurs the crucial distinction between science and technology. Of course, scientists may direct their thinking towards an extended understanding of the universe in such ways as seem to offer the greatest possibilities of material advance, although how they would choose their directions seems highly mysterious. But in so doing they would cease to behave as scientists. It is no function of the scientist to discriminate between what is commercially and materially useful and what is not, nor is there anything in scientific training or methods which enables the scientist to exercise that choice with greater skill and certainty than anyone else. It is even more obvious that attempts on the part of scientists, as scientists, to set up rules as to the uses to which scientific discovery could or should be put, leads them away from the position in which they have any right to speak with authority.

Secondly, it is assumed that the web-like unity of technical progress *can* be broken down into pieces disparate enough to be examined independently, and that common observation provides reliable grounds for assuming that some inventions are more significant than others. In the steady and continuous accretion of ideas and methods, refinements, observations and original combinations of bits of existing knowledge, some of the steps taken are bigger than others, some of the individual contributions have more to do with the final result than others, and even when a group of individuals have been working together, the final result is not in the nature of a chemical reaction but a sum total to which the members will have contributed in varying degrees.

On this view, the part played in invention by individuals cannot be looked upon as of little or no account. It is the practice of some writers to present a fuzzy picture of invention as 'a social process'; to suggest that, if one inventor had not done what he did when he did, someone else would have done it; that inventions have come to ripeness at a predestined point in time influenced little, if at all, by human will, courage and pertinacity; that if Watt and the other great names had never been heard of, the world would have been much the same. But this attitude – that nothing can be understood

[1] Following here Professor M. Polanyi, who has developed his views in full in *Science and Freedom*, pp. 36–46.

unless all is understood, that by piling one unresolved enigma upon another some all-comprehending solution is made the more likely – involves the error of 'seeing depth in mere darkness', as Sir Isaiah Berlin once put it.

It is just as important to be clear about what is here being accepted as about what is being challenged. It is not to be doubted that any and every invention, if the story of it be probed long and deeply enough, can be traced back almost indefinitely. The history of an invention has almost unlimited dimensions in space and time. Every inventor, however original he may appear to have been, is laying bricks upon a building which has long been in the course of construction from innumerable and mainly unknown hands. This is, of course, no new idea. Samuel Smiles[1] (whose analysis of the nature of invention has been so oddly neglected and many of whose findings have in recent years been re-discovered and re-presented to us by sociologists, although they speak in much less straightforward language than his) wrote in 1863 of how 'the living race is the inheritor of the industry and skill of all past times; and the civilisation we enjoy is but the sum of the useful effects of labour during the past centuries'.

It is, however, one thing to concede that historians can often do little more than emphasise the highlights of the past, and that we are, in truth, all beneficiaries from the work of our ancestors. It is quite another to view the history of invention as a remorseless unrolling of a pattern in which no one man has played a part of his own choosing or one for which there were not immediately available a full supply of equally competent understudies. It seems misleading to write down or write off the fertilising effect of the big jump in ideas or the outstanding and recognisable influence of the seminal mind. For 'there is the hand which lights the lamp, but other hands must keep it burning and carry it on'.[2]

Even if the great inventors of the past century and a half had never been born, but some powerful, dynamic, impersonal push had ultimately produced the same results, the delay in itself would be significant. Timing here is of the essence of the matter. If invention is a continuing process, any loss of time is a permanent loss, the whole course of technical progress is set back permanently, the time lost is never made up. But, in fact, individuals have made a permanent impress on the shape of things. It is true that the rapid advances in the past 150 years seem to give meaning to such phrases as 'the march of science', 'the march of technology' and to create the impression of movements pressed along by forces independent of the will, the decisions, the struggles of individuals. In that period one genius trod closely upon the heels of another. But to describe an epoch especially rich in outstanding inventors as one in which individual contribution can no longer be identified is a curious inversion of facts, and as great a misapprehension as it would be to talk of the 'march of painting' in the

[1] *Industrial Biography: Iron Workers and Tool Makers*, pp. 167–8.
[2] L. J. Ludovici, *Fleming, Discoverer of Penicillin*, p. 152.

Netherlands or 'the march of drama' in the England of the sixteenth and seventeenth centuries.

The third presupposition is that a useful working distinction can be made between 'invention' and 'development'. Just as a distinction is made between science and technology, so technology itself can be divided into these two parts. Invention is something which comes before development. The essence of invention is the first confidence that something should work, and the first rough tests that it will, in fact, work.

When Watt conceived of his engine and then, years later, made the first satisfactory model; when Cartwright, with the help of mechanics, put his ideas of a power-loom into a shape which operated; Whitney built his first primitive cotton gin; Perkin produced the first aniline dye; Goodyear his first batch of vulcanised rubber; Gilchrist-Thomas lined a steel converter with a basic material; Bell transmitted speech over a short wire; Cross produced viscose on a small scale; Diesel found a way of producing ignition in a cylinder by compression; Baekeland, ignoring traditional ideas, manufactured bakelite; the chemists of Tootal Broadhurst Lee imparted to a piece of cotton cloth crease-resisting properties; Lawrence constructed his first rickety cyclotron; Whittle ran his first turbo-jet; Poulsen magnetised a wire and recorded sound on it; Carothers drew through an improvised spinneret a fibre now known as nylon; Fleming watched the action of penicillin on bacteria; Farnsworth conceived of his image dissector tube for television and constructed his first crude model to demonstrate the soundness of his notions; Whinfield first saw the fibre subsequently named Terylene; Kroll, in his private laboratory, produced tiny quantities of ductile titanium; the chemists at Imperial Chemical Industries found a white plastic material, now known as Polythene; Carlson, working with the principles of photoconductivity and electrostatics, devised xerography, a novel system of copying documents: in all these cases an invention, subsequently proving of great importance, had been made.

Development is a term which is loosely used in general discussion to cover a wide range of activities and purposes, but all these activities seem to satisfy three conditions. One, development is the stage at which known technical methods are applied to a new problem which, in wider or narrower terms, has been defined by the original invention. Of course, it may happen that in the course of development a blockage occurs, existing technology may provide no answers, and then, what is strictly another invention is called for to set the ball rolling once more. Two, and consequentially, development is the stage at which the task to be performed is more precisely defined, the aim more exactly set, the search more specific, the chances of final success more susceptible to measurement than is true at the stage of invention. Invention is the stage at which the scent is first picked up, development the stage at which the hunt is in full cry. All the money in the world could not have produced nylon or the jet engine or crease-resisting fabrics or the cyclotron in 1900. At the time of writing it is possible to say that all the money in the world may not produce a cure for most forms of cancer, or lead to the discovery of economical methods of storing electricity on a large scale. Three, development is the phase in which commercial considerations can be, and indeed must be, more systematically examined, the limits of feasibility imposed by the market are narrowed down. As one moves from invention to development the technical considerations give way gradually to the market considerations.

This distinction is crucial because this essay is almost wholly concerned with invention and not with development.[1] Since little or nothing will be said about development until the end of the book, and since invention can be negatively defined by describing what it is not, it may be useful to dilate a little upon the process of development itself.

(*a*) The task of producing on a large scale may be different in kind or highly different in degree from that of producing on a small. In the chemical industry especially the manufacture in quantity of what has already successfully been produced in a test tube in a laboratory may confront the developers even with different chemical reactions arising out of the scale of operation. At the present time, results which can be produced on a minute scale by the use of solar or atomic energy cannot simply be multiplied arithmetically to obtain useful results. A rotating wing can easily be constructed to carry its own weight, the problem is to use it to carry more than that. It is easy to imagine that at some stages what is here described as development may involve something which it is almost impossible to distinguish from invention. And, at the other extreme, since the final proof of how a big apparatus embodying the ideas derived from a small will really work must often be settled by constructing a large apparatus – a pilot plant or something even bigger – development merges into the function of investment and risk-taking.

(*b*) A small model may be constructed of materials which would not be pure enough or durable enough or sufficiently susceptible to manipulation to small enough margins to enable a larger machine with a practicable working life to be constructed. The history of mechanical invention is strewn with cases where a search had to be made for better materials (which might itself involve invention) or for new uses for materials already known. The wooden cylinder was not good enough for the steam-engine; the progress of the jet engine depended crucially upon metals which would withstand abnormal heat and strains; in the development of the transistor, particularly the silicon transistor, one of the important lines of search was for materials in a particularly pure form.

(*c*) Where the inventions are of new materials such as those produced on a vast scale in the chemical industry in recent years, these materials may have properties which render them impossible to handle with the existing mechanical devices. So the need arises for new types of machines, new

[1] But see Chapters VIII and X for some comments on Development.

mechanical processes to manipulate, transform and transport the new stuffs. Or new machines may be needed for producing queerly shaped or exceptionally finely adjusted equipment. The development of the safety razor, the zip fastener, the jet engine, barbed wire, are cases in point.

(*d*) It seems proper to include in development the search for the reasons why something 'works'. If a process is largely empirical, interest in the exploration of it may be purely intellectual curiosity, which is science. But there may be good commercial reasons for organising an investigation. Where a process works but it is not known how or why, that process is not strictly under control. If something goes wrong, if the experience is apparently not repeated, it becomes the more difficult to determine the essence of the divergence from expectation and to seek a remedy. There are some instances where an invention has been totally lost for these reasons. Again, systematic improvements are more difficult to make where there is complete dependence upon traditional and half understood methods.

Although, therefore, the attempt to draw a very sharp line between invention and development would be as foolish a waste of time as to seek to determine whether twilight is day or night, yet there seems to be a fundamental distinction between the two.[1] The one is a beginning, without which the other is of no avail; is in the nature of things largely unpredictable; arises more immediately out of technical preoccupation and the combination of that with the sense of artistry, craftsmanship and making things fit which is characteristic of much invention. The other will be informed at each step by economic calculation, each successive step becoming more deliberate and specifically designed to a defined end.

The history of nylon is one good illustration: in 1935, after seven years' work of varying fortune and many disappointments, work which might have led anywhere or nowhere, W. H. Carothers, in the laboratories of the du Pont Company, produced the first nylon fibre and du Pont undertook to translate it into a marketable product. By 1939 large-scale production of nylon hosiery had commenced. Thus, in a matter of four years of development, du Pont had reached its appointed goal. Estimates put the total cost of the early stages of research and development at over $1 million; at that time 230 technical experts were engaged in the work. What precisely was

[1] Much subtlety has been devoted to the establishment of distinctions between 'invention' and 'discovery' and between different types of 'discovery', such as 'discovery for' and 'discovery of'. Thus the jet engine might be said to have been invented, since it had never existed before, while penicillin might be said to have been discovered, because it had been there all the time and had now been first found. The former was the result of an original manipulation of materials, the latter was the first recorded observation of a natural phenomenon. But these distinctions appear to raise more perplexities than they resolve. In this essay the word 'discovery' is applied as a non-discriminating term to anything new, whether the novelty is of a scientific and inventive, a development, or an improvement type.

involved in the development undertaken after Carothers's initial discovery?

First it was necessary to find ways of producing on a large scale the intermediate constituents of nylon which, up to that time, had been made only on a small scale. The two important materials were adipic acid and hexamethylenediamine. Adipic acid had been manufactured in Germany for some time but there had been no commercial exploitation of it in the United States. The German processes were not readily adaptable to the plants of du Pont and it became imperative to develop a new catalytic technique for this purpose. Hexamethylenediamine posed even greater difficulties; it was merely a laboratory curiosity and had never been manufactured on a commercial scale before. Success here required the discovery of new catalysts and the proper handling of heat transfer problems. Next, a great deal of work had to be done at the stage where the materials react to form the long chain molecules of the nylon polymer. The first polymers were made in glass equipment in Carothers's laboratory, but glass equipment was completely unsuitable for commercial manufacture and metal equipment had to be designed. Methods of controlling the degree of polymerisation had to be evolved, since a failure to stop the reaction at precisely the right time resulted in the production of different and far less useful polymers than nylon. The technologists had to learn how to make one batch of the product exactly like another.

At the next stage of manufacture the flakes of the polymer had to be melted and some means found to transfer the molten mass to the spinning machines. Only pumping gave the filaments adequate uniformity, but unfortunately there were no existing pumps suitable for the task. A new type of pump was required embodying new alloys capable of withstanding the heat of the molten polymer. At the next stage of spinning, the machinery had to be specially designed for the task, since nylon could not be spun in the same manner as cotton, wool, viscose or cellulose acetate. The winding and the cold drawing processes also confronted the developers with problems which were novel and for which specially designed machines were required. Thus at each one of these stages – the mass production of what had formerly been made only on a small scale, the maintenance of unusual degrees of purity, the flexible controlling of the chemical processes and the devising of mechanical aids for handling materials with novel properties – the developers were confronted by one hurdle after another. It was only when the process reached the stage of knitting and weaving that existing and familiar techniques could be called in to help. But at every stage workers knew what they were looking for, and, with varying degrees of certainty, they knew it could be found.

Another apposite illustration is that of penicillin. Here fifteen years elapsed between Fleming's discovery in 1928 and the commencement of large-scale production; for seven of these, from 1932 to 1939, no work on its development was being done. The main phase of the development was concentrated into the four years 1939–43 and this was itself divided into two periods: that of the discovery of a method of isolating penicillin and of the evidence of its value as a chemotherapeutic agent by the Oxford team under Sir Howard Florey from the autumn of 1939 to the summer of 1941; and that of the development of methods of large-scale production in the United States. It is this second period after 1941 which constitutes the development proper, although even in the first period there were times when the character of the problem to be solved and the dependence for success upon skill in manipulation closely approached the tasks of development.

The early part of the story of penicillin is very familiar. Fleming had found penicillin unstable and in consequence difficult to work with and, in the absence of a biochemist to work on the final stages of isolation, had abandoned the attempts to do so. Raistrick, who came near to success, was able to grow the mould in a synthetic medium and to store it for longer periods. The Oxford team took up from this point and by 1941 had found ways of making penicillin and of storing it, and had produced clinical evidence of its effectiveness. Up to that point the cost of work was small: the special grants to the Oxford team for penicillin research amounted to £20,000 over the six years 1939–45.

The real beginning of the development period perhaps commences with Florey's visit to America in 1941 for the purpose of encouraging American manufacturers to produce penicillin on a larger scale. No more inherently intractible task has ever been put to manufacturers: to produce a mould on an industrial scale under absolutely sterile conditions. The first strides to final success were made by the staff of the Research Laboratory of the United States Department of Agriculture at Peoria. There the suggestion was made for deep culture fermentation which had great advantages for large-scale production since it obviated the use of small bottles, each of which needed to be injected with the spores of the mould by hand and then handled again to have the fluid drawn off. Another important discovery made at Peoria was that the use of corn steep liquor in the fermentation medium greatly increased the yield of the mould.

The value of penicillin for military use led the American Government to support the development of the new production process as a matter of urgency. In many universities and government laboratories a search went on for better strains of the mould. Numbers of pharmaceutical companies built plants for large-scale production. But large-scale manufacture posed increasingly serious problems of bacterial contamination. Sterile zones had to be created in the factories, which called for techniques for sterilising large volumes of air in order to prevent bacteria reaching the culture fluid while it was being drained from the vats and going through the extractive processes. Gradually, by care and thoroughness, experience was accumulated which turned novelty into routine.

The study of invention must necessarily be qualitative, selective and impressionistic, more historical than scientific. This may at first sight seem paradoxical, for there are perhaps no records so complete and voluminous as those appertaining to the activity of invention. The Patent Office of every industrial country contains in its files of specifications the story of industrial originality, unique in its completeness, in its detail, in its accessibility. Yet the frustrating and exasperating fact remains that these monumental heaps of documents cannot be reduced to any form which gives a sharp picture. Of course, patent statistics are not without relevance for some purposes. But, unfortunately, although every patent presumably involves an invention, not every invention involves a patent. The patent has a legal meaning but, as a unit, its economic meaning, for reasons which will be enlarged upon at a later stage, remains tantalisingly obscure.

So, too, the writings on invention are of extraordinarily mixed quality. There seems to be no subject in which traditional and uncritical stories, casual rumours, sweeping generalisations, myths and conflicting records[1] more widely abound, in which every man seems to be interested and in which, perhaps because miracles seem to be in the natural order, scepticism is at a discount. Perhaps no one can hope entirely to escape the mild mesmerising influence of the subject. But by comparing the records of one authority against another; by interviewing wherever possible the inventors themselves; by discussions with scientists, technologists, patent lawyers, Patent Office officials, businessmen, financiers and research administrators, an endeavour has been made here to identify the points at which the systematically ascertainable facts and collective wisdom seem to support each other and thereby give credibility to generalisations about the fascinating and elusive power of the human mind to originate.

[1] One intriguing aspect of this subject is how many different countries claim the credit for the same invention and tend to ignore the contributions of other countries. It was usually found that the story of an invention, when based on American records, gave greater credit to the American contributors and less to British contributions than the same story based on British records. It is not only the Russians who believe that everything has been invented within their national boundaries!

CHAPTER II

MODERN VIEWS ON INVENTION

All philosophers who find
Some favourite system to their mind,
In every point to make it fit
Will force all nature to submit.

THOMAS LOVE PEACOCK, *Headlong Hall*

THERE is one type of mind which finds it tempting to stress the contrast between the world of today and that of yesterday and to think of change as a series of big fresh starts: there is another type congenitally disposed to believe that there is nothing new under the sun, that all that has been said and done has happened before. As between these two extremes, both likely to give a distorted perspective, there can be little doubt that the greater part of modern writing about invention and technical progress strongly inclines to the view that we live in a new world in which thinking of the present or the future in terms of past experience is largely irrelevant, and that our ideas must be recast and our institutions reformed to fit fresh surroundings. Social scientists are now tending to speak with more confidence about the scale on which inventions will be made and the sources from which they will arise. There have, indeed, been some odd switches of thought since the end of the First World War. In the early 1930's it was widely believed that technical progress would normally be so swift and disturbing that a high level of 'technological' unemployment would be usual and inevitable. In the later thirties, due mainly to the failure of the American economy to continue to expand at an unbroken rate, the view gained currency that technical progress would usually be too sluggish to create sufficient profitable investment openings for the savings arising under full employment: secular stagnation and chronic unemployment were inevitable, in the absence of public intervention, because technical progress would never be on a large enough scale.[1]

The period following the Second World War with its general shortage of capital, its full employment and the impact upon the public mind of the

[1] By much more cautious and scientific methods Professor Simon Kuznets (*Economic Change*, chap. 9) has suggested that in each industry taken separately, technical progress will tend to slacken after a time. The suggestion, of course, carries with it no support for theories of secular stagnation of whole economies. And it seems that, by extending the cases beyond those chosen by Mr. Kuznets, exceptions to his tentative hypothesis are to be found. The whole subject, however, is well worth more study than it has received since the publication of the original stimulating article in 1929.

discovery of atomic energy, has brought forth a fresh crop of generalisations which, compared with the pessimistic views of the inter-war period, are oddly sanguine in tone.[1] Unbroken and rapid technical advance, it is thought, can now be taken for granted; the causes of it, the institutions which best foster it, are understood, and understood so well that society may be within measurable distance of the power deliberately to control it or, failing that, to predict with a high degree of certainty what the future holds.

This new and fashionable doctrine, subscribed to by many scientists, technologists, economists, statesmen, business men and popular writers, does not seem anywhere to have been expounded in final or authoritative shape. Nor is it difficult to pick out inconsistencies between its varying formulations. The mainspring is sometimes held to be the growing power of science, sometimes the increasing skill and eagerness with which technologists pick up and use scientific knowledge. By some, inventions are now thought to be easier to make than formerly; there is a gathering momentum, an 'autocatalytic process', driving things on. Others believe that inventions have now become more difficult to make because all the easy inventions have already been made. 'Necessity', to some is still 'the mother of invention'; to others, inventions are thought to pour out almost automatically and prodigally in an ever-increasing stream in the richer industrial communities where needs are least urgent. Some hold that there never was a time when inventions were more eagerly seized upon by industry for commercial exploitation; others that there is still a great deal of 'resistance to change'.

Whatever the discrepancies and dissensions, there appears to be a broad area of agreement among these who adhere to the spectacular view of modern invention. A few quotations, drawn from a vast literature, will serve to indicate the gist of the doctrine.

James B. Conant:[2]

> As theory developed in physics and chemistry and penetrated into practice, as the degree of empiricism was reduced in one area after another, the inventor was bound to disappear. Today the typical lone inventor of the eighteenth and nineteenth centuries has all but disappeared. In his place in the mid-twentieth century came the industrial research laboratory and departments of development engineering.

J. K. Galbraith:[3]

> A benign Providence . . . has made the modern industry of a few large firms an almost perfect instrument for inducing technical change. . . .

[1] Although alarms have been sounded about the unemployment which might arise from 'automation'. These fears seem similar to those which arose in the 1920's about 'rationalisation' and in the 1930's about 'technocracy'.

[2] Speaking in May 1951.

[3] *American Capitalism*, p. 91.

There is no more pleasant fiction than that technical change is the product of the matchless ingenuity of the small man forced by competition to employ his wits to better his neighbour. Unhappily, it is a fiction. Technical development has long since become the preserve of the scientist and the engineer. Most of the cheap and simple inventions have, to put it bluntly, been made.

W. R. Maclaurin:[1]

We have now reached a stage in many fields where inventions are almost made to order, and where there can be a definite correlation between the number of applied scientists employed (and the funds at their disposal) and the inventive results. But one really gifted inventor is likely to be more productive than half a dozen men of lesser stature.

Walton Hamilton and Irene Till:[2]

Most discoveries patented today can be anticipated. . . . For the most part, technicians are not self starters. The bulk of them in fact are captives; those in corporate employ are told by business executives what problems to work on. . . . The solo inventor's real opportunity is to seize or blunder upon a pioneer idea; as a technology foliates from its base, his self reliance is hardly a match for a bevy of experts who can be kept on the job. . . . A captive technology offers no chance to invent except to those already in control, or to others on such terms as those in control dictate.

W. B. Kaempffert:[3]

It is not difficult to predict the effect of industrial group research on invention. As organised invention and discovery gain momentum the revolutionist will have no chance in explored fields. He will have to compete with more and more men who have at their disposal splendidly equipped laboratories, time and money, and who may work for three or four years before producing a noteworthy result. Only the exceptionally brilliant trained scientist will be able to meet these explorers on their own ground. Possibly Edison may be the last of the great heroes of invention.

J. D. Bernal:[4]

Many intelligent non-scientific people still think of science as it appeared to be in the nineteenth century, as the product of individual efforts of men of genius, instead of, as it now is, a highly organised new profession

[1] 'The Sequence from Invention to Innovation', *Quarterly Journal of Economics*, Feb. 1953.

[2] *Law and Contemporary Problems*, vol. 13 (1945), p. 252.

[3] *Invention and Society*, p. 30.

[4] *Science and Industry in the Nineteenth Century*, p. 4.

closely linked with industry and government. . . . It is almost as difficult in an age of vast engineering and chemical factories, each furnished with its own research department, to recall the intimate traditional and practical character of the old workshops and forges from which the modern giants are descended.

A. Hunter:[1]

The days when one individual's inventiveness and enterprise could transform an industry are in the past. In this context the big firm again shows to advantage. Perhaps we are labouring the obvious. These are all well known facts of economic life. . . .[2]

The pith of the modern view is, therefore, that in the nineteenth century most invention came from the individual inventor who had little or no scientific training, and who worked largely with simple equipment and by empirical methods and unsystematic hunches. The link between science and technology was slight. Manufacturing businesses did not concern themselves with research. In the twentieth century the characteristic features of the nineteenth are rapidly passing away. The individual inventor is becoming rare; men with the power of originating are largely absorbed into research institutions of one kind or another, where they must have expensive equipment for their work. Useful invention is to an ever-increasing degree issuing from the research laboratories of large firms which alone can afford to operate on an appropriate scale. There is increasingly close contact now between science and technology, both through the closer association of the workers in the two fields and because the border-line between the two formerly separate functions is becoming obliterated. The consequence is that invention has become more automatic, less the result of intuition or flashes of genius and more a matter of deliberate design. The growing power to invent, combined with the increased resources devoted to it, has produced a spurt of technical progress to which no obvious limit is to be seen. It will be noted that statements of the kind quoted above assert facts, imply causes and express satisfaction as to results. Something is actually occurring in the world of technology which is heartily to be welcomed, or is inevitable, and which is bringing about improved standards of living.

[1] 'The Control of Monopoly', *Lloyds Bank Review*, Oct. 1956.

[2] Even at the risk of repetition the following ought to be quoted: 'It has become generally established that scientific progress is the result of well-organised research teams. The day of the garret scientist, working alone in a near-bare loft by the flickering light of an oil lamp is almost past. For the scope of knowledge in any one field is so vast that few individuals can fully master it. In addition, an individual effort is dwarfed by the large scale attack on the frontiers of our technical knowledge by incalculable numbers of scientific workers in many great laboratories with unlimited facilities. Even in purely theoretical contributions, the facilities available in these million dollar laboratories are almost indispensable to original work; in experimental investigations the facilities of large laboratories are even more essential.' (A. Coblenz and H. L. Owens, *Transistors, Theory and Application*, p. 1.)

This, then, is the sharp contrast drawn between an earlier heroic age of clumsy individual pioneering and a modern age in which highly trained, closely organised teams of technologists, fortified by an easily accessible and constantly expanding body of scientific knowledge, move forward with deliberation to results which can largely be predetermined. Not all modern writers, however, share this view about nineteenth-century invention. Whitehead once wrote[1] that –

> The greatest invention of the nineteenth century was the invention of the method of invention. . . . The whole change has arisen from the new scientific information. . . . It is a great mistake to think that the bare scientific idea is the required invention. . . . One element in the new method is just the discovery of how to set about bridging the gap between the scientific ideas and the ultimate product.

That is to say he claimed for the nineteenth century those very achievements which other writers attribute uniquely to the twentieth.

It is perhaps even more significant that some writers in the nineteenth century did not picture their own century as twentieth-century writers now often describe it. In 1808, in a paper presented to the Manchester Literary and Philosophical Society,[2] Ewart challenged the view that invention was, or could effectively be, carried on without scientific knowledge and contacts. A knowledge of the principles of mechanics, he argued, must help the inventor and assist him to distinguish real from illusory improvements. Nor would he allow that the history of mechanical discoveries supported the contrary opinion. Both Huygens and Hooke, he pointed out, had been scientists and, if they had not, they might not have invented the balance. Smeaton had used theoretical knowledge in his inventions. Watt also was a man of scientific attainments, but for which he might not have invented his improvements to the steam engine. Charles Babbage in his *Decline of Science in England*, published in 1830, stressed the importance of science in technical progress and alleged that the failure of official bodies to support science was endangering industrial expansion. Professor Tyndall, in a lecture on the electric light[3] in 1879, quoted with approval Cuvier as saying:

> Your grand practical achievements are only the easy application of truths not sought with a practical intent. . . . Your rising workshops, your peopled colonies, your vessels which furrow the seas; this abundance; this luxury, this tumult, all come from discoveries in science.

[1] *Science and the Modern World*, pp. 120, 121.

[2] 'On the Measure of Moving Force', *Memoirs of the Literary and Philosophical Society of Manchester*, Second Series, vol. 2. Ewart was a well-known engineer who had been apprenticed to Rennie and then worked for Boulton and Watt, becoming a lifelong friend of Watt. In 1792 he became Samuel Oldknow's partner in his cotton business; the following year he set up as a cotton spinner in Manchester on his own. In 1835 he left this business to become Chief Engineer and Inspector of Machinery in H.M. Dockyards.

[3] *Proceedings of the Royal Institution*, vol. 9 (1879–81), pp. 22, 23.

It is clear that at this time the potential use of scientific discoveries, and the manner in which the inventor could base his work upon them, were recognised in many quarters.

Some experts in the nineteenth century were even prepared to argue that invention had already become 'a social process' in which the contribution of no one individual could be crucial. Engineers such as I. K. Brunel and Sir William Armstrong made this their chief argument for the abolition of the patent system; all inventions, they claimed, were merely improvements or adaptations of existing knowledge. And, in developing their case, they employed arguments that modern writers have employed about the twentieth century: that most inventions were made simultaneously by several people; that inventions are called for by the existence of a need and that, since so much knowledge was already available, the need would always be met and probably by more than one person.

It is, therefore, of more than ordinary interest to try, as a first step, to put into proper perspective invention and the inventors of the nineteenth century; the next chapter is devoted to this subject.

CHAPTER III

INVENTORS AND INVENTION IN THE NINETEENTH CENTURY

For we are like the Chinese in reverse.
Our feeling for the future's so prodigious
We might be termed descendant-worshippers.
MARTYN SKINNER, *The Return of Arthur*

THERE has been much writing about the many nineteenth-century inventors, from which it seems possible to obtain an accurate general picture of the manner in which they lived. It is much less easy to be certain about how they worked and thought. The inventor's mind must always be a matter of some mystery, although it ought to be possible to discover how far he based his work on scientific knowledge, how methodical was his approach to a subject, and whether he was disposed to seek the help of scientists. In fact, many contradictory answers have been given to these questions, especially to the second, and usually the more extensive the writings about a particular inventor the more numerous the contradictions. In such cases all that can be done is to draw attention to the conflicting opinions, although, wherever an inventor's own statements are available, they have been given because they sometimes resolve the conflicts and should, perhaps, be treated with more respect than those of writers whose information is second-hand.

I

THREE INVENTORS OF PRIME MOVERS

A preliminary glance at this subject may be taken by comparing three British inventors who have been responsible for great changes in the uses of power: Watt, Parsons and Whittle.

Watt, the son of a Greenock carpenter and merchant, was educated at the local grammar school, where he showed a talent for mathematics. After working for a time in his father's workshops, he decided to become a mathematical instrument maker. He served his apprenticeship in London, then returned to Glasgow, where he met Dr. Black, the famous chemist, and John Robison. These connections obtained for him the post of mathematical instrument maker to the university. Watt's friendship with Black grew and with it his interest in, and knowledge of, the scientific developments of the age. It was by learning from Black's scientific methods that Watt was able to understand why the Newcomen type of steam-engine was so inefficient, and to invent a means of making it more efficient through the use of a separate condenser and a steam-jacketed cylinder.

The development of the engine was slow. Watt had conceived his idea in 1765; ten years later he made the first satisfactory engine. He does not seem to have had the ability to develop an invention rapidly. Cautious and uncertain of success, he was unable to see in advance which was the most hopeful course. He tried all possible alternatives until he found the right one, but always remained dissatisfied with any solution which he thought could be improved. H. W. Dickinson has written that Watt's attitude of mind was more that of the scientist than of the craftsman.[1] Boulton was the ideal partner for Watt; he pressed him to produce results and made him design the engines of higher power, able to give rotary movement, then being demanded by manufacturers.

In Birmingham as well as in Glasgow, Watt spent much time in the company of scientists: as a member of the Lunar Society, he knew Joseph Priestley, William Small, Erasmus Darwin and Thomas Wedgwood. He helped Priestley in his experiments leading to the discovery of the constitution of water, and Wedgwood with 'sun pictures', the ancestors of photographs. Thus Watt spent much time in the company of the most distinguished scientists of his day, and from their teaching he was led to make perhaps the most important invention of the industrial revolution.

Sir Charles Parsons was the son of the third Earl of Rosse, himself an astronomer and President of the Royal Society. Few inventors can have been more suitably educated. As a child Parsons was taught by tutors at home; his father helped with his scientific education, while he was able to gain practical knowledge of engineering in the workshops where his father's telescopes were made. Later he went to Trinity College, Dublin, and Cambridge, where he read the Mathematics tripos, for that in Engineering had not yet been founded. From Cambridge he went to work in various engineering firms, but he had already invented his 'epicycloidal' engine, in which the cylinders revolved around the crankshaft. It was several years later, in 1884, when he was about thirty, that he made his first steam turbine. He said later[2] that he had started work on the steam turbine because calculations from known data, based on the analogy between the flow of steam under small differences of pressure and the flow of water in a hydraulic turbine, showed its practicability.[3]

Some writers have stressed the apparently intuitive nature of Parsons's mental processes. Dr. Stoney, who worked with him, said[4] that he had an extraordinary intuition in all matters connected with design, and seemed able to solve problems instinctively, without the use of formal mathematical

[1] H. W. Dickinson, *James Watt, Craftsman and Engineer*, p. 40.
[2] *The Steam Turbine*, Rede Lecture, Cambridge, 1911, given by Sir Charles Parsons.
[3] H. W. Dickinson, *A Short History of the Steam Engine*, p. 194.
[4] *Ibid.*, pp. 19–78.

reasoning. One of his greatest gifts was that of obtaining accurate results from experiments conducted with apparently crude apparatus: for example, in designing the *Turbinia*'s hull he experimented with models drawn across a pond by a fishing-rod and obtained results which were later confirmed by tank testing. Dickinson has suggested that there is a close parallel between Parsons and Watt: but Parsons seems to have lacked Watt's cautious approach, to have had more confidence in his ideas and a better realisation of the range of their possible applications. One thing, however, he did have in common with Watt – interest in science and a hobby of scientific research. Where Watt had studied the composition of water and 'sun pictures', Parsons attempted to make artificial diamonds, but without success.

Sir Frank Whittle had none of the advantages of upbringing possessed by Parsons; his background more closely resembles that of Watt.[1] Whittle's father owned a small engineering business, and in the workshop Whittle gained his first practical experience of mechanics. A passion for aircraft led him to make models and then to join the R.A.F., where he obtained the normal technical education of a cadet. At this time Whittle became interested in jet propulsion as the best solution to the problem of high-speed flight at high altitudes. Later, in 1929, he realised that the most efficient jet engine would be one in which the jet was generated by a gas turbine.

Up to the time of his invention Whittle had had less contact with scientists than either Parsons or Watt, although it should be borne in mind that the increased number of scientific and technical publications now available make personal contacts less essential as a means of obtaining knowledge. Whittle's invention, like Watt's, was not based on a recent scientific discovery. After his invention, Whittle took a further R.A.F. course in engineering and then went to Cambridge to read the Mechanical Sciences tripos, winning first-class honours. Before his invention he had received a less thorough education than Parsons although a more specialised training in mechanical engineering than Watt. In the eighteenth century a formal education in that subject could only be obtained by an apprenticeship to an engineer, of whom few existed. Whittle, though he had never specialised in the study of their design, was also less ignorant of engines than Watt. His methods of work contrast with those of Parsons in that he relied more on theoretical calculation than most engineers, and less on intuition.

The histories of these inventors, therefore, do not give the impression that a fundamental change in the methods and characteristics of inventors has taken place in the present century, though the technical environment in which they live has changed, and with this the facilities for keeping in touch with scientific developments.

[1] Whittle had a more systematic technical education than F. B. Halford, one of the most successful aircraft engine designers, who had no formal technical education at all, apart from what he received at school.

II

THE HIGH-PRESSURE STEAM-ENGINE

After Watt, the most important improvement in the steam-engine was the introduction of the high-pressure engine by Richard Trevithick in England and Oliver Evans in the United States. Their inventions cannot be said to have been based on scientific discoveries. Yet both men were in contact with the scientists of their day, while Evans was a man of some scientific attainments.

Trevithick's father was an engineer in charge of the pumping engines at Cornish tin mines, where Watt's and Newcomen's engines had first been used. Trevithick was educated at the village school, but at the age of twenty he was considered enough of an expert to be put in charge of the engines at a mine. We do not know how the idea of using high-pressure steam came to him: it had indeed first been suggested by Leupold in 1720, and Watt had included a reference to it in his steam-engine patent, but no practical use had been made of the idea. Watt was strongly opposed to its use on the ground that the boilers would explode; he once said that Trevithick deserved hanging for introducing the high-pressure engine. Whether or not Trevithick knew of these previous suggestions is unknown.

The best source of information on Trevithick's life is his letters to Davies Gilbert, later President of the Royal Society, and Gilbert's recollections of him. Gilbert, a patron of young men of scientific leanings, was also a Cornishman; when the idea of using high-pressure steam had occurred to Trevithick he enquired of Gilbert what

> would be the loss of power in working an Engine by the force of Steam raised to the Pressure of several Atmospheres, but instead of condensing, to let the steam escape. I, of course, answered at once that the loss of power would be one Atmosphere. . . . I never saw a Man more delighted.[1]

Gilbert became the confidant and friend of Trevithick; when the latter went to London in 1802 to enrol his patent on steam-engines and steam-carriages, Gilbert gave him an introduction to Davy, who was greatly interested in the invention. Through Davy, Trevithick met Count Rumford, the founder of the Royal Institution. Rumford also was interested in Trevithick's ideas and suggested a design of fireplace for his steam coach. The evidence all suggests that Trevithick, despite his lack of education and his possession of a mind which bubbled over with ideas, some of which Gilbert wrote 'were so wild as not to rest on any foundation at all', was an inventor who enjoyed the friendship and respect of some of the leading scientists of his time.

Trevithick's high-pressure steam-engine was built some years before that of Oliver Evans, but Evans's conception dates back some twenty years earlier. In 1800 he had explained his principle to Robert Patterson, who encouraged

[1] Letter from D. Gilbert to J. S. Enys, 1839. Quoted in H. W. Dickinson and A. Titley, *Richard Trevithick: The Engineer and the Man*, pp. 43–4.

him to go ahead with the construction of an experimental engine. Before that year there was no ready market for, and indeed little general interest in, such engines in the United States, but in 1800 Evans wrote that he was now determined to build a high-pressure engine in case he should die before he had done so. Evans was a farmer's son and little is known of his education; the knowledge and high level of literacy he possessed suggest that it must have been better than was normal at the time. Despite the lack of contact with scientific thought in Europe, he was able to gain a surprisingly good knowledge of science. According to his most recent and thorough biographers,[1] his knowledge of thermodynamics was well in advance of the time; his practical experience with steam-engines may have helped to give him a better realisation of the truth than scientists then possessed.

He seems to have first realised the power of high-pressure steam during the 1770's when he heard of some youths filling an old musket with water and placing it in a fire, where it went off as if filled with gunpowder – an incident suggesting that knowledge of the attributes of steam was fairly general. He had never seen a steam-engine, but when he came across a book on the subject he was astonished to find that steam was only used at two or three pounds' pressure, and condensed to form a vacuum under the piston so that it was driven down by atmospheric pressure. He then decided that he would design an engine to use high-pressure steam, which would be lighter than the atmospheric engine and could be used to propel carriages.

At first Evans could get no financial support for his schemes for steam-carriages, but his invention of flour-milling machinery established him in business. When he started to manufacture steam-engines about 1803 his prosperity increased, despite his continued difficulties with infringers of his patents. In 1815 he even published a small book of advice, *Oliver Evans to his counsel*; he employed fifteen counsel at one time. For most of his life he was in contact with other men of scientific leanings. In the 1780's he had corresponded with John Fitch, the unlucky steamboat pioneer, who also lived in Philadelphia. Later he met and corresponded with men such as Robert Livingston and Fulton, and engaged in a controversy with Colonel Stevens over the originality of their respective inventions. Evans's writings in this controversy show that he had a fairly sound knowledge of the history of the steam-engine and of the work of Savery, Black and Dalton. The Bathes write:[2]

> Early biographers of Oliver Evans seem to have regarded him as just a poor inventor who was everlastingly engaged in battle and litigation . . . and who died in poverty, broken-hearted. . . . However, in spite of many obstacles, . . . his financial strength grew and at the time of his death he was a man of property with staunch friends and a few bitter enemies which must always be expected by a successful man.

[1] G. and D. Bathe, *Oliver Evans: A Chronicle of Early American Engineering*, p. xvi.
[2] G. and D. Bathe, *Oliver Evans*, pp. xv, xvi.

III

THE STEAMSHIP AND THE STEAM-LOCOMOTIVE

The application of the steam-engine to the ship and the railway employed the energies of many inventors, with a consequent multiplicity of conflicting claims to priority. One interesting fact about the inventors is the extent of the contacts between them.

There was a web of connections between the pioneers of the steamboat: Rumsey, Fitch, Miller, Fulton, Stevens, Symington, Bell and Livingston. James Rumsey had demonstrated a steamboat, driven by a jet, on the Potomac in 1785 and John Fitch ran a paddle-boat on the Delaware in 1787. Fulton and Stevens knew of these experiments. Patrick Miller built the first steamboat in Britain in 1788 using engines designed by Symington. This proved impractical, but Symington built Britain's first successful steamboat in 1801. Stevens experimented with steamboats at Hoboken from 1802 onwards. Fulton was also experimenting with steamboats in France in 1802, where he met Livingston, Stevens's brother-in-law. In 1804 Fulton visited Scotland and met Symington and Henry Bell, who later built the *Comet*. When Fulton returned to America he was in correspondence with Bell about the details of Miller's boat.

The development of practical steamboats was thus the work of men who profited from the ideas and the mistakes of others. None of them worked in isolation: some of the pioneers received encouragement from their scientifically minded contemporaries. For example, a Rumsean Society, of which Franklin was a member, was founded in Philadelphia in 1787 to help Rumsey.

The development of the railway locomotive provides us with a similar picture. The first locomotive to be tried on rails was that of Trevithick, which was run on the Pen-y-darran iron-works railway in 1804; but it was nearly ten years before locomotives were used to draw coal from the Tyneside collieries to the river. George Stephenson was the outstanding figure there, but Blackett and Hedley of the Wylam colliery had built and used a locomotive before him.

All three had known Trevithick. In 1805–6 Blackett, who had heard of the Pen-y-darran experiment, ordered a locomotive from Trevithick which was built in Newcastle but not used, its weight probably being too great for the rails. Trevithick himself seems to have visited Newcastle at this time and met George Stephenson. Much later, in 1827, Trevithick was described to Robert Stephenson as 'your father's friend and fellow-worker'.[1]

George Stephenson, Blackett, Hedley and other locomotive builders, such as Blenkinsop and Murray, knew of one another's work; they lived within a short distance of one another. George Stephenson, according to Smiles, regularly visited Wylam to see the engine there and also went to see the locomotives of Blenkinsop and Murray. It was from this careful study of

[1] H. W. Dickinson and A. Titley, *Richard Trevithick*, pp. 204–5.

the work of others that Stephenson was able to make the most successful engines.

The later development of the railway engine, after 1825, owed less to George Stephenson than to his son Robert and to Timothy Hackworth, who reduced its weight and increased its power. Robert Stephenson had received a thorough education; his father, who had had to struggle to educate himself, was determined that his son should not have to do so.

IV

TEXTILE INVENTORS

The inventors of labour-saving machinery in the textile industry are generally thought of as untutored men of genius; and uneducated, in the formal sense, most of them were. But many of them were closely interested in science and were friends of scientists, while their education, if self-administered, was better than might be expected. Nor do they seem to have worked in isolation from one another: Samuel Smiles, in his *Industrial Biography*, wrote that mechanical inventors worked on each other's views and suggestions, and their own experiences. Hence many inventions were made more or less simultaneously.

Little is known about the earlier inventors of cotton machinery, such as Kay, Wyatt and Paul, and Hargreaves, but they mostly seem to have been men of some social standing and good education. More is known about the later inventors: Samuel Crompton – who evolved the mule from Hargreaves's jenny and Wyatt and Paul's and Arkwright's spinning by rollers – was the son of a skilled craftsman whose early death forced Crompton to start work, spinning at home, when he was fourteen. Up till then he seems to have attended the local school; from then until he was nineteen he studied mathematics at evening classes. In his spare time he was working on the mule and wrote of his inventive efforts:

> The whole of my powers both of body and mind were concentrated in one continued endeavour to accomplish the object of my pursuit; which was that every thread of cotton should be (as near as possible) equally good.[1]

He succeeded in this aim, but, in the absence of patent protection, he did not make his fortune. A shy and retiring man, the battle for his rights was uncongenial to him, and he continued to earn his living by spinning. He even refused an interesting offer from the first Sir Robert Peel to work in Peel's engineer's shop, which must have been one of the earliest attempts to get an inventor to do industrial research.

Crompton, after 1800, when a subscription raised £500 for him, ran a spinning business. In 1812 he was voted £5000 by Parliament, but he lost this in business and was living in poverty at the time of his death. Of his

[1] H. C. Cameron, *Samuel Crompton*, p. 30.

character, his most recent biographer[1] has written that he was a man of some culture and education, a good mathematician and a musician with a wide knowledge of classical music.

Dr. Edmund Cartwright is an interesting contrast. He was a Fellow of Magdalen College, Oxford, and unusual among the amateur scientists of that time in that he invented machinery. He was said to have invented the power-loom after being shown over Arkwright's mill and learning that weaving had become the bottle-neck in the production of cotton cloth. As he had no knowledge of machinery, his first efforts to produce a power-loom were failures; but with the help of experienced workmen he succeeded in devising a reasonably successful loom. He then set up a factory in Doncaster, but this venture did not prove a success; fortunately he possessed a country living, so he was not reduced to poverty. He continued to mix in scientific circles in London, where Fulton was one of his friends.

Cartwright invented a wool-combing machine and also a closed-cycle alcohol engine, which he thought might serve to power Fulton's ships, but neither was successful. He next became manager of the Duke of Bedford's experimental farm. In 1810 he was granted £10,000 by Parliament on the grounds that his loom had come into use after his patent had expired. Whereupon he bought a farm and spent the rest of his life dabbling in agricultural experiments.

V

MACHINE TOOLS AND INTERCHANGEABLE PARTS

Henry Maudslay, one of the first and the greatest of the many British machine-tool inventors of the nineteenth century, started work in 1783, at the age of twelve. At fifteen he worked in a smithy; at eighteen, having established a reputation for ingenuity, Bramah sent for him to help to devise machinery for lock-making. In 1797 he left Bramah and set up a shop of his own; he had already, in 1794, invented the slide-rest lathe and he proceeded to specialise in the production of machine tools of increased accuracy, which for the first time made such parts as screws and bolts interchangeable. He was not merely a mechanical genius wholly ignorant of science; astronomy and the manufacture of telescopes for his own use was his chief hobby. Faraday was his close friend and a frequent visitor to his works. Like many inventors of this period, scientific discoveries were of as much interest to Maudslay as his practical achievements were to the scientist.

Eli Whitney, as well as being the inventor of the cotton gin, was Maudslay's contemporary in the introduction of interchangeability of parts and the beginnings of the modern technique of mass production. He was educated at Yale and lived amongst the intellectual society of New England. He had no contact with men of similar inventive ideas simply because there were

[1] *Ibid.* p. 116.

no such men in America. A recent biography[1] remarks that he had to work alone, whereas his British contemporaries could rely on their predecessors' work, or turn to one another for help.

Just as Whitney made muskets on mass-production lines, one of his contemporaries, M. I. Brunel, designed block-making machinery for the navy during the Napoleonic war, although he was able to turn to Maudslay for help in making the machinery to mass-produce the blocks. Brunel was a French naval officer who had fled his country after the Revolution. He had been well educated, having a particular flair for mathematics. In England Brunel became a well-known figure in scientific circles and a Vice-President of the Royal Society.

VI

THE TELEGRAPH AND TELEPHONE

The electric telegraph was one invention of the early nineteenth century which was based on scientific discoveries and which was largely the work of scientists or of men working in collaboration with scientists. The discoveries at the beginning of the nineteenth century of the characteristics of electricity, and of the Voltaic battery, a means of producing and storing it, enabled electric currents to be employed for such purposes as the telegraph. Scientists, Ritchie in England, Schilling in Russia, Gauss and Weber in Germany and Joseph Henry in America, first devised telegraphs as a matter of scientific interest and entertainment. These, however, were crude systems; much had to be done before the telegraph could be used commercially. This work was also carried out by scientists or scientifically minded men: Wheatstone and Cooke in England, Morse in America and Siemens in Germany.

Charles Wheatstone was a professor at King's College, London, for most of his life, though his chief claim to fame is his work on the telegraph and the dynamo. He was, in fact, more concerned with applied than pure science. Despite his scientific standing he had not had any scientific education, in which he resembled most other scientists of the time, such as Davy and Faraday. Being interested in science, Wheatstone began to write articles on acoustics and other subjects for scientific periodicals, gaining enough of a reputation to be made Professor of Experimental Physics at King's College in 1834. He started experiments there on the rate of transmission of electricity along wires; he then met W. F. Cooke, who had spent some time studying in Germany and had there constructed a telegraph based on that of Gauss and Weber. Cooke and Wheatstone between them developed a practical system which was adopted for use in Britain.

Joseph Henry had made a telegraph in America in 1831. In the same year Morse visited Europe and, stimulated by what he heard there of the telegraph, had conceived his system on the return voyage. Morse was a painter, and his knowledge of electricity was limited to hearing some lectures at

[1] J. Mirsky and A. Nevins, *The World of Eli Whitney*, chap. XIII.

Yale. His first attempt to execute his ideas failed. He was then Professor of Painting at New York University and, on the advice of Professor L. D. Gale of the University, he studied the published work of Joseph Henry and later consulted him in person. His eventual success, while it owed much to his own persistence and the work of his colleague Vail, would not have been possible without the help of the scientists.

The telegraph was an invention which revealed to physicists an application of their work and also provided new phenomena for them to study. For this reason the improvements to the telegraph and its extension to undersea uses were largely the work of scientists. Lord Kelvin was consultant to the company which laid the Atlantic cable; his work, and the invention of the siphon recorder, were essential to its success. The invention of the telephone, however, was not the work of academic scientists, though both Bell and Gray were men of wide scientific knowledge and Gray was the chief engineer of the American Western Electric Company.

VII

THE VULCANISATION OF RUBBER AND EARLY CHEMICAL INVENTIONS

Although a number of chemical discoveries of practical value were made by scientists during the nineteenth century, one of the most important, that of the vulcanisation of rubber, was made by Goodyear, who was not a scientist. The depiction of Goodyear as the comic figure of nineteenth-century invention is, however, misleading and unjust. He realised the difficulties of what he was trying to do and the element of chance that must exist in such a purely empirical search. He made no pretensions to chemical knowledge, but found that chemists could give no assistance to him, all their attempts to rid rubber of its stickiness having failed.

> The inventor was, however, encouraged in his efforts by the reflection that that which is hidden and unknown, and cannot be discovered by scientific research, will most likely be discovered by accident, if at all, and by the man who applies himself most perseveringly to the subject, and is most observing of everything relating thereto.[1]

The United States Commissioner of Patents, in his report for 1858, wrote that Goodyear had made himself so much a master of the subject of rubber that nothing could miss his attention; rubber and sulphur might have been heated together before, and a thousand times afterwards, without the world benefiting. Goodyear was not entirely out of touch with scientists. Professor Gale carried out for him experiments with his earlier method of curing rubber by using nitric acid; Silliman of Yale certified to the effectiveness of Goodyear's vulcanisation process, and Professors Booth and Boye of Philadelphia reported on it in 1844.

[1] C. Goodyear, *Gum Elastic*, 1855, p. 101.

The discovery, in 1856, of mauve, the first aniline dye, was made by W. H. Perkin. At the time he was a student, eighteen years of age, at the Royal College of Chemistry. He was experimenting with compounds, discovered by Faraday and Hofmann, along lines suggested by Hofmann, and while trying to synthesise quinine he discovered the dye. There was, therefore, perhaps an even greater element of chance in this discovery than in that of Goodyear, despite Perkin's more scientific background.

The inventions of the Leblanc and Solvay processes for producing soda were different in character. N. Leblanc, surgeon to the Duke of Orleans, discovered in 1789, apparently after reading of a process incorrectly stated to do so, that sodium sulphate, limestone and charcoal heated together produce soda. Other soda processes tried in Britain and France at this time were based on already known reactions and proved unprofitable. Once the raw materials used in the Leblanc process became available commercially, and its development problems were solved, it began, after 1810, to supplant the production of soda from plant ash.[1]

The Solvay ammonia-soda process, which replaced the Leblanc in the latter part of the century, was patented in 1861 and considerably improved in 1872. It represented the industrial application, after many previous attempts, of Fresnel's discovery in 1811 that sodium bicarbonate was precipitated from salt saturated with ammonia when carbonic acid was passed into the solution. Professor Graham, the Glasgow University chemist, seems to have encouraged the use of this reaction industrially during the 1830's. A patent on the process was taken out in 1838 by Dyar and Hemming, and many unsuccessful attempts to use the process were made before Solvay succeeded.[2] Solvay was not a scientist, but he was the son of a quarry owner and salt refiner, so that he had early practical experience of chemical processes. This, combined with his inventive ability, enabled him to succeed where others had failed.

VIII

STEEL

New processes for the production of cheap steel revolutionised the use of metals in the second half of the century. Sir Henry Bessemer, Sir William Siemens and Sidney Gilchrist-Thomas were the inventors mainly responsible. Bessemer, the first to succeed in producing cheap steel, was an inventor by profession. Whether it was known at the time that the carbon in iron could be burnt out by exposing it to a blast of air, or whether Bessemer made this discovery, is a matter which is not entirely clear.

Bessemer's own description is that he noticed, during experiments with melting pig-iron in a reverberatory furnace, that a lump of pig-iron which

[1] Charles C. Gillespie, 'The Discovery of the Leblanc Process', *Isis*, June 1957.

[2] See D. W. F. Hardie, *A History of the Chemical Industry in Widnes*, pp. 43–8, for a description of the British attempts.

had remained incompletely melted had been transformed into wrought iron when he directed a blast of air onto it. Another explanation has been given by James Nasmyth, who maintained that the discovery was inspired by his own patent for the use of steam in puddling iron. The crucial discovery once made, Bessemer invented the converter to make it practically usable. When he ran up against difficulties associated with the presence of phosphorous, he took advice from Dr. Henry, a well-known English chemist.

Bessemer was more an inventor than a scientist, averse, like many other inventors, to the influence of accepted ideas. His father owned a type-foundry in Hertfordshire and sent his son to the village school. Bessemer, however, preferred working with his hands and he soon went to work in the family factory. His invention of a method of producing bronze powder by machinery, which he kept secret for forty years, gave him financial security and he spent the twelve years before his invention of the Bessemer process working on various inventions of his own and acting as a consulting engineer.

After he retired from the steel business with a fortune of nearly a million pounds, he devoted his time to his favourite hobby, astronomy, and experiments with a solar furnace. Though more an engineer than a scientist, and without a formal scientific education, he was not ignorant of science nor of the help scientists might be able to give him. He wrote in his *Autobiography* that, when he started his work on steel, his knowledge of metallurgy was only what an engineer observes in the foundry or smith's shop, but that this was an advantage because it meant he had nothing to unlearn.

Siemens, in contrast, was more of a scientist than an inventor. Sir Frederick Bramwell claimed[1] that his scientific attainments were so great that he was more of a physicist than an engineer. Siemens said of himself that it was his disposition to seek for new applications of first principles to accomplish his objects. In the regenerative furnace Siemens's object had been to secure higher temperatures and greater efficiency in the use of fuel by employing the heat of the outgoing fumes to heat the incoming air. Like the other members of his family, Siemens had settled into a career as an engineer and inventor early in life, starting by assisting his elder brother, Werner, by whom he was sent to England to sell inventions.

Gilchrist-Thomas was also a scientist by inclination, but he was self-taught. A London police court clerk, his interest in science led him to attend evening classes. Hearing a lecturer mention the problem of phosphoric ores in the Bessemer converter, he determined to try to invent a means of overcoming it. After carefully studying steel-smelting and all that had been done before to try to solve the problem, he saw that the solution lay in lining the converter with a basic material which would absorb the phosphorus. His cousin, Percy Gilchrist, was a chemist at a Welsh ironworks and carried out the necessary experiments for him. At first Gilchrist-Thomas was ignored when he announced his invention; but its success led to its recognition and

[1] Presidential Address to the British Association, 1888.

widespread use, especially in Germany, for the invention made it possible to utilise the vast beds of phosphoric ores there.

IX

THE HEROIC AGE OF INVENTION IN THE UNITED STATES

The last third of the nineteenth century saw inventions proliferate in the United States; it is indeed often regarded as the great age of 'heroic' invention. Yet it was essentially a period in which scientific knowledge found practical application and, together with the same period in Europe, it constitutes an excellent example of the bridging of the gap between scientific ideas and ultimate products which Whitehead stressed as one of the greatest inventions of the nineteenth century. Much knowledge of physical, and especially electrical, phenomena had been accumulated during the first two-thirds of the century by the work of such scientists as Davy, Faraday, Henry and Helmholtz, and this was freely available to inventors in the scientists' published works and the scientific textbooks. Faraday's *Electrical Researches*, for instance, was the bible of Edison's formative years.

The invention of an economic means of generating electricity in the form of a practical dynamo provided the incentive to invent electrical equipment. Itself the result of a long evolution from Faraday's original discovery of electromagnetic induction in 1831, to which scientists and scientifically minded inventors such as Pixii, Wheatstone, Werner von Siemens and Pacinnotti had contributed, the dynamo began to be produced commercially in 1870 by the Belgian carpenter, Zenobe Gramme. With it, the age of electricity could begin.

Edison, of course, was the greatest figure amongst these American inventors. Much has been written about him and much of it is conflicting: he has been described as possessing a wonderful power of going straight to the practical end, never equalled since Leonardo;[1] as inventing by the method of trying anything, which method has consequently become known as the 'Edisonian';[2] as not being the sort of experimenter to look for a needle in a haystack;[3] as always making a careful study of available information to get a clear idea of the existing state of the art and then trying the ideas which occurred to him.[4]

Much of this confusion is perhaps due to the fact that Edison invented in both the mechanical and the chemical fields and methods tend to be less empirical in the former than in the latter.[5] Late in his life Edison said:

> When I am after a chemical result that I have in mind, I may make hundreds or thousands of experiments out of which there may be one

[1] H. S. Hatfield, *The Inventor and His World*, p. 35.

[2] *Encyclopaedia Britannica*, 14th edition, article on Invention.

[3] C. Matschoss, *Great Engineers*, Trans. H. S. Hatfield, p. 331.

[4] A. E. Kennelly, 'T. A. Edison', *American National Academy Biographical Memoirs*, vol. xv, p. 298.

[5] See pp. 63–4, 128–9.

> that promises results in the right direction. . . . There is no doubt about this being empirical; but when it comes to problems of a mechanical nature, I want to tell you that all I've ever tackled and solved have been done by hard, logical thinking.[1]

It is certain he co-operated with scientists: before he built his Menlo Park laboratory he used the Princeton University laboratories where the Henry Professor of Physics, Dr. C. F. Brackett, assisted him in his early work. Afterwards, he usually had well-educated young scientists as his assistants, while his library of scientific works was extensive.

Alexander Graham Bell came of a family of distinguished elocutionists. His grandfather had been Professor of Elocution at London University, while his father invented 'visible speech', a means of teaching speech to the deaf. Bell thus had a hereditary interest in the nature of speech which was strengthened by his own scientific interests. He spent a year at Edinburgh University and two at University College, London, studying anatomy and physiology. Helmholtz's analysis of sound waves and demonstration of how vowel sounds could be built up by a group of tuning-forks, described in *Sensations of Tone*[2] in 1863, naturally interested Bell and provided the knowledge of the nature of sound waves which was essential for the invention of the telephone. Helmholtz's ideas were the basis of Bell's attempt to transmit several messages over the same wire. Bell's experiments with this 'harmonic telegraph' were not a success, but in the course of them he discovered a method of transmitting speech over a wire by causing the voice to create electrical waves identical to the sound waves produced by the voice.

In 1870 Bell emigrated to Canada and in 1873 was appointed Professor of Vocal Physiology at Boston University. He had contacts with scientists; when working on the harmonic telegraph he met scientists of the Massachusetts Institute of Technology, who told him that Joseph Henry had anticipated some of his work and Bell consulted Henry in Washington. The success of the telephone brought a fortune to Bell, who then settled down to a life devoted to various aspects of scientific research and the encouragement of science in America. He founded the Volta laboratories in 1880 and the magazine *Science* in 1883.

Emile Berliner, one of the generation of inventors whose early efforts were encouraged by Bell's success, was a less well educated man. He had come to America as a youth from Hanover; he worked for some time as laboratory assistant to Dr. Fahlberg, the discoverer of saccharine, and spent his evenings studying at the Cooper Institute in New York or reading an old textbook of physics. In 1876 he produced his first invention, a loose-contact microphone; this led to his joining the Bell company for a time to

[1] H. Ford, *My Friend Mr. Edison*, p. 44.

[2] This work of Helmholtz was a development of that done earlier by Wheatstone and Willis.

develop a microphone transmitter which would be suitable for long-distance telephony. In 1888 Berliner vividly described the atmosphere of this period.

> There appeared about that period something in the drift of scientific discussion which ... foretold the coming of important events. . . . Prominent daily papers commenced to publish weekly discussions on scientific topics; series of scientific books ... were eagerly bought ... popular lectures on scientific subjects were sure of commanding enthusiastic audiences . . . scientific periodicals were expectantly scanned for new information; and the minds of both professionals and amateurs were on the *qui vive*.[1]

Berliner clearly regarded himself and his fellow-inventors as scientists; it was recognition by professional, academic scientists that he most desired. He was delighted when, on a visit to Germany in 1889, Helmholtz visited him and congratulated him on his invention of the gramophone.

Two other American inventors of this generation in the electrical field were C. F. Brush and Professor Elihu Thomson; the success of both can be explained by their sound knowledge of science. Brush attributed the superiority of his dynamo over other designs to his understanding of the principles of magneto-electricity. He had a sound technical education before he became interested in electricity, having taken a degree in engineering at University of Michigan. Thomson had been educated at Philadelphia High School but his scientific knowledge was largely self-acquired. The Franklin Institute in Philadelphia, a meeting-place for those with scientific interests, was of great help to him. After he had spent some active fifteen years in invention, he began to withdraw from practical work and to devote himself to scientific research to which he, like many other inventors, was drawn by inclination.

C. M. Hall, who invented, almost simultaneously with P. Héroult in France, the electrolytic process for the production of aluminium, was another inventor who, although he did most of his experiments in a garden shed, had had a good education in science. He studied chemistry at Oberlin College, Ohio, under Professor F. F. Jewett.

Sir Hiram Maxim had many of the characteristics traditionally attached to inventors of the more erratic sort, and so might be thought to have been unmethodical in inventing and ignorant of science. This is presumably the reason why conflicting descriptions of his methods have appeared. Lord Moulton wrote:[2] 'He rarely, if ever, took the trouble to ascertain the existing state of knowledge in any subject which he took in hand.' Yet P. F. Mottelay wrote:[3]

[1] F. W. Wile, *Emile Berliner*, 1926, pp. 45–6.

[2] In his Introduction to P. F. Mottelay, *The Life and Work of Sir Hiram Maxim*, 1920, p. xvi.

[3] In the Foreword of P. F. Mottelay, *Life and Work of Sir Hiram Maxim*, p. xii.

He always said he left no stone unturned in order to become expert at anything he had to do, and often spoke of the 'glorious' period, when all his leisure time was given to studying such books as he could procure on the subject that happened to engage him at that moment.

Maxim claimed in his autobiography:[1] 'I have always gone in very strongly for what is known as book learning: so whenever I took up any new thing I read everything I could find on the subject.' The *Dictionary of American Biography*, in its article on Maxim, states that his success as an inventor was the consequence of his combining mechanical skill with a knowledge of physics and chemistry.

X

ELECTRIC LAMPS AND ARTIFICIAL SILK

A group of amateur and professional scientists and inventors were responsible for two of the chief British inventions of the last quarter of the nineteenth century. They were Sir Joseph Wilson Swan, C. H. Stearn, F. Topham and C. F. Cross. Swan invented a carbon filament incandescent lamp and a squirting process to make nitrocellulose fibres for lamp filaments; Stearn and Topham helped him to produce the lamp, Stearn being an expert on the production of high vacua and Topham the glass blower who made the glass globes. Stearn and Topham later invented the means of producing a textile fibre from cellulose, which was made by the viscose process discovered by C. F. Cross.

Swan was an example of the self-taught amateur scientist, a type which was common in England during the nineteenth century. He was early apprenticed to a druggist, but in his spare time he studied science and became an expert on photography. He had begun thinking of a carbon filament incandescent lamp in 1845 when he heard of J. W. Starr's U.S. patent on a lamp. For the next thirty-four years the idea of making such a lamp never entirely left Swan, and he watched the inventions relating to electricity and electric lighting in this period for signs of progress. After experimenting with carbonised filaments, he made a lamp in 1860 in which a carbonised strip was rendered incandescent; but he could not obtain a good enough vacuum, so that the filament burnt away. It was not until he saw Crooke's radiometer in 1875, and heard how a near-perfect vacuum could be obtained by the use of a Sprengel mercury pump, which had been invented in 1865, that he revived his experiments.

Stearn was a Liverpool bank clerk and, like Swan, an amateur scientist. He had made a special study of high vacua; consequently he was able to give Swan the assistance he needed in producing glass lamps with a near-perfect vacuum. He conducted the experiments for Swan, Topham doing the glass-blowing; by the end of 1878 they produced a lamp with a filament

[1] Sir Hiram Maxim, *My Life*, p. 180.

which did not waste away. Swan wanted to get better filaments: in 1880 he made one of cotton treated with sulphuric acid and in 1883, seeking a means of producing a more even thread, he obtained a filament by squirting a nitro-cellulose solution through small holes. Though he realised that it had possibilities as a textile fibre and he exhibited samples at the 1885 Inventions Exhibition, Swan did not concern himself with its development for this use.

A synthetic textile fibre was not developed in England until nearly twenty years later; it was made from cellulose produced by the viscose process, which C. F. Cross had invented in 1892. Cross was a consulting chemist, who had been educated at King's College, London, Zürich University and Owens College, Manchester. Consulting chemists at that time were expected to do some original research to establish their reputations. He studied the chemistry of cellulose and, in the course of becoming the accepted expert on this subject, he invented the viscose process. Its value was that it enabled pure cellulose to be obtained from wood pulp, the cheapest raw material.

At first Cross and his partner Bevan had little idea what to do with viscose; they do not seem to have considered using it to produce a textile fibre. At the time Stearn was running an electric-lamp factory; he was a friend of Cross and he made some experimental lamp filaments from viscose instead of from the then widely used cellulose reduced by the chloride of zinc process. Samples of these viscose filaments were shown to Sir James Swinburne when he visited Stearn; Swinburne commented that they should make a good textile fibre. This started off Stearn's work which ultimately led to the introduction of viscose rayon. One critical problem was that of obtaining a thread fine enough for textile use from the fibres formed when the solution was squirted into the precipitating bath. Stearn told Topham, then also running an electric-lamp factory, of his difficulties and Topham invented the 'spinning box', the key to the successful production of viscose rayon.

The viscose rayon industry, therefore, arose directly out of the work of a consulting chemist, a former glass blower and a former bank clerk.

XI

THE INTERNAL COMBUSTION ENGINE

The internal combustion engine was another important European invention of the later nineteenth century. The gas and petrol engine was a French and German invention; the heavy-oil engine a British and German invention. The idea of a gas-engine had been widely discussed during the nineteenth century but it was not until 1859 that Lenoir, a practical mechanic, produced such an engine similar to the steam-engine in design. The essential principles of a successful design were soon established: the pre-compression of the mixture to the greatest possible extent, and the four-stroke cycle. The idea that compressing the mixture would increase efficiency seems to have been widespread at this period; who first combined this idea with that of the four-stroke cycle is a matter of controversy. Some writers maintain that Alphonse Beau de Rochas, who published a full description of the working and advantages of compressing the mixture and the four-stroke cycle in 1862, was the inventor; others that it was Otto, who is said to have made a model engine incorporating these features at the end of 1861.

Beau de Rochas was a railway engineer, who hoped that the gas-engine would replace the steam-engine on locomotives for use in hilly country. Otto worked for a Cologne merchant; Lenoir's work aroused his interest in the gas-engine. Otto had left school at sixteen, but while there he had learnt some science and now he began a study of all available information on the gas-engine culminating, it is said, in his first model engine.

Whatever the priorities, Otto was responsible for making the first four-stroke gas-engine to be sold commercially, in 1876. But it was not until it had been improved by Daimler, Maybach and Benz that the motor-car became a practical possibility.

Daimler was a baker's son, who had spent two years at the Stuttgart Polytechnic after being apprenticed to a gun-maker. After spending several years abroad, he met Maybach who possessed great mechanical skill and wished to become a machine-builder. From then on Daimler and Maybach worked together, spending ten years at Otto's factory, from 1872 to 1882. They subsequently set out to make the petrol-engine lighter and more powerful so that it could be used to provide power for vehicles. In this Daimler succeeded; but his concept of the use of the engine as a source of power that could be temporarily attached to vehicles had to give way to Maybach's desire to design cars as a whole. The culmination of Maybach's work was his design of 'the first modern car', the Mercedes of 1900.

Benz had also benefited from the German technical education system, having spent four years studying engineering at the Karlsruhe Polytechnic. Here his attention was first turned to the gas-engine because his professor held that the steam-engine was too big and heavy. Though he thought of the possibility of a car before Daimler and concentrated more exclusively on creating one, Benz's designs were cruder than those of Daimler and Maybach and he was averse to changing them. His designs, therefore, rapidly became out-dated and his business as a manufacturer suffered.

The heavy-oil engine, now best known as the diesel engine, was developed more or less simultaneously with the petrol-engine. The first engine to operate without a vaporiser or ignition was that invented in 1890 by Herbert Ackroyd-Stuart, who owned an iron foundry at Bletchley, where he conducted his experiments with the advice of Professor William Robinson of University College, Nottingham. His lack of capital prevented him from carrying on his experiments, and he sold in 1891 a licence to manufacture his engines to the company now known as Ruston and Hornsby.

Rudolph Diesel was responsible for taking the final step of developing an engine in which ignition took place solely by the heat generated by compression. He had had a good technical education at the Augsburg School of Industry and the Munich Technical High School. The suggestion in a lecture that a higher thermal efficiency than that of the steam-engine was possible was, he said, the beginning of his life's work: 'From that time, the desire to realise the Carnot cycle ruled my life'.[1] His problem was to construct an engine which would withstand the pressures involved in compressing air to thirty or thirty-five atmospheres; after four years of experiments, he produced an engine in 1897.

Inventors of the latter part of the nineteenth century in Germany mostly enjoyed the benefit of the excellent technical education system. It is doubtful, however, whether they were any better fitted to invent than their British and American counterparts who had acquired their scientific knowledge by self-education. But it is interesting that Germans in the nineteenth century were as sceptical as are many modern writers of the merits of nineteenth-century inventors who claimed scientific attainments without having had an academic scientific education.[2]

XII

AERONAUTICS

The history of the aeroplane began with Sir George Cayley. By 1799 he had conceived and, it would appear, tested, a model of a flying machine that gained lift from a fixed wing, with longitudinal stability and steerage provided by a cruciform tail. Propulsion he suggested, could be provided by paddles. For more than fifty years after this invention, Cayley intermittently pursued experiments with model and full-scale gliders, which culminated in the first flight by a manned glider in 1853. At the same time he was working on the principles of aerodynamics. His paper 'On Aerial Navigation', published in *Nicholson's Journal* in 1809–10, laid the foundations for future studies of aeronautics.

His concern with the need for a prime mover lighter than the steam-engine, to make powered flight possible, led Cayley to invent alternatives. In 1807 he described the hot-air engine – better known as the Stirling engine, though Cayley has priority as its inventor – and an internal combustion engine using gunpowder as a fuel. During his life he frequently returned to the design of these and alternative engines, though without much practical progress. Outside aviation, his inventions included a form of caterpillar tractor, artificial limbs, and instruments.

This remarkable man, a Yorkshire baronet and landowner and, from 1832 to 1835, the representative of Scarborough in the House of Commons, was an important member of the scientific establishment of the time. He was a founder-member of the British Association and a founder of the Regent Street Polytechnic. Among his friends were I. K. Brunel, Bryan Donkin,

[1] C. Matschoss, *Great Engineers*, Trans. H. S. Hatfield, p. 300.

[2] F. W. Wile, *Emile Berliner*, 1926, pp. 203–4. Berliner had to overcome this attitude when he visited Germany in 1889.

Sir William Fairbairn, Sir Goldsworthy Gurney, George Rennie and P. M. Roget, secretary of the Royal Society.

W. S. Henson, the next significant figure in Britain, added some sound engineering improvements to Cayley's concepts in his 'aerial steam carriage' patented in 1842. He suggested screw propellers for propulsion, which Cayley inexplicably neglected, and he improved methods of construction. But a model that Henson built with John Stringfellow during 1844–7 failed to fly.

The first model aeroplane to fly, a monoplane with tractor propeller and cruciform tail, powered by clockwork, was built by a French naval officer, Félix du Temple, in 1857–8. He later built a full-sized aeroplane to the same basic design with a steam engine, which became the first to hop into the air, after a downhill run, in 1874. He seems to have carried out his aeronautical experiments in private, if not in secret. Alphonse Pénaud had more influence on aviation in France for he was the first to demonstrate the flight of a model aeroplane in public in 1871 and his introduction of the elastic band as the power unit of models made experimenting cheap and easy. In 1876 he patented an aeroplane which combined all his ideas: an amphibious monoplane, with a retractable undercarriage, variable-pitch propeller and a joystick for control.

The most important work in Britain at this time was the studies of aerofoils made by F. H. Wenham and Horatio Phillips, using the first wind tunnels. Wenham was a marine engineer by profession, but with scientific interests which extended from aerodynamics to microscopy and photography. When the Aeronautical Society was founded in 1866 he gave the first lecture, and through a meeting with Octave Chanute in 1875 he aroused in this famous American engineer an interest in flying. Chanute, in turn, seems to have passed on information about the British work to the Wright brothers, which influenced their choice of the biplane. Phillips was also a member of the Aeronautical Society; the son of a gunmaker, he spent several thousand pounds on building an elaborate wind tunnel in the 1880's for tests of aerofoils, and on a full-sized machine which, when tested on rails, lifted at 40 m.p.h.

The possibility of powered flight was brought close in the 1890's by the work of Chanute, Lilienthal and Pilcher. Chanute's book, *Progress in Flying Machines*, published in 1894, assembled the information he had acquired in his wide and constant travel. This, together with Lilienthal's book, *Bird Flight as the Basis of Aviation*, published in 1889, became the would-be aviators' bible.

Lilienthal's regular gliding flights aroused interest all over the world. Although his death in one of his gliders and the apparent crudity of its design give the impression that he was an intrepid birdman rather than a scientific experimenter, his research was more methodical than might appear. He had graduated in engineering from the Technical Academy in Berlin; and, after working as a designer with a manufacturer of mining machinery, achieved independence and earned funds for his aeronautical experiments by inventing a tubular boiler and establishing his own business to exploit it.

Percy Pilcher carried on this work with gliders and was also killed while flying. He was the son of wealthy middle-class parents; after studying naval architecture at London University, he became assistant lecturer in Glasgow University in 1892. The head of his department, Professor (later Sir John) Biles encouraged his aeronautical experiments, although Lord Kelvin, while giving practical help, thought Pilcher would break his neck. Pilcher later worked with Sir Hiram Maxim, who had made abortive experiments with a steam-powered aircraft, and in 1898 he set up as a consulting engineer in partnership with W. G. Wilson (who later invented the Wilson epicyclic gearbox). Their business was successful and broad in scope; it included the designing of cars. Before his death in 1901, Pilcher was negotiating to establish a company to finance his further aeronautical experiments.

'Garret inventor' is thus the least appropriate description to apply to these early pioneers of the aeroplane. They were men of social standing and often of wealth; and, although most of them worked on their own, they were far from ignorant of the work of other inventors. If these inventors were isolated or oppressed, it was because their ideas seemed so wild and improbable to their contemporaries.

XIII

SCIENTIFIC EDUCATION IN THE NINETEENTH CENTURY

It may well be that the disposition of many modern writers to regard nineteenth-century inventors as uneducated and empirical in their methods is a direct outcome of the difficulty which academically educated people often have in understanding the possibilities of self-education. But as there were so many examples of great scientists who were self-taught, such as Faraday, Davy, Sturgeon and Wheatstone, there seems no reason to believe that inventors could not also teach themselves science. The facilities to do so certainly existed; self-education was a most highly regarded occupation in the Victorian era. Mechanics' Institutes, or similar organisations, were common, while books enjoyed a wider sale than might be imagined today. Even at the beginning of the century, standard science text-books existed.[1]

Scientific periodicals were also available. The Royal Society began to issue its *Philosophical Transactions* in 1665, providing a source of information on the work of the leading English scientists. The *Philosophical Magazine* was founded in 1798 and provided a wide field of information. The Society

[1] Professor Eaton Hodgkinson gained his early knowledge of science from the works of Emerson, a village schoolmaster, and Simpson, a weaver, which were said to reflect eighteenth-century scientific literature, and to have been written for the mass of mankind. See *Memoirs of Manchester Literary and Philosophical Society*, Third Series, vol. 2. R. Rawson, *A Memoir of Professor Eaton Hodgkinson*, pp. 151–2.

of Arts published *Transactions*, as did the Royal Institution after 1851, where the lectures attracted wide attention. There were other less famous journals devoted to scientific topics, such as Walker's *Annals of Philosophy* in which Wheatstone published his earliest work, *Nicholson's Journal* and the *Mechanics' Magazine*. The second half of the century saw the foundation of technical magazines; *The Engineer*, one of the earliest, was established in 1856.

In America, the first scientific periodical seems to have been the *Transactions of the American Philosophical Society*, publication of which began in 1771. Other societies followed this example by publishing their *Transactions*; but the foundation of the *American Journal of Science* by Benjamin Silliman of Yale in 1818 provided the first national scientific journal. Silliman was more interested in applying science to practical use than in pure science, and the object of his Journal was to provide a general source of scientific information for the benefit of scientists and others. It did much to help the progress of American science, having considerable influence until the Civil War. The Smithsonian Institution, founded in 1846, was a source of scientific publications, so was the Franklin Institute, founded in 1824. Later in the century, the spread of interest in science was shown by the success of the *Popular Science Monthly*, founded in 1872, which was read by the class of amateur inventors which had appeared during the Civil War, and in the last two decades of the century many journals on engineering and electrical subjects appeared.

Not only was self-education encouraged by writers such as Smiles in the nineteenth century, but facilities grew up to make it possible. The Mechanics' Institutes in England, and their equivalents in America, were the most important of these. Professor John Anderson of Glasgow was the pioneer of the provision of scientific education for working men: he started evening classes in 1760 and left a fund to continue them. This was used to found the Andersonian Institution, where J. B. Neilson, the inventor of the hot blast, was one student. This served as a model for the Mechanics' Institutes, which, by 1850, numbered over six hundred, with over a hundred thousand members. Organisations such as the Manchester Literary and Philosophical Society were interested in improving scientific and technical education; a college of Arts and Sciences was founded by Dr. T. Barnes to give evening lectures in 1783, though it proved an over-ambitious venture. In a paper to the Society[1] Dr. Barnes said that the sole object of the plan was the improvement of manufactures by the improvement of the arts on which they were based – Chemistry and Mechanism. A similar interest was shown by the Society of Arts and the Society for the Diffusion of Useful Knowledge, both of which helped the Mechanics' Institutes.

Mechanics' Institutes were created in America as well as in England by the interest of working-men in the inventions and discoveries that were

[1] R. Angus Smith, *A Centenary of Science in Manchester*, p. 167. *Memoir of the Manchester Literary and Philosophical Society*, Third Series, vol. 9.

changing their methods of work. The Franklin Institute of Philadelphia was the best known and most important. It was founded as a meeting-place for tradesmen, businessmen and scientists to discuss the practical application of science and to encourage it by granting premiums to inventors. Elihu Thomson and Edwin Houston were two inventors among its members.

XIV

CONCLUSIONS

The information drawn together in this chapter constitutes, of course, only a fragment of the vast records of technology in the nineteenth century: but a glance has been given at some of the personalities most closely linked with the appearance of the steam-engine, the steamship, the steam-locomotive, textile machinery, machine tools, the electric telegraph and telephone, dyes and artificial fibres, cheap steel and the aeroplane; at some of the outstanding names in the great burst of invention in the United States late in the century; and at three of the most important fields of invention at the end of the century. The conclusions may now be assembled.

Did scientists and inventors have but few contacts with one another in the nineteenth century? It is certainly true that a tradition grew up that science should be pursued for its own sake, but this does not seem to have meant that scientists felt that they should have no contact with industry or inventors. The links, indeed, were numerous and close: of Watt with Black and Priestley; Trevithick with Gilbert; Morse with Gale; Goodyear with Gale and Silliman; Bessemer with Henry; Edison with Brackett; Bell with Henry; Hall with Jewett. There were many distinguished scientists who were also inventors and interested in the application of the discoveries, among them Kelvin, Joule, Davy, Dewar, Hofmann, Bunsen, Babbage and Playfair. The scientists who joined the British Association and the American Association or took part in the proceedings of the Lunar Society or the Manchester Literary and Philosophical Society can hardly have remained isolated from the industrialists and inventors who formed the rest of the membership. And it was frequently true that those inventors who were not formally trained scientists showed a high respect for scientific knowledge and an anxiety to acquire it.

Did the inventors of the nineteenth century work in conditions of comparative isolation as contrasted with the 'team work' of later days? It appears that in general they were a more closely knit society than is popularly supposed. From Newcomen onwards, small groups are to be found working together, perhaps in very little different a fashion from that now described as teamwork. In some cases the passage of time may have left one inventor remembered and others associated with him forgotten, so that the extent to which inventors worked together has been overlooked. In other cases, where we cannot tell which member of a group was 'the' inventor, there must have been co-operative effort. Wyatt and Paul; Watt and Murdock; Watt and Wedgwood; Trevithick, Gilbert and Vivian; the Siemens family; Bessemer, Allen and Longsdon; Sidney Gilchrist-Thomas and Percy Gilchrist; Swan and Stearn; Edison at Menlo Park; Thomson and Houston; Otto and Langen; Daimler and Maybach; might today all be regarded as teams of inventors.

What truth is there in the view that, in the past, invention was 'merely empirical' and that it has since become 'scientific'? Much turns here on what is meant by 'empirical'. The word is sometimes used to suggest that nineteenth-century inventions and discoveries were accidental; something was found which was not being looked for, although close and systematic observation might have been employed. Examples are the discovery of mauve, the first aniline dye, by W. H. Perkin; of synthetic alizarin by Caro or of the vulcanisation of rubber by Goodyear. But there are examples in the twentieth century to match these, such as the discovery of polyethylene, tetraethyl lead, Freon refrigerants, Duco lacquers, Neoprene, where chance played a large part even in the work of large industrial research laboratories. Here, indeed, the important distinction is not so much that between the centuries as that between invention and discovery in different fields. Discoveries in chemistry and chemotherapeutics are still largely 'empirical' in the sense that the peculiar properties of compounds have still often to be found by testing in the hope that unpredictable virtues will thereby be revealed. It is not surprising that, up to now, there continue to be so many fortunate and 'accidental' discoveries in this field. Mechanical invention, on the other hand, is not so likely to be favoured by accident. A mechanism has to be thought of from the beginning as a system, and designed as a whole; and whilst it is true that, in the process of getting a mechanism to work, trial and error may produce results which were unexpected, yet there is a general sense in which mechanical invention is, and chemical discovery is not, theoretical.[1] This is why there have probably been many more 'accidental' and empirical inventions in the twentieth century than there were in the nineteenth.

There is, indeed, one feature of modern research which, while not altogether without parallel in the nineteenth century, may be regarded as peculiar: large-scale systematic search in a field where results are *a priori* likely. When penicillin was discovered, it was natural that a world-wide search should be organised to discover whether moulds other than penicillin possessed chemotherapeutic properties. Streptomycin was found by S. A. Waksman after the examination of ten thousand soil cultures. There is nothing in the nineteenth century which quite matches the *scale* of these

[1] It is interesting to find that writers on invention in the nineteenth century, who would be thinking mainly of mechanical invention, were accustomed to stress the 'theoretical' nature of invention. Thus Herbert Spencer, *An Autobiography*, vol. 2, p. 436: 'Evidently constructive imagination finds a sphere for activity alike in an invention and in a theory. Indeed, when we put the two together, we are at once shown the kinship; since every invention is a theory before it is reduced to a material form'.

efforts. On the other hand, the *method* is just that which Goodyear pursued in his search for a way to 'cure' rubber, Edison in his efforts to find better materials for lamp filaments and many chemists in their attempts to discover new dyes after Perkin's original work.

Sometimes, however, 'empirical' is used to connote that the invention was not based on prior scientific knowledge. There is, of course, more such knowledge now than there used to be and, in the present century, more inventions, or even a higher proportion of all inventions, may have been based upon it than in earlier periods. But to speak of this century as 'the age of science' and of the nineteenth century as the 'age of empiricism' is far too bald a statement to contain the whole truth. The electric telegraph, the electric generator and perhaps even the methods of steel-making were based on prior scientific knowledge. Conversely, in the twentieth century it is difficult to see any specially close connection between science on the one side and, on the other, the invention of the turbo-jet engine, the improvements in aero-engine fuels, the early work of Farnsworth on television, of Armstrong on radio or of Wankel on the rotary piston engine. The extent to which the technical progress of the past half-century has been directly dependent upon science can easily be exaggerated,[1] for practical advances have frequently been the result of empirical experiment and not of scientific methods.

Be that as it may, the task of the inventor may remain much the same in the one case as in the other. The high-pressure steam-engine was not based on pure science, there was no scientific knowledge available on the subject when the engine was introduced: the invention of the engine led to the enunciation of the laws of thermo-dynamics. On the other hand, Diesel's attempt to apply these laws led to the diesel engine. But in both cases the main task of the inventor was the same: how to design the equipment which would stand the stresses involved – in the former, the stresses of high steam pressures and temperatures, in the latter, the strains of high compression ratios.

Distinctions between the present and the past, therefore, must not be drawn sharply. History is not simple and cannot be made to seem so without distortion of facts. The modern view about invention in the nineteenth century seems to have arisen out of a misunderstanding. It is based on the assumption that science and the spread of formal scientific education have transformed the methods of invention, whereas they have only modified them; and that, in consequence, no previous inventors can have worked in the same manner as modern inventors, whereas they could and did.[2]

[1] See J. R. Baker, *The Scientific Life*, pp. 113–24.

[2] It would take us too far afield in this volume to search for the sources of the myth of the 'heroic age' of invention. But one suggestion may be worth while following up. Views about invention and inventors in the nineteenth century were greatly influenced by the long string of highly popular industrial biographies produced by Samuel Smiles. But Smiles seems to have been much more interested in unsuccessful, or at least unrewarded, inventors than in successful. His passionate concern was with the struggles of those

CHAPTER IV

SOME RECENT IMPORTANT INVENTIONS

The human understanding is most excited by that which strikes and enters the mind at once and suddenly, and by which the imagination is immediately filled and inflated. It then begins almost imperceptibly to conceive and suppose that everything is similar to the few objects which have taken possession of the mind, whilst it is very slow and unfit for the transition to the remote and heterogeneous instances by which axioms are tried as by fire.—FRANCIS BACON, *Novum Organum*, Book I.

I

IN the nineteenth century there was apparently a stronger link between science and invention and a much more closely knit society of man concerned with technical progress than is frequently supposed. The next obvious step is to try to ascertain what has been happening in the present century and, for this purpose, a study has been made of some seventy inventions which can reasonably be considered modern. In these case histories no attempt has been made to enter upon detailed technical matters; the purpose has been to identify the individual or individuals who appear to have made the greatest contribution to ultimate success, to determine the conditions under which the work was carried out and generally to try to isolate the factors which contributed to, or impeded, the advance. The information for each case was collected from printed records, which are plentiful but scattered and often conflicting; from inventors themselves wherever possible or those closely connected with them; or from scientists and technologists whose views and judgments should carry most authority.

It might at first sight be supposed that, by collecting facts and opinions in this way, it ought to be possible to assemble a body of knowledge sufficient

who, as lone workers, pursued their discoveries through adversities and obstacles. His recent biographer (A. Smiles, *Samuel Smiles and His Surroundings*) makes this clear. In the most successful period of his life, when he was inundated with invitations to produce biographies of successful men, he chose to write the stories of Thomas Edward, a poor cobbler and amateur naturalist of Banff, and of Robert Dick, a Thurso baker with a passion for geological studies. In this period, it is true, he wrote a biography of one George Moore, a highly successful merchant prince, but it is reported that he 'hated doing it' for the story was 'unromantic'. It is true also that he later produced the life-story of James Nasmyth, the inventor of the steam hammer, 'not one of the usual Smiles heroes, but a man of education and background', but here again it was not of his own original choice. Nor was he enthusiastic about writing the biography of Josiah Wedgwood. Smiles's predilection for the under-dog may well have given a twist to the interpretation of the history of technology in the nineteenth century.

to test any general statement about the circumstances under which modern inventions have arisen. Yet, although the evidence seems to support, often strongly, some general inferences, the element of personal judgment that has necessarily gone into them should not be overlooked.

The cases which are chosen for examination are listed below; from this list fifty-six of the most interesting histories, including the ten new ones, are printed in Part II and Part III.

Acrylic Fibres: Orlon, etc.
Air Conditioning
Air Cushion Vehicles
Automatic Transmissions
Bakelite
Ball-point Pen
Catalytic Cracking of Petroleum
'Cellophane'
'Cellophane' Tape
Chlordane, Aldrin and Dieldrin
Chromium Plating
Cinerama
Continuous Casting of Steel
Continuous Hot-Strip Rolling
Cotton Picker
Crease-resisting Fabrics
Cyclotron
DDT
Diesel-Electric Railway Traction
Domestic Gas Refrigeration
Duco Lacquers
Electric Precipitation
Electron Microscope
Electronic Digital Computers
Float Glass
Fluorescent Lighting
Freon Refrigerants
Gyro-Compass
Hardening of Liquid Fats
Helicopter
Insulin
Jet Engine
Kodachrome
Krilium
Long-playing Record
Magnetic Recording
Methyl Methacrylate Polymers: Perspex, etc.
Modern Artificial Lighting
Moulton Bicycle
Neoprene
Nylon and Perlon
Oxygen Steel-making
Penicillin
Photo-typesetting
'Polaroid' Land Camera
Polyethylene
Power Steering
Quick Freezing
Radar
Radio
Rhesus Haemolytic Disease Treatment
Rockets
Safety Razor
Self-winding Wrist-watch
Semi-synthetic Penicillins
Shell Moulding
Silicones
Stainless Steels
Streptomycin
Sulzer Loom
Synthetic Detergents
Synthetic Light Polariser
Television
'Terylene' Polyester Fibre
Tetraethyl Lead
Titanium
Transistor
Tungsten Carbide
Wankel Engine
Xerography
Zip Fastener

All these cases can be held to belong to the twentieth century; the year 1900, that is to say, has been taken as the dividing line between old and modern inventions. It is, of course, arbitrary to slice history into neat sections in this way. But there are some reasons for adopting this division. The turn of the century witnessed the rapid growth of new technologies, especially in the chemical field; and it was then that the first large-scale industrial research laboratories, so often taken as a symbol of a new era in technology, began to make their appearance. A number of inventions originating in the last twenty-five years have been included.

The choice of what might be regarded as 'important' inventions has also had to be largely arbitrary. Every case chosen has been a commercial success, shows promise of becoming so, or is obviously a discovery of great public value. Inventive virtuosity has not been accepted for its own sake. Common observation suggests that many of the items could hardly have been omitted from any list. Some of the inventions have transformed ways of living: radio, television, the jet engine, nylon, quick freezing, plastics, stainless steel, computers – to name only a few. Some take a high position among the great medical discoveries of the age: the penicillins, insulin and the treatment for rhesus-haemolytic disease, for example. There are many new techniques and tools of production, new instruments for research and new materials, less well known and with a less direct impact on the final consumer, which nevertheless have had final effects difficult to ignore: the continuous hot-strip rolling of steel, shell moulding, tungsten carbide tools, the catalytic cracking of petroleum, the cyclotron, the electron microscope, Neoprene, the cotton picker, oxygen steel-making and photo-typesetting serve to illustrate this group. There are many new consumer goods, often too lightly dismissed as 'gadgets', which have contributed much to the convenience of life judged by the ready way the consumer has taken to them, such as the safety razor, the zip fastener, the self-winding wrist-watch, the long-playing record, the ball-point pen and the Moulton bicycle.

Some of the cases range over more than one invention. It was sometimes necessary, in order to present a coherent story of advances popularly regarded as single inventions, to lump together several significant inventions. Radio, for instance, did not emerge suddenly as the result of one invention; there are several crucial as well as a multitude of lesser inventions properly falling under this head.

II

Taken together the items chosen, with certain exceptions such as atomic energy and the electronic devices employed in 'automation',[1]* seem to

[1] Those who are disposed to assume that the important ideas and devices embodied in modern electronic equipment were the product of large research groups and of team-work should recall that perhaps the most fertile worker in this field was Norbert Wiener, an American mathematician, whose writings reveal his highly individualistic outlook and

constitute a representative cross section of the technical progress of this century. The really important omission is, of course, atomic energy. It has been excluded partly because it is far too large a subject to be dealt with in anything short of a volume to itself, partly because it is still surrounded by secrecy which perhaps makes it premature for the outsider to try to write its history.[1] It is commonly assumed that the discovery of the means of creating an atomic explosion and of using atomic energy for peace-time purposes came about under conditions different in kind, as well as degree, from those associated with any other earlier inventions, even in the twentieth century; that this is final and conclusive proof of the merits of gigantic scale research, organised in systematic fashion, with scientists and technologists pacing and stimulating each other in the achievement of a plan laid out and firmly in the grip of skilful administrators; that this unique experience renders obsolete all prior experience as to the best way of organising for technical advance, even for peace-time purposes.

No judgment will be made in this volume on these sweeping claims. But when the time comes for a full assessment of them, the following points would have to be borne in mind.

The fundamental discoveries that led to atomic energy were made by academic scientists with fairly simple tools.[2] Perhaps the greatest of these scientists is reported as saying that 'we could not afford elaborate equipment, so we had to think'. The funds for the early work did not come from government sources; up to the end of 1940 in the United States virtually all research was supported by private foundations and the universities.[3] Almost up to the moment that the first bomb blew off, there were doubts as to its possibility. As for peace-time uses, outstanding scientists up to 1937 doubted whether the atom could ever provide significant supplies of energy. Atomic energy came into existence, as did Newton's discoveries,[4] not through the pursuit of clearly perceived social utilities, but through the private compulsions of a few men of genius.

The development of the bomb would seem to have no lessons to teach

---

temperament. He has pungently expressed his views concerning industrial research laboratories: 'Of course the large laboratory can make a limited case for itself. However it is perfectly possible for the mass attack by workers of all levels, from the highest to the lowest, to go beyond the point of optimum performance, and to lose many really good results it might obtain in the unreadable ruck of fifth-rate reports. . . . The great laboratory may do many important things, at its best, but at its worst it is a morass which engulfs the abilities of the leaders as much as those of the followers.' (N. Wiener, *I am a Mathematician*, p. 364.)

* [In Chapter X something is said of these electronic devices and one of the new case histories is that relating to Electronic Digital Computers.]

[1] One recent book, however, by Arthur H. Compton, *Atomic Quest*, throws a flood of light upon the conditions under which, and the methods by which, the atom bomb was produced.

[2] O. R. Frisch, *The Times Supplement*, Oct. 17th, 1956.

[3] Arthur H. Compton, *Atomic Quest*, p. 28.

[4] M. 'Espinasse, *Robert Hooke*, p. 33.

us about the best ways of exploiting new ideas for peace-time purposes. It was magnificent, but it was not economics. Time was of the essence of the matter, and economic and scientific resources were in free supply. Four separate methods for making fissionable material were pursued in parallel. Everything was tried as soon as thought of. The first bomb seems to have cost about 1500 million dollars. How much of all this represented wasted effort, which did more to delay than to expedite the final success, because of the sheer dead weight of the organisation and the dispersion of energy in the absence of the discipline of cost, we may never know. But success was achieved, the bomb blew off. If it had not done so the whole exploit would have represented one of the most massive misdirections of resources in history.

The story of the development of the bomb appears to provide some extremely interesting material about the difficulties of administering research, arising out of the conflicts between the different standpoints and methods of thought of scientists, technologists and administrators. Even when these several groups were drawn powerfully together by the feeling of national patriotism and the fear of common dangers, these tensions made themselves apparent.[1]

The peace-time development of the uses of atomic energy has been associated with two remarkable features: extraordinary optimism about the economic advantages of atomic energy[2] and the control of research by governments. This combination has produced results upon which it is still too early to pass judgment. The absence of ordinary commercial tests and the inevitable mistiness of public accountability has meant that the major government decisions have been accepted as acts of faith. But in any full-scale enquiry into this case, it would be relevant to ask whether the progress in the peace-time uses of atomic energy has not been slowed down by the failure in nearly all countries to broaden the base of participation in the work; whether excessive secrecy, restrictions upon the use of radio-active material, the granting of contracts only to larger firms and the restrictive employment of patents by public bodies have not been the blocks to advancement that they would certainly have been in other industries.

With that has gone most ambitious schemes, especially in Great Britain, for public economic planning over unprecedently long periods. Whether

[1] Arthur H. Compton, *Atomic Quest*, p. 113, is most interesting on this point. He made these remarks at one stage to General Groves: 'We (scientists) don't know how to take orders and give orders. But a scientist, if he is a responsible man, has a different kind of discipline. It is not possible for anyone to tell a scientist what he must do, for his proper course of action is determined by the facts as he finds them for himself. Then he needs a different kind of discipline. He needs to be able to make himself do what he sees should be done without having anyone to tell him to do it.'

[2] It has been doubted whether the theories regarding nuclear energy will ever have the same impact upon material change as did the structural theory of organic chemistry in the nineteenth century. (F. Greenaway, 'The History of Science and the Common Reader', *The Listener*, Nov. 10th, 1955.)

these schemes in the long run prove to be wise or not, they are in the nature of gambles. It is not yet clear that electricity can be produced more cheaply by the use of atomic energy furnaces than by the use of modern equipment fired by the more traditional fuels.[1] Large-scale investment in an industry where techniques are still in an embryonic state and are rapidly changing clearly involves the risk of miscalculations.[2] Optimistic ideas of the extent to which and the speed at which electricity generated by atomic energy can replace that produced in the older ways[3] have resulted in some countries in a drain upon the supply of scientists and technicians, thereby contributing to an existing shortage, and have tended to place at a discount the study of methods of improving the production and utilisation of other forms of fuel.[4]*

III

Turning now to the seventy case histories. Each was an intricate skein which refused to shake out into simple lines and which tended to become the more

[1] Calculations of costs must be based upon the price allowed for the plutonium by-product, and assumptions about the load factors of the atomic energy stations, which of course will depend upon the extent to which such stations replace those of the traditional type.

[2] In the British White Paper, *A Programme of Nuclear Power*, published in 1955, it was assumed that the first stations would produce 100–200 megawatts of electricity. In 1956 it was announced that they would produce 200–300 megawatts. This, fortunately, was an error in a favourable direction.

[3] In view of the optimistic general impressions on this subject, it comes as something of a shock that, even by the end of this century, it is estimated that less than one-half of the total electrical output of important industrial countries will come from the use of atomic energy.

[4] See the O.E.E.C. Report, *Europe's Growing Needs of Energy, How Can They be Met?*, 1956.

* [These comments were made ten years ago. Many of the doubts we then expressed seem to have been well founded. Looking back, it seems probable that British economic growth in the last decade would have been faster if there had been no nuclear power programme and if the scarce man-power resources devoted to it had been available elsewhere. The Atomic Energy Authority now estimates that nuclear energy will not be cheaper than electricity generated in stations burning untaxed oil until the late 1970's. These estimates imply the writing off of the £330 million which appears to have been spent by AEA on research and development up to 1966; they are also conditional upon the spending on R and D by the AEA of about another £180 million. (See Report of the Select Committee on Science and Technology, Session 1966–7, p. 473.) In recent years the AEA has been employing about 3,000 qualified scientists and engineers on research and development, about six per cent of the total so employed in the whole country. The position would seem to be that *if*, contrary to past experience, the estimates of the AEA about future costs of nuclear energy are not over-optimistic and *if* their estimates of alternative ways of generating electricity are not over-pessimistic and, accepting as a fact the very large sums already spent on the research and development of nuclear energy, then it may be the right thing to go forward with a large nuclear power programme. The whole story is but poor testimony to the value of government intervention in technical innovation.]

complicated the more thoroughly it was examined. Some of the histories, indeed, appear more sharp cut than others. Firmer contours, for instance, can perhaps be given to the story in the case of nylon, penicillin, insulin, bakelite, titanium, the transistor, the jet engine, float glass, oxygen steel-making and semi-synthetic penicillins than in the case of stainless steel, the electron microscope, chromium plating, fluorescent lighting, the acrylic fibres, or radio. But, at every point, the contributions of even the most outstanding workers are clearly bound up with the speculations, reasoning, guesses and mistakes of others.

Inventors have often started from different points and reached much the same results or, starting with a common problem, adopted widely varying means of solving it. Emerging ideas have proved intractable, been dropped and, when apparently forgotten, been subsequently picked up and used to effect. Inventions in one field have lain dormant until some ingenious inventor has seized the old idea, combined it with a notion gleaned from another apparently unrelated field and produced a new and fruitful combination. Inventions have been born before their time; either the public refused to accept them or a practical model could not be built because the rest of the technical art was backward. Identical or nearly identical ideas have emerged independently many years apart. Inventions not infrequently have been simultaneous or so near together in time that the determination of priorities has stretched the power of judgment even of the Courts with all the facts at their disposal. Useful products and processes have arisen from reasoning which subsequently proved to be fallacious. The simple route to the solution of a problem has sometimes seemed to be almost perversely overlooked by the cleverest inventors and the most tortuous path followed to a point to which there has all the time been an easy and direct line. Inventors, groping for solutions along complicated and expensive roads, have missed the target completely, while an individual entering the field with a fresh approach, crude equipment and a generous smattering of common sense has achieved success along a path which, in retrospect, looks perfectly simple. Chance and accident seem often to have played an important part in discovery, but so, too, has brilliant and disciplined thinking and a persistence amounting almost to obsession.[1]

Anyone who has handled material of this kind quickly becomes aware of certain odd characteristics, intriguing in themselves, of the literature regarding discovery and invention. There is, for instance, a strong propensity to simplify and to idealise the stories, to present them as a steady and logical march towards a final goal, to interpret them as the results of deliberate

[1] H. S. Hatfield, *The Inventor and His World*, p. 40, has vividly described the impression left with him. 'Their success [the pioneering inventors] is that of an ill-equipped and fanatical horde, storming positions before which nine-tenths of the attackers perish while the tenth that wins through is generally robbed of the fruits of victory. But it would be wrong to attribute the success of those who survive to pure chance, to the fact that numbers must triumph. The fanatical enthusiasm is the outcome of a creative drive, the success is the result of certain essential elements of character.'

planning. Whereas the reality is more often a series of stops and starts, of desperate frustrations and back-trackings, of logical steps intermixed with blind shots and of final success when it seems the most unlikely and the least hoped for. Successful inventors themselves often contribute to the romantic aura surrounding their final success; it is normally much more agreeable for them to think of their achievements as the outcome of a flawless chain of brilliant deductions than as the result of groping about among uncertainties. And subsequent writers, possessing more complete records of the lucky strokes than of the more numerous failures and searching for a tidy story rather than a muddled one, carry on the building up of legends. A second point which strikes the independent observer is the absence of agreement among the experts as to where the emphasis of the story should be laid. Inventors are disposed to belittle the technical novelties of the development carried on by firms; firms are frequently reluctant to give credit to pioneers outside their organisation. Scientists are inclined to be slightly contemptuous of the contributions made by inventors, inventors to be grudging about the guidance and leads they have derived from prior scientific advance.

It would be too much to hope that, in the cases collected and presented here, over-simplification has been avoided or the right balance always struck. Even with a body of fact universally accepted, others may be disposed to place upon it an interpretation different from that given here. For this reason fifty-six of the more interesting cases have been printed in Parts II and III (in a form which is necessarily a compression of the material collected) and references provided for each case.

From what has been said, it was not to be expected that all, or even most, of the cases would fall into neat and watertight groups. All that can be indicated, in each instance, is where the balance lies between autonomous and institutional research, between the work of research institutions of different types and different sizes, between the achievements of large teams and of individuals.

Before attempting to strike these balances, and in anticipation of matters considered in more detail in the next chapter, a word must be devoted to the definition of 'individual' inventor and invention. If, at the one extreme, an individual chooses the field of ideas in which to work, employs his own resources or acquires them from others who exercise no control over his work, stands to gain or lose directly from his inventive success or failure, works with limited resources and with colleagues subject to his guidance and leadership, then we clearly have an 'individual' inventor. These, very broadly, were the conditions under which the jet engine was invented. If, at the other extreme, an invention emerges in the laboratory of a firm where the research workers have been engaged on a 'set' problem and are salaried employees who would not normally gain directly from the invention, where the co-operation of a team has been involved and where considerable sums have been made available for the research, then equally clearly we have a case of 'institutional' invention. These were the conditions, very broadly, under which the crease-resisting process, float glass and semi-synthetic penicillins were discovered. Most cases, however, fall between these two extremes; it is precisely in determining whether a case lies nearer to the one extreme than the other that the element of judgment arises.

IV

More than one-half of the cases can be ranked as individual invention in the sense that much of the pioneering work was carried through by men who were working on their own behalf without the backing of research institutions and usually with limited resources and assistance or, where the inventors were employed in institutions, these institutions were, as in the case of univerities, of such a kind that the individuals were autonomous, free to follow their own ideas without hindrance. Into this group it seems proper to place:

Air Conditioning; Air Cushion Vehicles; Automatic Transmissions; Bakelite; Ball-point Pen; Catalytic Cracking of Petroleum; 'Cellophane'; Chromium Plating; Cinerama; Cotton Picker; Cyclotron; Domestic Gas Refrigeration; Electric Precipitation; Electron Microscope; Gyro-Compass; Hardening of Liquid Fats; Helicopter; Insulin; Jet Engine; Kodachrome; Magnetic Recording; Moulton Bicycle; Penicillin; Photo-typesetting; 'Polaroid' Land Camera; Power Steering; Quick Freezing; Radio; Rhesus Haemolytic Disease treatment; Safety Razor; Self-winding Wrist-watch; Streptomycin; Sulzer Loom; Synthetic Light Polariser; Titanium; Wankel Engine; Xerography; Zip Fastener.

Brief references may be made to some of these cases (more fully described in Parts II and III) where the designation of individual invention can be applied with a large measure of confidence. The jet engine was invented and carried through the early stages of development almost simultaneously in Great Britain and Germany by men who were either individual inventors unconnected with the aircraft industry or who worked on the airframe side of the industry and were not specialists in engine design; the aircraft engine manufacturers came in only after much pioneering had been carried on. The gyro-compass was invented by a young man who was neither a scientist nor a sailor but had some scientific background and was interested in art and exploration. The process of transforming liquid fats by hardening them for use in soap, margarine and other foods was discovered by a chemist, working in the oil industry, who pursued his researches, and his efforts to get the process adopted, single handed. The devices which made practicable the hydraulic power steering of motor vehicles were primarily the work of two men, one of whom worked strictly on his own while the other was the head of a small engineering company. The foundations of the radio industry were laid by scientists; but the majority of the basic inventions came from individual inventors who had no connection with established firms in the communications industry or who worked for, or had themselves created, new small firms. In the case of magnetic recording, the early crucial invention came from an independent worker, as did a number of the major inventive improvements; the interest of the companies arose much later. The first successful system for the catalytic cracking of petroleum, which opened up the way for many later advances, was the product of a well-to-do engineer who was able to sell his ideas for development to the oil companies. The history of the evolution of the cotton picker reveals two main lines of progress; in each case individual inventors working with limited resources were able to take their ideas to the point where large firms were prepared to buy or license their patents for subsequent development. Bakelite, the first of the thermo-setting plastics, was produced by a brilliant solo investigator. The first, and still the most important, commercially practicable method of producing ductile titanium was conceived of by a metallurgist working in his own laboratory. In the application of automatic transmissions to motor vehicles the credit for mechanical novelty has to be shared between individual inventors and companies, but the former should probably rank above the latter; actually the ideas of a shipbuilding engineer lie behind much of the modern progress but both in Britain and the United States inventors working single-handed have contributed a great deal to the present-day mechanisms. Up to 1938 only one large aircraft manufacturer had taken much interest in the helicopter and even that only as the result of the personal interest of the head of the firm; the progress was made by the enthusiasm of individual inventors, usually with limited resources, obtaining backing in unlikely quarters in a manner which would parallel the many stories of 'heroic' invention in the nineteenth century. The groundwork for the successful Kodachrome process was laid by two young collaborators, both musicians, whose ideas were taken up by a large photographic firm; the safety razor came from two individuals who struggled through financial and technical doldrums to great success; the zip fastener came from the minds of two engineers and was only taken up for large-scale production many years later; the self-winding wrist-watch was invented by a British watch repairer. The key invention in the history of air cushion vehicles came from an electronic engineer turned boat-builder. The successful rotary piston engine was the work of a man in his own research institute. Photo-typesetting machines have in large measure been the creation of independent inventors.

By virtue of the conditions under which the work was carried out, the individual invention group contains some important inventions and discoveries made in university laboratories. The cyclotron, invented by an American scientist, is a classic instance of the combination of scientific knowledge and mechanical ingenuity pushed forward, with the use of simple equipment at the early stages, to a brilliantly successful conclusion. With penicillin and streptomycin, not merely were the first crucial observations made in a university laboratory but some part of the early development occurred there. With insulin, a general practitioner came into a university to test out his ideas. Electric precipitation was made possible by two university scientists, one of whom, in Britain, showed that the method could be used and the other, an American, who found how it might be applied on an industrial scale. Although there is much dispute as to the real inventive source of methods of chromium plating, the ideas clearly arose in universities. The discovery of a method of preventing rhesus haemolytic disease came from university workers.

The inventions, some of them of outstanding importance, which seem to have had their origin largely in the research laboratories of manufacturing companies are:

Acrylic Fibres; 'Cellophane' Tape; Chlordane, Aldrin and Dieldrin; Continuous Hot-Strip Rolling; Crease-resisting Fabrics; DDT; Diesel-Electric Locomotive; Duco Lacquers; Float Glass; Flourescent Lighting; Freon Refrigerants; Methyl Methacrylate Polymers; Modern Artificial Lighting; Neoprene; Nylon; Oxygen Steel-making; Polythylene; Semi-synthetic Penicillins; Silicones; Synthetic Detergents; Television; Terylene; Tetraethyl Lead; Transistor.

This list can be subdivided in a number of ways. Some of the inventions have arisen in the research organisations of very large firms; others have been produced by much smaller firms. A few have been the outcome of 'directed' research where the target has been set for the research workers by the firm; others have had an element of accident about them in the sense that important and unpredictable discoveries have been made in the course of basic research work where no specific problem had been set. In some instances the firm must take the whole of the credit, in others individual inventors have independently been in competition with the firms and have made important contributions. In some cases, although the discovery was made in a firm, it was not a firm in the industry where such a discovery might normally have been expected.

Outstanding successes by very large concerns may first be mentioned. Nylon was discovered by a small research group, headed by an outstanding chemist, in the laboratories of du Pont; the firm made itself responsible for the development of this first truly synthetic fibre. Slightly later another very large firm, I.G. Farbenindustrie, produced and developed a similar fibre, Perlon. Several firms in Germany and the U.S. have devised methods of producing successful acrylic fibres; they were all very large firms. Freon refrigerants and tetraethyl lead were both produced in General Motors by small groups under Midgley and Kettering; the cases are interesting in that a motor engineering firm made these two important contributions in the chemical field and in that their discovery involved a strong element of chance. In the story of television, one outstanding figure was an employee of the Radio Corporation of America, but a number of the crucial inventions were made by a second American inventor who worked independently; and the first complete system for television broadcasting was created for the British Broadcasting Corporation by a British firm of modest size. The transistor was produced in the Bell Telephone Laboratories, a case which comes nearer than most to research directed towards a predetermined result. Polyethylene was discovered, in the course of some very broad scientific studies and as the immediate outcome of a fortunate accident, in the laboratories of Imperial Chemical Industries and developed by them; but methods of producing polyethylene at low pressures were later discovered at about the same time in one of the Max Planck Institutes in Germany and by American companies. In the discovery of the methyl methacrylate polymers known variously as Perspex, Lucite and Plexiglas, two large firms were primarily involved, I.C.I. and Röhm & Haas; but an independent research student appears to have made an important contribution. The diesel-electric locomotive probably embodied less inventive effort than many of those mentioned above; it represented the development by European and American firms, and especially by General Motors in the United States, of nineteenth-century inventions. The recent remarkable growth in the use of silicones represents the discovery of practical applications for compounds, produced by a British university scientist, the usefulness of which was first realised by scientists in an American company. The discovery of Neoprene is a romantic story in which a priest, occupying a chair in chemistry in an American university, was responsible for observations which were taken up by a large chemical firm and carried much further by them to a successful conclusion. Float Glass and Semi-Synthetic Penicillins were the results of costly decisions, courageously pursued, by large firms.

The list contains several important inventions emerging from firms much smaller than those mentioned above. Terylene was discovered by a small research group in the laboratory of a firm which had no direct interest in the production of new fibres. The continuous hot strip rolling of steel sheets was conceived of by an inventor who might well be considered an individual inventor and perfected in one of the smaller American steel companies. The crease-resisting process emerged from a medium-sized firm in the Lancashire cotton industry. 'Cellophane' tape was the product of what was virtually a one-man effort in a then small American firm. The virtues of DDT were found by a Swiss chemical firm which, for that industry, was of modest dimensions. Small firms played a vital role in the discovery of Chlordane, Aldrin and Dieldrin; Computers and Oxygen Steel-making.

Some cases seem to defy classification. Thus the long-playing record, where an institutional engineer, working on a side-line almost as an individual inventor, discovered a system which was taken up and developed by a sister corporation. Stainless steels had been known before an individual inventor and one working in the research laboratory of a company almost simultaneously recognised their virtues and prescribed more exactly the limits of their constitution. The origin of tungsten carbide tools seems to lie with a research worker in the electric-lamp industry whose discovery had very much wider repercussions than he could have expected. Success in the continuous casting of steel came only after continuous casting of non-ferrous metals had become firmly established; the main interest of this case lies in the fact that it was a small number of individuals, working mainly outside the steel companies, who were largely responsible for the successful pioneering. Of the remaining cases in the list, radar emerged from the work of government research stations, radio companies and scientists in the universities. Early interest in the long-range rocket, both on the practical and theoretical side, came from individual inventors, and enthusiasm was maintained by amateurs; but it was not until the German military authorities took up the subject during the Second World War that reliable rockets were built. Shell moulding was invented by the proprietor of a German foundry after a long personal search for improved methods of casting metals, and has subsequently been taken up by many large and small companies throughout the world.

Even where inventions have arisen in the research laboratories of firms, the team responsible for it seems often to have been quite small. It is usually found that one outstanding figure has been surrounded by, and has stimulated, a few devoted colleagues in an intimate relationship with his manner of thought and speculations.

Individual inventors seem to be less commonly found in some industries than in others. In the chemical field, in particular, there appear to be few counterparts to outstanding individual performers such as Whittle, Farnsworth, Armstrong or Kroll in other industries. This fact is intriguing for more than one reason. Individual chemical invention was not unknown in the nineteenth century – Goodyear, Perkin, Mercer and Cross are famous in that connection. It is still to be found in industries which may be regarded as peripheral to the chemical industry such as photography, metallurgy, textile finishing and chemotherapeutics. And at least some of the great chemical inventions of recent years, such as those of Carothers, Whinfield and Midgley, although made in industrial research laboratories, were produced by small groups operating with relatively inexpensive equipment. The reasons for the comparative absence of individual inventors in certain industries, and the consequences of it, will be speculated upon in later pages.

Finally, it is noticeable in the cases studied that no one country has a monopoly of invention. The outstanding names and groups are widely spread over the industrial countries. There is certainly no evidence here to support the view that invention occurs only in those countries where research is most highly organised in large institutions or corporations.

V

To sum up: A significant proportion of twentieth-century inventions have not come from institutions where research will tend to be guided towards defined ends. There are many similarities between the present and the past century in the type of men who invent and the conditions under which they do so. Many of the twentieth-century stories could be transplanted to the nineteenth without appearing incongruous to the time or the circumstances; far too many, indeed, to render tenable the idea of a sharp and complete break between the periods. The novelty of this century appears to lie in the relative decline of invention of the individual type in certain industries, industries in which the large industrial research laboratory is most commonly to be found.

CHAPTER V

THE INDIVIDUAL INVENTOR

Great floods have flown
From simple sources; and great seas have dried
When miracles have by the greatest been denied.
Oft expectation fails, and most oft there
Where most it promises; and oft it hits,
Where hope is coldest, and despair most fits.

All's Well that Ends Well.

I

THERE is a sharp and intriguing conflict of opinion between those who hold that the day of the individual inventor is done[1] and those who consider that not merely is he very much with us but that, from present appearances, he will continue to play an active part in technical progress. The views of the first group have been set forth in Chapter II.[2] Against them can be set the notions of some eminent scientists, technologists, individual inventors and students of invention. A few of many possible quotations follows.

Sir Edward Appleton:[3]

> Many people nowadays are inclined to think that, because of the rapid development of technology, the day of the inventor is over. I do not agree. . . . The world of the inventor is being invaded by the technologist; and this is a process which we may expect to continue. But happily, I think, there is still scope for the people who will not allow their objectives

[1] Sometimes the suggestion is that the individual inventor has already passed from the scene, sometimes that he is in process of so doing. His final disappearance has certainly been anticipated. Edison was often described as the last of the great heroes of invention, but when E. H. Armstrong, the highly inventive radio engineer, died in 1954, he was described as 'one of the last great individualists in the art of invention'.

[2] To these may perhaps be added the statement, in their Report for 1954, of the British National Research and Development Corporation: 'Our five years' experience requires us to report that . . . outside the field of light engineering and instrument manufacture, the isolated individual rarely appears to have any serious contributions to make to the advancement of technology. In the various fields of physics, chemistry, biology, medicine and the non-mechanical branches of the engineering science no meritorious proposals have reached us from such persons.' [In its Report for 1966 the N.R.D.C. stated that it had sifted and assessed more than 6,000 inventions from private individuals. Only about one per cent of these had been accepted for further assessment or development. It is, however, worth remembering that one of the outstanding cases of assistance in subsequent development made by the Corporation was that of the Hovercraft, which was certainly an independent invention.]

[3] Second Reith Lecture, *The Listener*, Nov. 22nd, 1956.

and ambitions to be influenced by the theoretical limitations of the day – people who are sceptical when they are told that something cannot be done.

Dr. Frank B. Jewett:[1]

I think it is inevitable that the great bulk of what you might call the run-of-the-mine patents in an industry like ours will inevitably come from your own people. . . . I think it is equally the case that those few fundamental patents, the things which really mark big changes in the art, are more likely to come from outside than from the inside. . . . There are certain sectors where the independent inventor cannot operate. . . . There are certain sectors where . . . the chances are ten to one that the fundamental ideas are going to come from outside big laboratories simply because of the nature of the things.

Dr. Vannevar Bush:[2]

In addition, there is the independent inventor, whose day is not past by any means, and who has a much wider scope of ideas and who often does produce out of thin air a striking new device or combination which is useful and which might be lost were it not for his keenness. . . . New ideas are coming forward with as great frequency today as they ever have, and while a great research laboratory is a very important factor in this country in advancing science and producing new industrial combinations, it cannot by any means fulfil the entire need. The independent, the small group, the individual who grasps a situation, by reason of his detachment is oftentimes an exceedingly important factor in bringing to a head things that might otherwise not appear for a long time.

Charles F. Kettering:[3]

He deplored the common impression that the public has got from much of the information given out about atomic bombs. The belief that science has passed beyond the need for the individual inventor is harmful. However important organised research may be, it still must have the aid of individual persons with the insight to recognise problems and needs, and perseverance in working their solutions.

Joseph Rossman:[4]

We may safely conclude that the majority of the most important inventions and achievements in industry are still being made by the individual inventor, particularly by the industrial inventor.

[1] Temporary National Economic Committee, Parts I–IV, pp. 971, 976.
[2] Temporary National Economic Committee, Parts I–IV, pp. 871, 872.
[3] *New York Times*, March 12th, 1950.
[4] *The Psychology of the Inventor*, p. 33. See also H. S. Hatfield, *The Inventor and His World*, p. 139, where he argues that there is still room for the individual even in such fields as chemistry.

Philo T. Farnsworth:[1]

> We must not lose track of the fact that inventions as such, important inventions, are made by individuals and almost invariably by individuals with very limited means.

W. J. Kroll (the discoverer of the method of making ductile titanium):[2]

> My example should refute all those who claim that today individual research ... is dead, and that it had been replaced by the collectivist planned work of teams ... who operate ... in private or government laboratories in hopes of discovering something under the command of a director who, although usually removed from the observation of the experiments, is supposed to do the thinking. ... The free individual who accepts risks ... can still today defy and outsmart the captive salaried research performed at extravagant cost in palatial laboratories where nothing else is wanted but ideas.

II

What is meant by the 'individual' inventor[3] and by the statement that he is disappearing or has disappeared? In one extreme sense every inventor is an individual inventor and every invention is an individual invention: since all human minds function independently, a new idea must arise in one brain. Teams do not think; corporate institutions do not have ideas. At the other extreme, no inventor conforms, or ever has conformed, to the caricature that has been drawn of him; a man who technically knows little or nothing, has virtually no contact with scientists, technologists or other inventors, is almost wholly without financial resources and has no links with the outside world of manufacture and commerce.

The adjective 'individual' must obviously apply to the conditions under which the inventor does his work: whether he is self-employed or works as an employee under contract for some other individual or institution; whether he is free to do what he wishes or is under agreement to think and work within prescribed lines; whether he works in a large team or a small; whether, within the team, he is one of many subject to the control of others or is the head of a group following his instructions and providing his ancillary services. At the one extreme are the Krolls, the Whittles, the Armstrongs of the world – inventors certainly not without scientific training and contacts with like-minded specialists but essentially prime movers who would find it difficult to work except under conditions of the utmost freedom. At the

[1] Temporary National Economic Committee, Parts I–IV, p. 988.

[2] 'How Commercial Titanium and Zirconium were Born', *Journal of the Franklin Institute*, Sept. 1955.

[3] There are other terms commonly and loosely employed in the same connection: private inventors, lone inventors, isolated inventors, inventors without industrial connections, amateur inventors, heroes of invention, etc.

other are the Midgleys, the Carothers, the Langmuirs, who apparently find it possible to work as employees in institutions, subject to some limit on the range of their activities.

As a first working approximation, therefore, the distinction between individual inventors and others can be taken as that between inventors who work on their own and those who are employed in an institution of some kind set up for the purpose of invention. Two qualifications must, however, immediately be added. One is that, despite many exceptions, inventors are very much a type whether they work inside or outside an institution. No one who has studied their lives or who has had contact with them can seriously doubt this statement.[1] The inventor is absorbed with his own ideas and disposed to magnify their importance and potentialities. He tends to be impatient with those who do not share in his consuming imagination and leaping optimism. He runs to grievances and feels them sharply. More often than not he would make a bad businessman. And so through a wide range of qualities of steadily narrowing generality. But, although of course there are not a few exceptions, his crucial characteristic is that he is isolated; because he is engrossed with ideas that he believes to be new and therefore mark him out from other men, and because he must expect resistance. The world is against him, for it is normally against change and he is against the world, for he is challenging the error or the inadequacy of existing ideas.[2] It is precisely because of these eccentric qualities that society has always found it so difficult to fit the inventor into its scheme of things. Gifted so highly in many rare ways, he is often oddly devoid of worldly knowledge and thereby more than ordinarily in need of assistance. Yet that assistance must be given largely on his own terms and in a manner which does not frustrate or destroy his originating powers. He is capable of self-deception yet he can be right when most others are wrong. If he is left free to his own devices as an individual inventor, he runs the risk of being neglected. If he tries to work in an organised institution, he may waste his energy battering against the ordered arrangements surrounding him.

Further, it would seriously misrepresent the realities to picture the whole of inventive activity in terms of the extremes – at the one end the independent inventor working on his own or with a small team of assistants and at the other the inventor in a large and organised research laboratory working under the guidance of others. There are very many, perhaps an increasing number, of different possible relations between the inventor and the outside world. There is the type often described as the industrial consultant. He may at some times and for some purposes be under contract to a large industrial

[1] Studies of the type will be found especially in J. Rossman, *The Psychology of the Inventor*, and H. S. Hatfield, *The Inventor and His World*, chap. 2.

[2] Anyone who doubts the feeling of isolation that inevitably surrounds the innovator should try to put himself in the position of Leonardo da Vinci when, more than one hundred years before Galileo, he wrote in his note-book in large letters 'The sun does not move'.

firm but simultaneously be following ideas for which there is no immediate sponsor. University teachers may act as consultants to firms, they may supervise rather narrowly specified research within their own laboratories with assistants and facilities paid for by firms who would expect the exclusive rights to any discoveries; or they may conduct research in more broadly defined fields supported from outside but with no restrictions upon the general publication of results. An inventor may be employed in quite another capacity in a firm and his inventive bent may be found within or outside the main stream of interest of his employer. An inventor may be a person who has been temporarily seconded by his firm to conduct research with greater freedom in an outside research institution or a university; he may be a university teacher working for a time in an industrial research laboratory. Within a large industrial research laboratory, the degree to which a research worker is controlled may vary greatly. He may be completely free for a part of his time to follow his own bent; the range of his choice of work may vary enormously. A man may be an inveterate inventor or he may invent once in a lifetime. He may provide his own capital or he may be backed by sponsors who partly support him as an inventor.

The possible gradations, of which only a few have been mentioned, between these limits are therefore determined by the combination of three conditions: who chooses the field of ideas in which the inventor works, who provides the resources for his work, and who stands to gain directly from his inventions. On the counting of these matters here Edison, Farnsworth, Armstrong, Whittle, Fleming, for instance, would be considered as individual inventors; Carothers, Langmuir, Shockley, Brattain and Bardeen would not.

It is now possible to analyse more exactly the meaning of the phrase 'the day of the individual inventor has gone'. It may mean either:

(i) that, in fact, the inventive ability of the community is being increasingly absorbed into organised research institutions, so that inventors are less free to pursue their own ideas, less dependent upon their own resources, less directly the recipients of the gains to be made from their inventions, and/or

(ii) the inventive ability of the community can only be put to the best use if it is increasingly set to work in organised research institutions.

The first statement is supposed fact. The second is a judgment of how best one kind of ability in the community should be employed. It is open to anyone to believe that both statements are true: that what is happening is indeed desirable. But the first is not, of itself, sufficient proof of the second. Yet the argument is frequently met with, that because the individual inventor is disappearing this constitutes adequate evidence that his abilities can best be employed in other ways, or even that society would do well to prevent him from pursuing his independent courses.

III

IS THE INDIVIDUAL INVENTOR DISAPPEARING?

The Qualitative Evidence

The evidence of the continued survival of the individual inventor is simply that many men who have lived in this century – numbers of whom are still, or were until recently, alive – fall into this category and by their genius have added enormously to the stock of useful ideas and to standards of living. Here is a list of some twentieth-century individual inventors with an indication of the field in which they made their major contributions:

H. Anschütz-Kaempfe (Gyro-Compass); E. Armstrong (Radio); L. H. Baekeland (Bakelite); F. G. Banting (Insulin); W. B. Barnes (Automobile Overdrive); L. Biro (Ball-point Pen); J. E. Brandenberger (Cellophane); S. G. Brown (Gyro-Compass); A. Campbell (Cotton Picker); C. Carlson (Xerography); W. H. Carrier (Air Conditioning); W. Chalmers (Plastic Glass); J. de la Cierva (Autogiro); Sir C. Cockerell (Hovercraft); F. G. Cottrell (Electric Precipitation); S. W. Cramer (Air Conditioning); J. Croning (Shell Moulding); F. W. Davis (Power Steering); Carleton Ellis (Chemicals); P. Farnsworth (Television); H. Ferguson (Tractors); R. Fessenden (Radio); C. G. Fink (Chromium Plating); A. Fleming (Penicillin); L. de Forest (Radio); K. Gillette (Safety Razor); L. Godowsky (Kodachrome); J. H. Hammond Jr. (Electronics); J. Harwood (Self-winding Watch); H. F. Hobbs (Automatic Transmissions); E. J. Houdry (Catalytic Cracking of Petroleum); S. Junghans (Continuous Casting); W. J. Kroll (Titanium); E. H. Land (Synthetic Light Polariser); F. W. Lanchester (Aeronautics); E. O. Lawrence (Cyclotron); L. Mannes (Kodachrome); G. Marconi (Radio); A. E. Moulton (Moulton Bicycle); C. Munters (Gas Refrigeration); W. Nickerson (Safety Razor); J. A. Nieuwland (Neoprene); W. Normann (Hardening of Liquid Fats); E. J. de Normanville (Automobile Transmissions); H. von Ohain (Jet Engine); R. P. Pescara (Helicopter and Free Piston Gas Generator); V. Poulsen (Magnetic Recording); R. Rossmann (Sulzer Loom); J. and M. Rust (Cotton Picker); P. M. Salerni (Automatic Transmissions); G. J. Sargent (Chromium Plating); F. Seech (Ball-point Pen Ink); H. Sinclair (Fluid Flywheel); A. Steckel (Steel Rolling Process); G. Sundback (Zip Fastener); E. A. Thompson (Automobile Transmissions); S. A. Waksman (Streptomycin); F. Wankel (Wankel Engine); Sir F. Whittle (Jet Engine); K. Ziegler (Polyethylene).

Space forbids further mention of more than a few of these.

Edwin H. Armstrong (1890–1954), a student and later a Professor of Electrical Engineering at Columbia University, worked independently during the whole of his career and was able to sell his patents on the feedback circuit, the superheterodyne and the super-regenerative circuit for very large sums to the bigger radio corporations. He was also responsible for the introduction of frequency modulation. His ideas were often belittled or neglected, but he overcame much opposition by his undisputed technical genius and his keen sense for business. 'He was never so happy as when flying in the face of some accepted theory or confounding an unimaginative engineer.'[1]

S. G. Brown (1873–1948), an English inventor, educated in science at the University of London, took out 235 important patents on equipment for telegraphy, telephony, radio and the gyro-compass. Although he owned a company he had little interest in business affairs, preferring to be free to work on his inventions and their development. What he himself regarded as one of his outstanding successes, an all British gyro-compass, brought him no profit and much litigation, but the patent on it was in 1930 extended for ten years, which was said to have been unprecedented since the time of Watt. Brown once said, 'If there were any control over me or my work every idea would stop'; he never accepted financial aid for experimental work, or for the cost of producing an original device.[2]

Carleton Ellis (1876–1940), one of the few successful twentieth-century individual inventors in chemicals, took out over 800 patents. He was educated at the Massachusetts Institute of Technology and thereafter quickly began to invent, but was much harassed by infringers of his patents. He later set up research laboratories of his own from which flowed a stream of inventions, particularly in the field of plastics, paints and lacquers, which appear to have brought him substantial returns.

Philo T. Farnsworth (1906–) is usually regarded, along with Zworykin, as one of the two outstanding early pioneers in television. His education was a broken one, but he pursued some scientific studies at Brigham Young University. He continued to develop his early conception of an electronic system of television as an independent worker with some outside financial backing. He has always emphasised his preference for research on a small scale and with simple equipment. Business interested him little and, although the commercial organisation based on his work and his patents has since become large, he has mainly devoted himself to technical research and its administration.

Dr. Siegfried Junghans (1887–1954) did more than any other man to make the continuous casting of metals a reality. The son of a watch and clock manufacturer, he served in the German Navy from 1906 to 1918. After taking a course in metallurgy and analytical chemistry at Stuttgart Technical High School, he was given control of a brassworks owned by his family. Since he lacked the normal foundry equipment, he wanted to produce castings simply and cheaply; by 1927 he had formulated his ideas on continuous casting and had begun experiments. When his firm was taken over by Wieland-Werke A.G. in 1931, positive results had been obtained and, after the success of the test plant erected in 1932, commercial production soon began. Herr Wieland told Junghans: 'If you had been a brass-

[1] L. P. Lessing, 'The Late Edwin H. Armstrong', *Scientific American*, Apr. 1954.

[2] Sir Richard Gregory, *Obituary Notices of Fellows of the Royal Society*, vol. 7, p. 322.

founder you would never have risked this work, for only an outsider could do it'.[1] After being made technical director, Junghans left Wieland in 1935 to work privately on the application of his process to all metals. Despite financial difficulties he was successful, and in 1936–9 he sold licences for various non-ferrous metals at home and abroad. The war interrupted his experiments, although he was supported by the German Government and built a production plant. In 1948 he restarted work on steel, a subject he had been considering since 1936, and in March 1949 the first cast was produced. Despite further financial difficulties he carried on his experiments, making agreements with four steel firms from 1950 onwards. Before his death in 1954, pilot plants had been built. Steel is now being continuously cast commercially.

Edwin H. Land (1909–) the inventor of the first practical synthetic light polarising sheet and later of the 'Land' camera, had, by the time he left Harvard University, gone far to the perfecting of his first invention. Later he and a Harvard physics instructor set up a consultant business and put the final touches to his polarising material, for which there proved to be much demand. Subsequent inventions brought him considerable commercial success, but, despite his activities as a businessman, he has continued as an active and versatile inventor.

Sir Frank Whittle (1907–), joined the R.A.F. in 1923 to be trained as a rigger for metal aircraft; in 1926 he was appointed as a cadet; and in 1928 he joined a fighter squadron. At this time his mind had already turned to the idea of novel types of aircraft engines and in 1930 he applied for a patent on a turbo-jet engine. Between then and 1939, when the Air Ministry became responsible for financing his work, he struggled along without any encouragement either from the Air Ministry or from aero-engineering firms and, in 1935, he actually allowed his patent to lapse. After 1935 he obtained limited backing from a London finance house and he was able to devote all his time to his research, although still an Air Force officer, first at Cambridge University and later on the Air Force Special Duty List. When the first jet aircraft flew successfully in Great Britain in May 1941, it could be said to be the product of Whittle's genius and the help of a small, but highly enthusiastic and extremely devoted, group of collaborators.

Juan de la Cierva (1895–1936), the inventor of the autogiro, was building gliders and was interested in aircraft before spending six years at the Civil Engineering School in Madrid. He later pursued privately his study of mathematics and aeronautical theory. His first seven machines were privately financed; the next, built in 1924, was financed by the Spanish Government. It was only in 1925 that he was able to devote his full time to aeronautics, and in that year, as a result of an invitation from the British Air Ministry, he brought his latest machine to England, where his subsequent work was done. Apart from the encouragement of the Air Ministry, he received financial backing from Lord Weir and his family. In 1926 the Cierva Auto-

[1] Quoted in communication from Frau Junghans, Nov. 1956.

giro Company was formed and in the 1930's a number of autogiros were built by G. & J. Weir Ltd., the marine engineers. The autogiro has, in fact, given place to the helicopter, but Cierva's practical and theoretical work created a body of knowledge on the attributes of rotating wing aircraft which paved the way for the helicopter.

F. W. Lanchester (1868–1946), one of the most versatile inventive geniuses of his age, made his greatest contributions in aerodynamics and mechanical engineering. In the course of his life he took out over 400 patents. In 1894 he gave the first explanation of the vortex theory of sustentation in flight, ideas which came to be fully accepted only in the 1920's. Lanchester's motor car, developed from first principles between 1894 and 1899, was far in advance of its contemporaries. He studied at the Royal College of Science for three years, although he never graduated, and at the same time attended evening lectures in engineering and workshop tuition at Finsbury Technical College. From 1909 to 1930 he was consultant to B.S.A. and Daimler. His life was embittered because many of his inventions were used long after he conceived them, so that the fame and rewards often went to others. He once wrote: 'If I were to say what I think to be the salient feature of my career, I think it would be to point to the fact that my work has been almost wholly *individual*. My scientific and technical work has never been backed by funds from external sources to any material extent.'

Francis W. Davis (1887–) is an independent inventor who made important, if not indeed the greatest, contributions to the system of hydraulic power steering for passenger cars. He studied engineering at Harvard, was associated with the Pierce-Arrow Motor Company for twelve years and in 1922 set up as a consulting engineer. Most of his experiments were conducted in a small garage in Waltham, Massachusetts. Davis has taken out more than twenty patents on power steering-gears and pumps, under which some of the largest American motor-car firms have acquired licences.

Lee de Forest (1873–1961), one of the great radio inventors, whose name will always be associated with the three-element vacuum tube, was inventing long before he went to Yale, where he took a doctor's degree in mechanical engineering. He found it difficult to work under conditions short of complete autonomy. This, combined with his lack of business judgment, exposed him to many vicissitudes. In his autobiography he speaks of

> my very isolation and poverty of opportunity for . . . experiment in those early formative years which forced my mind to create its own resources of contemplation, imagination and wonder; to find within myself a resourcefulness and ingenuity to make the utmost from next to nothing, and so enabled me later on to overcome great and genuine difficulties in the path to achievement.

W. J. Kroll (1889–), the inventor of the process of producing ductile titanium which goes under his name, was born in Luxembourg and studied metallurgy at the Technical High School at Charlottenburg. He had a number of significant metallurgical discoveries to his credit before 1923, when he established his own laboratory in Luxembourg which he ran with the help of one labourer, a mechanic and a secretary. There he conceived of his titanium and zirconium processes which have subsequently been developed and widely used. In 1940 he fled to the United States and worked as a private consultant and experimenter.

Felix Wankel (1902–), the inventor of the Wankel rotary piston engine, was deprived of a university education when his family's fortune was lost in the German inflation of 1919–20. After working for a scientific publisher, he set up a car repairing business in 1924 but soon started designing rotary piston engines, the concept that was to preoccupy him for the next thirty years. In 1928 he tried out some of his ideas in the form of a rotary valve for reciprocating engines. This type of valve had aroused more interest than his ideas on rotary engines. By 1934 BMW were supporting his work for aircraft engines and in 1936 the German Air Ministry put him in charge of a research institute. After 1945 Wankel was compelled to carry on his work on rotary engines, valves and other ideas – including a hydrofoil boat – at his home, but in 1951 he was able to establish a small research institute. He gained the support of NSU for the development of his rotary valves for motor-cycles and, by 1953, when his concept of a rotary piston engine was nearing its final form, he collaborated with the Company in the development of the engine. His main concern, as he himself has put it, is with the fundamentals of engine design:

> 'I am rather like the majority of physicists, who have little interest in the practical application of a process discovered or invented by them.'

*The Quantitative Evidence: Patent Statistics**

It is indisputable that there still are notable individual inventors. But the qualitative evidence can tell us nothing of the possible trends. Patent statistics are usually assumed to provide strong evidence of the disappearance of the individual inventor for, at first glance, they suggest that inventions arising in corporations are not merely more numerous than individual inventions but constitute, in the twentieth century, a steadily rising proportion of the total.

In the United States the percentage of patents issued to corporations rose steadily from 18 at the beginning of the century to 58 in 1936. Since the end of the Second World War the figure has fluctuated; it reached a peak of 64 in 1946, fell to 54 in intervening years, in 1954 rose again to 61 and in 1955 was again at about the level of 1936. The percentage of patents issued to corporations was very much higher than average in some industrial fields. Thus in 'Chemicals and Related Arts' the corporation percentage rose from 34 in 1916 to 85 in 1945; in 'Radiant Energy, Signalling, Sound

*[On p. 198 later statistics on Patents are referred to and the reasons are given why it now appears that such statistics may well be more valuable than is suggested above.]

and Electricity' the percentage rose from 39 in 1916 to 72 in 1945. Conversely, of course, it was lower than the average in other industries. Thus in 'Aeronautics', in 1945, the corporation percentage was 48, in 'Internal Combustion Engines', 49.

In Great Britain, from sample enquiries made, the broad statistical picture appears to be much the same. Patent applications from corporations were about 15 per cent of the total in 1913, 58 per cent in 1938 and 68 per cent in 1955. For the group 'Electric Discharge Apparatus' the percentage of patents granted attributable to companies is now 95; the corresponding percentages for 'Synthetic Resins and Cellulose', 'General Organic Chemistry', 'Dyes and Dyeing', 'Calculating etc. Apparatus', 'Chemistry, Inorganic, Distillation Oils and Paints' and 'Electronic Discharge-Tube Circuits' being 92, 95, 89, 77, 88, 97.

If it were the case that patent statistics faithfully measured invention and if the distinction between corporation and individual patents could be relied upon, these figures would mean that individual invention was becoming relatively less important and that in certain branches of industry the work of the individual has largely ceased to count. Unfortunately, the relevance of patent statistics to what is really happening in the field of invention is very obscure. Indeed, were it not for the fact that no other statistical material exists the patent statistics would properly be ruled out of court as useless. But, because there is no other quantitative material, and especially because the patent statistics reveal a marked trend, it is natural enough to suppose that 'some figures are better than no figures'.[1]

The reasons which might lead the cautious statistician almost to despair of the value of the patent statistics are these:

1. All inventions are not patented. Presumably practically all individual inventions will be patented, but corporations may decide to rely upon secrecy. There is no evidence as to whether the dependence upon secrecy has been growing or declining.
2. Patents cover inventions of enormously varying importance. A large part of all patents have no commercial value at all. There is no way of determining whether the merit of the average individual invention is greater or less than that of the average corporation invention or whether the relation between the two has changed over time. In some industrial groups – such as oil refining and photography – where the statistics show a very high proportion of patents being issued to

[1] The use of patent statistics as a measure of inventions or of inventiveness has led to some very odd suggestions in other directions. Thus some writers, with Spenglerian pessimism, argued that the decline in the annual number of patents in most countries between the wars was evidence of a general decline in inventiveness, 'the beginning of the end of the scientific age in the West'. It would be interesting to know what they would make of the big jump in patents since 1945. Other writers, by comparing the patents per head of the population, have sought to place different countries in the order of the natural inventiveness of their people.

corporations, it is well known that some of the most important inventions are to be attributed to independent workers.

3. The standard of patentability may have changed over time (indeed it is often asserted that it has risen). That is to say, that ideas of a certain inventive level would have been accorded patent rights at earlier times which in fact would not, in most countries, be accepted as patentable now.[1] There is no way of determining how a more rigorous testing for patentability will have affected the individual-corporation ratio.
4. Patents may be taken out, especially by corporations, not with the intention of making use of the invention but simply for the purpose of blocking competitors or being in a position to make bargains with them. It is not known how far this has increased the number of patents taken out by corporations, but it seems likely that the blocking patent is most common in the chemical and electrical fields.
5. The methods employed for differentiating between individual and corporation patents (although the only methods employable) are unsatisfactory as a record of changes in time because the corporation form has become increasingly prevalent. A patent which in earlier days might have been issued to a non-incorporated business or a partnership is more likely in these days to be issued to a corporation. And patents formerly issued in the name of the head of the firm or inventor will now probably be issued in the name of the corporation.[2]
6. The methods may also be misleading in distinguishing between individual and corporation patents at any one time. In the American statistics, for instance, if an individual inventor had assigned his patent to a corporation before the issue of the patent this would be counted as a 'corporation' patent. The earlier and the more eagerly corporations take up the inventions of individuals, the larger would appear to be the corporation percentage.
7. Thirty or forty years ago Patent Offices were accustomed to handle and judge upon patent applications more quickly than they do now. There is therefore now a longer period before the final issue of the patent on the invention of an individual during which a corporation might buy up the rights to the patent (in which case it would be counted as a corporation invention). If, on top of this, it were also true that corporations now more vigorously seek out and buy up the inventions of individuals, then the proportion of corporation inventions would be still further exaggerated.

[1] It has been said that Edison's patent relating to his invention of the incandescent bulb would have been held invalid today.

[2] A further complication is the difference between different patent systems. In the American system a patent can never be issued to a corporation; the patent must be issued to the inventor. Where there is an assignment to a corporation the name of the inventor appears and then the assignment after it. In Great Britain patents can be sealed in the name of a company.

All kinds of inconclusive speculations can be based upon patent statistics. The deductions drawn here (admittedly influenced by the qualitative impressions gained in the course of the enquiry) are that corporations are certainly now accounting for a larger part of useful invention than formerly but that their part is probably exaggerated by the patent statistics. In some fields, notably chemistry and electronics, the dominance of the corporation is greater than elsewhere. But the fact that about a quarter of the patents issued in the United States and Great Britain are still issued to individuals ought not to be overlooked. In view of the palpable limitations to the value of patent statistics, it clearly will not do to reject the qualitative evidence merely by appealing to the quantitative.

IV

REASONS FOR THE DECLINE OF INDIVIDUAL INVENTION

The reasons usually put forward for the decline in the relative significance of the individual inventor fall into two groups. First, it is said that his powers to perform his functions as in the past are progressively weakening because he does not possess the financial means to test out his ideas, because he is not highly enough trained to bring modern knowledge to bear upon his tasks, and because he lacks contacts with like-minded workers and the stimulus arising from such co-operation. These arguments will clearly be more applicable to the individual private inventor than to autonomous workers in universities. Second, it is suggested that, as more institutions for directed research are created, tempting careers in them will be offered and accepted by a large proportion of the scientific and inventive minds of the age. The individual inventor will no longer be able to make a living.

In some fields, chemicals, and especially biochemicals, electronics and nuclear physics, complex and costly equipment is often undoubtedly essential. But there are many exceptions, among major and minor inventions, in relatively old and relatively new industries and in varied techniques, to any general rule. In the textile industries in the twentieth century most of the important inventions were conceived of without expensive equipment or heavy costs.[1] The fundamental ideas of the jet engine were taken a very long way by Whittle with the most meagre resources.[2]

E. O. Lawrence in inventing the cyclotron had no elaborate laboratory facilities.[3] In the invention of the transistor no expensive equipment was

[1] The important inventions were high draft spinning and automatic weft replenishment shortened processing, the control of tension during large package spinning, centrifugal spinning, double twisting, automatic winding and warping, the Sulzer and the Fayolle looms, F.N.F. knitting machines, tension control finishing machinery and continuous processing for bleaching and dyeing.

[2] Sir Frank Whittle, *Jet*, *passim*, but especially pp. 48–51. R. Schlaifer, *Development of Aircraft Engines*, p. 90, points out that in Great Britain the total cost of the work of the jet engine 1936–9 was about £20,000 and in Germany the cost of the corresponding work there was about the same.

[3] See pp. 248–9.

required.[1] In his path-breaking work in television, Farnsworth in the early stages worked with improvised equipment and, indeed, deliberately avoided expensive devices.[2] It was always said of Carothers that 'he worked with crude equipment' and Kettering has expressed his preference for simple equipment in many of his experiments. John Rust throughout most of the period of his ultimately successful work on the cotton picker[3] and Marconi, in his early days in radio,[4] are other cases where ideas went forward at moderate cost. Other inventions which would certainly fall into this list are: air conditioning, much of the work on automatic transmissions, bakelite, power steering, the triode tube, the synthetic polariser, xerography, Kodachrome, titanium, electric precipitation and the hovercraft.

Nor should it be overlooked that many individual inventors have in fact enjoyed a first-class scientific or technical education. Scientific and technical literature is extensive and easily and cheaply obtained, so extensive indeed that it is not possible for anyone to keep abreast of more than a fraction of it. Intellectual communication in these days is good and is improving. Those who are not employed in scientific and technical institutions may nevertheless be connected with scientific societies and associations and derive aid from industrial research associations or government research departments. A Patent Office is, after all, a public institution provided, on equal terms to everybody, for giving access to the details of current technical achievement. To suggest that, as a general rule, extensive and detailed records of existing scientific and technical knowledge are denied to the independent worker of calibre and purpose is surely an exaggeration.

The virtues and limits of team work will be examined in a later chapter and we pass on to the question of incentives. The inventor has got to live, and, with the dwindling of substantial personal fortunes in many Western countries due to higher levels of taxation, fewer men can now pursue a career of independence. The effect of this push is increased by the simultaneous pull into attractive paid employment in research institutions. How far is it the case that the individual inventor can no longer obtain rewards for his work?

It may be true that inventors are scurvily treated by society. But there is nothing new in this. Indeed it is one of the strongest traditions in the history of this subject that they rarely received fair treatment. Their laments run right through the ages for reasons which have remained strangely unchanged. There is room here for only a few of the very many cases which could be quoted.

In the sixteenth century Leonardo da Vinci was hitting out at his detractors:

> If indeed I have no power to quote from authors as they have, it is a far bigger and more worthy thing to read by the light of experience. . . .

[1] See pp. 317–19. [2] See pp. 307–10.
[3] See pp. 243–5. [4] See pp. 286–9.

They strut about puffed up and pompous, decked out and adorned not with their own labours but by those of others, and they will not even allow me my own. And if they despise me who am an inventor how much more should blame be given to themselves, who are not inventors but trumpeters and reciters of the works of others?[1]

In the seventeenth century Sir William Petty, not a man to be deterred by small difficulties, bitterly analysed his own experiences:

Although the inventor, oftentimes drunk with the opinion of his own merit, thinks all the world will invade and encroach upon him, yet I have observed that the generality of men will scarce be hired to make use of new practices, which themselves have not thoroughly tried, and which length of time has not indicated from latent inconveniences; so as when a new invention is first propounded, in the beginning every man objects, and the poor inventor runs the gauntlet of all petulant wits; every man finding his several flaw, no man approving it, unless minded according to his own advice: now not one out of a hundred outlives this torture, and those that do are at length so changed by the various contrivances of others, that not any one man can pretend to the invention of the whole, nor will agree about their respective shares in the parts. And moreover, this commonly is so long adoing, that the poor inventor is either dead, or disabled by the debts contracted to pursue his design; and withal ruled upon a projector, or worse, by those who join their money in partnership with his wit; so as the said inventor and his pretences are wholly lost and banished.[2]

At the end of the eighteenth century Eli Whitney, one of the fathers of the machine-tool industry and inventor of the cotton gin, was engaged in a desperate defence of his patent rights:

Politicians of the day, who have thought to enhance their favour with the people . . . have encouraged their trespass by crying out against monopolies, as leading directly to Aristocracy. The effect of this popular clamour was much extended by the direct dependence we were known to have on the Federal Government for the protection of a property which had been created by their legislative Acts.[3]

In the middle of the nineteenth century George Stephenson, despite all his successes, refused a knighthood and the fellowship of the Royal Society as a protest against what he regarded as unjust treatment, particularly over his part in the invention of the safety lamp.[4] And Herbert Spencer, whom

[1] E. MacCurdy, *The Notebook of Leonardo da Vinci*, p. 61.
[2] Quoted from *Sir William Petty*, E. Straus, p. 31.
[3] Quoted from *The World of Eli Whitney*, J. Mirsky and A. Nevins, p. 125.
[4] John Rowland, *George Stephenson*, chap. 14.

one might have supposed nothing would suppress, was sadly concluding:[1]

> My experience is, I suspect, very much the experience of most who have tried to make money by inventions. Non-success, due now to unforeseen mechanical obstacles, now to difficulties in obtaining adequate pecuniary means, now to infringement of patent rights, now to unfair treatment by a capitalist, is almost certain to result. Probably it is not too much to say that there is one prize to fifty blanks.

Speaking of his experiences in the latter part of the nineteenth and the early part of the twentieth centuries, Edison said:

> I have made very little profit from my inventions. In my lifetime I have taken out 1180 patents, up to date. Counting the expense of experimenting and fighting for my claims in court, these patents have cost me more than they have returned me in royalties. I have made money through the introduction and sale of my products as a manufacturer, not as an inventor. . . . We have a miserable system in the United States for protecting inventions from infringement. I have known of several inventors who were poor. Their ideas would have made them millionaires, but they were kept poor by the pirates who were allowed through our very faulty system of protection to usurp their rights. The usurpation is particularly apt to obtain in the case of some great epoch-making patent.[2]

And in the twentieth century we have Whittle complaining in these words:

> It has been the custom of certain individuals to treat me as a 'gifted amateur' inventor, etc. and talk of 'taking my child and sending it away to school', to say that I have no production experience, etc., and, I believe, to represent me as a somewhat difficult and temperamental individual. On the one hand a good deal of lip service has been paid to my achievements, but on the other it has been implied that I am fit only to have 'bright ideas' with the results that, as it seems to me, I have been regarded as being either too biased or too incompetent to make a good judgment or to give good advice on major matters of policy.[3]

Undoubtedly, although there are much brighter spots in the picture than would be suggested by these few quotations, society has treated the inventive genius clumsily and harshly, and he, in turn, has been baffled by the perversity of men in rejecting what was designed only for their good.[4] And if this is true of the giants in the history of invention how much truer must it be of a thousand smaller men with less to offer and less determined in their assaults on the barriers of custom and indifference.

[1] Herbert Spencer, *An Autobiography*, vol. 1, p. 321.

[2] *Saturday Evening Post*, Sept. 27th, 1930.

[3] Sir Frank Whittle, *Jet*, p. 203.

[4] Charles F. Kettering put the matter to us in a nutshell when he said 'Inventors must not bruise easily'.

But, to repeat, there is nothing new here to suggest that in recent times the relations of the individual inventor with the outside world have become altogether impossible. In some ways, indeed, his position may have been eased. The courts have become much more sympathetic to the small man appealing against what he considers to be unfair treatment on the part of corporations, which go to great trouble to avoid charges of injustice or exploitation. Moreover, the market for ideas has considerably widened. In the nineteenth century an individual inventor would probably have the narrow choice of either plunging into manufacture on his own account or disposing of his ideas to one or a few firms. In the twentieth century, although undoubtedly more firms produce their own inventions, even these are often anxious to use first-class ideas from outside. And there are now many more firms in total and more firms which do not conduct their own research and are prepared to rely upon outsiders for ideas. The disposition of businesses to diversify their products, in order to widen their markets, may well lead them to consider possibilities outside their existing interests and thereby create an increased number of outlets for the individual inventor. Apart from the firm or the individual financier, the inventor may obtain backing or find a customer for his wares in the non-profit-seeking research association of the Battelle type; or the specialised profit-seeking research organisation of the A. D. Little, Inc. type; or the industrial research association of the British type; or the specialised public financial institutions such as the Industrial and Commercial Finance Corporation or the National Research and Development Corporation of Great Britain; or organisations, deliberately devoting risk capital to small speculative inventions, of the kind which have now become numerous in the U.S.

Despite the strong tradition that inventors never make money, there are good reasons for believing that the financial inducements have been and are still considerable. Many inventors, perhaps the majority, do not cover their costs. Even so, it would by no means follow that the chance of gain had ceased to act as a powerful incentive. For a limited number of glittering prizes may be more effective in evoking effort than more moderate but widely scattered returns. Considerable numbers of inventors both in the nineteenth[1] and twentieth[2] centuries gained considerably from their patents and were not infrequently raised to levels of affluence.

Against all this has to be set the fact that the inventor in these days may

[1] Up to the beginning of the twentieth century some of the inventors who seem to have made money were: A. G. Bell, C. Benz, H. Bessemer, C. F. Brush, S. Colt, W. F. Cooke, G. Daimler, R. Diesel, G. Eastman, O. Evans, S. Z. de Ferranti, E. Glidden, C. Goodyear, C. M. Hall, P. Héroult, E. Houston, E. Howe, D. Hughes, Lord Kelvin, C. von Linde, H. Maudslay, H. Maxim, C. N. McCormick, J. Mercer, S. Morse, J. B. Neilson, N. A. Otto, C. Parsons, W. Perkin, M. Pupin, Werner von Siemens, William Siemens, I. Singer, E. Solvay, G. Stephenson, J. Swan, N. Tesla, S. Gilchrist-Thomas, E. Thomson, J. Watt, G. Westinghouse, C. Wheatstone, E. Whitney, J. Whitworth, J. Wilkinson.

[2] It would perhaps be invidious to quote names here. But see L. Hardern, *T.V. Inventors' Club*, chap. 23; *Some Rich Inventors*, for an account of some of these in Great Britain.

well find a congenial niche in a research institution, where he will work under comfortable surroundings with adequate equipment and assistance in routine tasks; the risks then are less, his personal security greater; the anxieties of selling his ideas no longer harass him; the opportunities for contacts with other inventors, with developers and producers are increased. On the other hand, he thereby forgoes the chance of large and spectacular personal successes, unless he is prepared to take them in the form of professional prestige, since his work belongs to his firm. His freedom of action will probably be narrowed. His security of income will be obtained by sacrificing the opportunities of pursuing his own interests.

In striking a balance it must not be overlooked that the psychology of the inventor is complex and that his motives are certainly not wholly economic; he is nearly always something of an artist or craftsman, prizing independence for its own sake. Nevertheless there must be many inventive persons who, had they lived in the nineteenth century, would have worked on their own, but will in these days choose to join an organised group.

V

THE CONTRIBUTION MADE BY THE INDIVIDUAL INVENTOR

If invention ever became the prerogative of full-time professional employees there are grounds for believing that it would be weakened in range, liveliness and fertility.

The Uncommitted Mind

The essential feature of innovation is that the path to it is not known beforehand. The less, therefore, an inventor is pre-committed in his speculations by training and tradition, the better the chance of his escaping from the grooves of accepted thought. The history of invention provides many instances of the advantages, if not of positive ignorance, at least of a mind not too fully packed with existing knowledge or the records of past failures. The comparative novice may, indeed, plunge wildly, wasting much time, but his initial lack of knowledge may not be pure loss for he may be less deterred by probable difficulties and the false steps of the past. Inventors who have struggled through to success have often confessed that 'if they had known the difficulties they would never have started'.[1] Nearly all the important inventions in steel production in the nineteenth century came from rank outsiders whose ignorance of what could not be done was their chief asset.[2]

[1] Henry Ford, *My Life and Work*, pp. 85–6: 'I am not particularly anxious for the men to remember what someone else has tried to do in the past, for then we might quickly accumulate far too many things that could not be done. . . . If you keep on recording all your failures you will shortly have a list showing that there is nothing left for you to try – whereas it by no means follows because one man has failed in a certain method that another man will not succeed.'

[2] J. D. Bernal, *Science and Industry in the 19th Century*, p. 94. Bessemer, in giving evidence before the Committee on Letters Patent of the House of Commons, 1871, said:

It has been said that Farnsworth may have benefited from his lack of contact with the outside scientific world and the work being pursued in television elsewhere.[1] He tells the story of how at one stage a professor gave him four good reasons why his ideas, subsequently successful, could not possibly work.[2] Shortly before the jump forward in solid state physics represented by the discovery of the transistor, scientific authority was being claimed for the belief that nothing more was to be known or found in this field. The first success in the use of short-wave radio for long-range transmission was achieved by sceptics who refused to be deterred by the formal proof of mathematicians that this was impossible. C. F. Carlson, the inventor of xerography, tells of how his technical friends looked with scorn upon his waste of time. The first use of the controllable pitch propeller evoked in Europe theoretical proofs that such complicated mechanisms had no advantages.[3]

The list of instances where a fresh and untutored mind has succeeded when the experts have failed, or have not thought it worth while trying, could be greatly extended. Some of them seem almost fantastic yet there is good authority for them; Gillette, the inventor of the safety razor, was a travelling salesman in crown corks. The joint inventors of Kodachrome were musicians. Eastman, when he revolutionised photography, was a bookkeeper. Carlson, the inventor of xerography, was a patent lawyer. The inventor of the ball-point pen was at various times sculptor, painter and journalist. The automatic telephone dialling system was invented by an undertaker. All the varieties of successful automatic guns have come from individual inventors who were civilians.[4] Two Swedish technical students were responsible for the invention of domestic gas refrigeration; a twenty-year-old Harvard student for success in producing the first practical light-polarising material. The viscose rayon industry was largely the result of the work of a consulting chemist, a former glass blower and a former bank clerk. An American newspaperman is credited with being the father of the parking meter. J. B. Dunlop, one inventor of the pneumatic tyre, was a veterinary surgeon.

Scientists and technologists trained in one field have often been responsible for inventions in a field relatively unknown to them. B. N. Wallis, an aircraft designer, was responsible for some of the most important innovations in bomb design during the Second World War although he had little or no knowledge of that subject when he first turned his interest to it. Midgley had been trained as a mechanical engineer, but his great achievements when with General Motors were in the chemical field. Von Neumann,

---

'I find that persons wholly unconnected with any particular business have their minds so free and untrammelled to view things as they are, and as they would present themselves to an independent observer, that they are the men who eventually produce the great changes'.

[1] G. Everson, *The Story of Television*, p. 12. [2] In a personal interview.

[3] R. Schlaifer and S. D. Heron, *Development of Aircraft Engines*, p. 629.

[4] Lt.-Col. Chinn, author of the authoritative history of the machine-gun, in an interview given to *The American Rifleman*, Feb. 1956.

a theoretical mathematician, made basic computer discoveries. Föttinger had been trained as an electrical engineer, yet revolutionised hydraulic power transmission.

During the last war men with very varying backgrounds were in Great Britain given a chance to devote themselves to new weapons of war and methods of operation and they produced a steady stream of ideas, many of which were highly successful.[1] Houdry was an engineer but he was responsible for the first practicable method of the catalytic cracking of petroleum. Goldmark, an engineer in charge of the Columbia Broadcasting System's colour television experiments, invented the long-playing record.

The outsider often has less inhibitions in challenging accepted ideas and in trying out simple solutions when those better informed are searching for results by more complex and scientific routes. Long practice in a profession tends to form a crust on the mind (not, of course, without its value for many purposes) which results in things being taken for granted, in assumptions becoming deeply embedded, in simple questions being asked the most rarely.[2] The growth of scientific specialisms may prove obstructive to innovation in two ways. First, specialist groups come to accept traditional interpretations of events difficult to break down except by the sceptical outsider who does not share the group loyalties. Second, the groups tend to become insulated from one another. For the specialists in one group, recognising how much there is to learn in their own corner, become reluctant to advance views outside their own immediate range of knowledge.

Outlandish Exploration

Even where the individual is well informed of what is being done elsewhere, he may deliberately devote himself to unorthodox ideas, recognising that his chance of making a mark is much less if he sets out to compete with larger organisations in producing mere improvements. The search for real originality, of course, reduces the chance of success, but the successes will be bigger when they come. Carlson in inventing xerography 'purposely steered away from anything resembling photography or other known chemical processes because he felt that this ground had already been fully explored by Eastman Kodak and other big companies in the field'.[3] E. W. Davis, who was largely responsible for the ultimate success of the use of taconite in providing supplies of iron ore, turned in this direction because it was known that the iron ore firms were exploring the conventional sources for ore.

[1] For an extremely vivid account of one such group see G. Pawle, *The Secret War, 1939–1945*.

[2] Whittle gives a most interesting illustration of this. In the design of turbine blades for the jet engine Whittle, who had no special knowledge of the subject, had taken for granted a phenomenon which the specialists of the British Thomson-Houston Co. had overlooked because they had unconsciously carried over into the new machines assumptions which had been legitimate only for more primitive types of turbines. 'The affair more or less left a permanent scar on the relationship between Power Jets and the B.T-H.' (F. Whittle, *Jet*, pp. 69–74.)

[3] In a personal interview.

Earl Thompson invented synchro-mesh when the gear companies were all searching for easy change devices but trying free wheels, constant mesh gears, etc., all of which were far less effective.

Indeed, nothing is more characteristic of the individual inventor than this disposition to fold his tent and quietly steal away to other territory when large-scale organised research comes into his field.[1]

The Importance of Observation

The exceptional and largely intuitive powers of individuals to identify unexpected variations have been the source of much individual discovery and invention. The more striking cases, as might be imagined, are to be found among those who have worked close to nature. The achievements of amateur botanists, plant breeders[2] and students of animal behaviour are well known. Amateur geologists and archeologists have been equally successful. And for anyone prepared to study practically any subject as a 'naturalist', with the insatiable and indefatigable energy of the inventor, the scope must be almost unlimited.

Technical search has in recent years been speeded up by recording machines which are quicker, more reliable, more accurate and more continuous than the personal observer could ever be and such equipment is naturally made most use of in organised research institutions, where resources are ample. But machines record variations only in those factors they are set to measure. Where new and unsuspected factors can intrude, fruitful observation depends upon subtle personal skills for which the machine offers no real substitute. Indeed, the very paraphernalia of research may militate against the chances of discovery. For excessive reliance upon the machine may lead to the atrophy of the power of personal observation, through lack of its exercise. And interest in complicated and expensive equipment, with the temptation to refine results beyond a useful limit, can easily lure the worker from his singleness of purpose and offer a seductive diversion from the painful road of original thinking.

The Advantages of Large Numbers

There is a school of thought which believes that every invention makes further invention more difficult and less likely and thereby narrows progressively the chances of the individual. If there were, in fact, a finite stock of possible inventions and if it could be assumed that the easiest would be

[1] Thus de Forest (*Autobiography*, p. 348), 'I had now pioneered in wireless and radio for nineteen years. The field was becoming somewhat crowded. The war had aroused lively interest in the new art on the part of such large concerns as the Telephone Company, General Electric, and Westinghouse. The Radio Corporation had been formed; and soon a hundred engineers would be making a smooth and beaten road out of what had been a wild and fascinating trail.'

[2] Luther Burbank, the great plant breeder, was no scientist, but to him 'every plant had a face' and his successes could be set down to an uncanny sensitivity to deviations from the normal, a gift he was unable to teach to others.

made first, then each invention would render the next step more complex and costly and less certain. Or again, if inventive progress is conceived of as a series of 'master' inventions, each giving rise to a limited family of smaller 'improvement' inventions, then ultimately the fecundity of the master invention would become exhausted. It is difficult to draw much support from past experiences for these fundamentally pessimistic assumptions. The conception of a fixed stock of inventive possibilities implies the existence of some ultimate barrier to knowledge. And even if the progeny of each major invention were limited, the number of master inventions, providing the jumping off point for still further inventions, might remain unrestricted.

It is a more realistic hypothesis to suppose that each new invention multiplies the possible combinations of existing ideas and thereby widens the scope for originality. A new invention may at one stroke give fresh value to older ideas up to then unutilised because of some unsurmountable obstacle. A vast accumulation of imperfect ideas is always lying dormant, lacking only some element which can bring them to life. 'Inventions that come before their time' must often await some further new idea, perhaps having a usefulness in itself but also acting as a fertilising agent in bringing older ideas to fruition. Thus recording machines were practicable by 1900 except for one drawback which was overcome by the introduction of electronic amplifiers in the 1920's. Inventions, each having an importance in its own right, may be combined together to produce a third invention. The modern jet engine was a combination of jet propulsion and the gas turbine; the diesel-electric locomotive of the diesel engine and electric traction. One invention may create a demand for a complementary one. If an invention speeds up only a part of a process, the unimproved part becomes a drag on output and points to a new need. The manner in which, in the eighteenth and nineteenth centuries, the cotton-spinning and weaving operations successively outstripped each other is familiar. As the speed of the aircraft increases, the long period spent in travelling from the airport to urban centres becomes absurd and turns minds to methods of vertical take-off. The 'engine' of the automatic machine-gun had to wait for the right fuel in the form of smokeless powder; the gun could not operate on black powder. The vast increase in the operating speed of calculating machines could not be properly utilised until methods of recording the results had been invented. The invention of the jet engine led to great forward strides in metallurgy, that of the transistor to improvements in methods of purifying materials. Many new inventions create the need for progress in ancillary manufacturing equipment – good illustrations are provided by the stories of penicillin, nylon, the zip fastener, the safety razor.

The discovery of a material with novel physical qualities may be tantamount to a reshuffling of all the technical cards in the pack. If, for example, a metal should be discovered lighter for a given strength than anything known, it might make practicable many devices hitherto frustrated by gravity, and it would have widespread influence upon the design of machines, the relative advantage of hand and natural power and of different types of natural power. Thus small air-conditioning units remained impracticable until the design of aerofin, an exceptionally compact, light surface of copper. And new hard alloys remained useless until, with the invention of tungsten carbide tools, methods were discovered of working these alloys. A new material may lead to a widespread search for possible uses and then the more numerous the searchers the better the chance of success. One extraordinary recent case of that is provided by the discovery of the silicones: within a short period new uses have been found for them in virtually every industry; the possibilities in the future seem to be limited only by the generality, energy and ingenuity put into the search.

The opportunity for invention therefore continually proliferates; what is discovered is a minute fraction of what is discoverable. A sudden mutation, a master invention, opens up new fields both through its cross fertilisation with older ideas and through its possible crop of improvement inventions, and it further clears the ground for other possible major innovations.[1] This helps to explain the well-known phenomenon of chance in discovery.

> Probably the majority of discoveries in biology and medicine have been come upon unexpectedly, or at least had an element of chance in them, especially the most important and revolutionary ones. . . . Perhaps the most striking examples of empirical discoveries are to be found in chemotherapy, where nearly all the great discoveries have been made by following a false hypothesis or a so-called chance observation.[2]

[1] There is, indeed, a very suggestive analogy between the multiplication of different types of plants, the variety arising partly through inter-specific hybridisation and partly through mutation, and the multiplication of inventions, deriving partly from usual combinations of existing ideas and partly through a sudden outcropping of master inventions. Thus although there are tens of thousands of different varieties of roses now grown, 95 per cent of existing rose species have not yet received attention by the hybridist. (Ann P. Wylie, 'The History of Garden Roses', *RoyalHorticultural Society Journal*, Feb. 1955.)

[2] W. I. B. Beveridge, *The Art of Scientific Investigation*, p. 31. One of the most interesting cases is that while the world was being ransacked for improved strains of penicillin mould, Dr. Raper of the Peoria laboratory, where most of the study of these moulds was being carried out, found one of the best strains on a mouldy cantaloupe in the local market.

S. C. Harland, 'Recent Progress in the Breeding of Cotton for Quality', *Journal of the Textile Institute*, Conference Issue, Feb. 1955, has made the following comment: 'Almost all the facts we can at present use for quality breeding have emerged as by-products from some apparently trivial and useless investigation, not only in cotton but also in other plants. Suppose, for a moment, that all genetical research on weeds or ornamental plants, had been discouraged and suppressed fifty years ago, and only work on economic plants permitted. Our present store of really useful knowledge in that event would be extremely sketchy. Earlier in this lecture I stressed the great importance of haploids for the future breeding of cotton for quality. The first known haploid was found and studied in a weed. . . . It was found in the same relatively small institute that gave us the most important tool in plant breeding – colchicine.'

A more recent and brilliant exposition of the same thesis is to be found in Arthur Koestler's *The Sleepwalkers*.

There was an element of chance in the discoveries of the first aniline dyes, Duco lacquers, polyethylene, the transistor, Teflon, Freon refrigerants, ethyl gas, Neoprene, Float Glass and semi-synthetic penicillins. The accidental element in invention has been commented upon elsewhere[1] and reasons given for believing that it is likely to be more important in the chemical than in the mechanical or the electronic field. But, even in the latter group, cases are to be found in which investigators have sought for one thing and found another, where a misunderstanding of a process or a chance observation or concatenation of circumstances has led to practical results.

It seems quite inconceivable that, in this constantly expanding universe of ideas, knowledge and techniques, the maximum results would be obtained if the harvesting were wholly left to those who were thought fit to be employed full time in all the research organisations ever likely to be set up. The pursuit of every interesting technical speculation, the testing of every feasible hypothesis, the making of all possible chemical compounds, the search without limit for all the special properties of each compound, could easily use up the time of every living person on the globe with any technical aptitude whatever. Quite apart from the question whether any particular individual would be more effective inside or outside a research institution, many of the potential free-lancers would be unorganisable – part-time inventors, inventors who have one good idea and never anything more. The whole of the inventive ability of a large community is not organisable in a formal way.

It is sometimes thought that invention must in the future come from restricted and organised groups because only within the research institution will needs be known. In one sense this is true. When, for example, Whittle had taken the invention of the jet engine to a certain point, only those aware of that progress would know that further advance depended upon the appearance of new alloys capable of standing extreme stresses and temperatures. More generally, where an invention is being developed, only the developers can possibly know of the next obstacles.

This, however, is not to prove that knowledge of need is always confined to a narrow circle. With consumers' goods, any or all of the consumers may have recognised the need; indeed they may be more conscious of it than producers who are at one remove from the consumers' mind. Most motorists must have felt the need for something to keep their windscreen clear in rain long before the windscreen wiper was invented. Every woman must have been aware of the need for a hairpin which would not slip out long before an individual inventor thought of putting a crinkle in the hairpin for this purpose. In the vast range of consumer goods, including all the durable consumer goods, domestic machines of every kind, office equipment, motor cars, etc., the knowledge of need is not restricted. A second group who become aware of need from experience and may thereby be led on to invent, are factory workers themselves. They are in a position to observe at first

[1] See pp. 63–4, 128–9.

hand how their machine works and what are its defects. Such observations, although unsystematic, are extensive and continuous.

VI

To sum up: a greater part than formerly of the inventive individuals of western countries now find their way into organised institutions where inevitably some restrictions must be placed upon their activities. At times they have little or no choice: the very machines which are the adjuncts of their thinking, especially those which generate extremes of physical phenomena such as cold and heat and pressure and penetrate to the heart of matter, are quite beyond their private means. At other times they have a choice, but they may choose the relative security of an organisation. Although it varies greatly from one industry to another, the trend is indisputable and it is reflected in the increased likelihood that new inventions will emerge from such organisations. But since invention has traditionally been so closely bound up with independence and since, even in this century, so many significant innovations have seen the light of day under these conditions, it may be asked whether the growing importance of the industrial research laboratory is an unmixed blessing. To put it in terms of policy, is the trend one which should be consciously encouraged? Or is it one to which we should seek to set limits by trying to make easier the lot of the inventor who prefers to choose a path for himself?

CHAPTER VI

RESEARCH IN THE INDUSTRIAL CORPORATION: I

Whatever is proposed to him (the man of genius) as the object of his application, can never fix him, unless it be that which nature has allotted him. He never lets himself be diverted from hence for any length of time, and is always sure to return to it, in spite of all opposition, nay sometimes in spite of himself. Of all impulses, that of nature, from whom he has received his inclinations, is much the strongest. – ABBÉ DUBOS, *Réflexions critiques sur la poésie et sur la peinture.*[1]

I

ALTHOUGH inventions still continue to emerge in the more traditional ways – from the individual inventor and the university laboratory – new sources have appeared in recent years such as the research organisation of the industrial corporation, the industrial research association of a whole industry, the specialised institution (profit or non-profit seeking) and the government research laboratory. Of these newer centres that which has attracted most attention and aroused the greatest hopes of further technical advance is the industrial research organisation owned by the firm itself, and especially by the large firm. This is a modern development. Its beginnings are to be found at the end of the last century but the momentum of its growth has increased within the past decade and it is perhaps not too much to say that it constitutes the most spectacular change since 1900 in the activities of the industrial corporation. Very sweeping claims are now made both as to the benefits which have already accrued from industrial research and as to its future potentialities. It is not infrequently suggested that the operation of a research organisation is a condition of survival – 'any firm which does not conduct research would quickly go bankrupt'. It is even supposed that there is a close correlation between the standards of living of different countries and the sums they expend in this way. Running through much contemporary discussion is a strong implication that the gains from industrial research are without limit, that no firm and no country ever devotes sufficient resources to these activities. And, as a derivative, the proposition is sometimes advanced that industrial research organisations, as they continue to grow and multiply, will challenge and even finally supplant all other sources of invention.

This and the following chapter are, therefore, concerned with three questions:

[1] Quoted from *The Life of David Hume*, C. E. Mossner, p. 71.

(*a*) Is the growing practice of combining research and manufacturing within a manufacturing corporation one which is likely to contribute greatly to the flow of inventions?

(*b*) Are there any limits to the contributions of the industrial research organisation and, if so, where would we expect to find industrial research most widely and profitably pursued?

(*c*) More specifically, are there reasons to believe that very large firms, or firms in a monopoly position, will be more able and more willing to exploit the benefits of research than smaller firms or firms operating in a competitive industry?

II

The distinction made earlier between invention and development remains here as important as ever. Some firms are engaged both in the search for invention and in development and even they themselves may often find it difficult to decide where the one ends and the other begins. It is all the easier to confuse the two because in nearly all available statistical records, as, for example, of research expenditure, they are grouped together. The difference may be put in terms of well-known events. Du Pont discovered nylon and they also developed it. They did not discover Terylene but, as Dacron, they developed it. I.C.I. developed both nylon[1] and Terylene but they were responsible for the discovery of neither. On the other hand, they discovered and developed polyethylene. Rolls-Royce, along with other firms, developed but did not invent the jet engine. Eastman Kodak developed but did not invent that system of colour photography known as Kodachrome. Sulzer Bros. developed but did not invent the diesel engine. And so on through a large number of possible cases, some of which are referred to in detail in other parts of this volume.

This distinction, of course, narrows down the scope of this chapter.[2] In the interest of brevity and euphony the word 'research' will henceforth be employed to cover the attempts on the part of firms consciously to organise for the purpose of increasing the flow of inventions.

III

Why should it be supposed that the combining within one firm of the apparently incongruous functions of manufacture and the search for invention will contribute to the success of both?

Research is always something of a gamble; attempts to make it systematic do not guarantee results. Chance has played a great part in the inventive achievements of industrial research laboratories, particularly in the chemical field. Why should firms ever think it worth while to embark on such gambles?

[1] In co-operation with Courtauld's through British Nylon Spinners, Ltd.

[2] For some comments on the subject of Development see Chapters VIII and X.

Why should they not content themselves with existing ideas and accepted principles and devote themselves exclusively to the tasks of manufacturing and selling known products as efficiently as possible? This would not, of course, preclude them from introducing a thousand and one minor modifications and improvements in their products and manner of operation; but it would stop short of deliberate and organised efforts to discover really new things, it would be a matter of polishing stones instead of searching for diamonds. Or, if the circumstances seemed to call for some revolutionary departure in their activities, why should they not rely upon outside sources for their ideas and buy these as they might buy raw materials or machines from outside? Why should invention not be a specialised occupation, normally divorced from manufacture and sale?

It might at first sight appear that the well established firm would have no special interest in inventions, for these often involve the scrapping of existing equipment and methods and generally create disturbances in the market, the final result of which no one can foretell. Indeed, not so very long ago it was a common enough doctrine that, wherever they could do so, firms, and particularly larger firms, would try to maintain the *status quo* by sterilising patents and suppressing inventions. The popular view now seems to be almost the opposite one: that firms, even those oligopolistic firms which are supposed to be reluctant to engage in price competition, will light-heartedly engage in a costly, competitive race for technical leadership.

There was never much concrete evidence to back up the older scarifying stories of sinister and deliberate suppression of inventions; the more closely they are examined the less authentic they usually prove to be. On the other hand, the grounds seem flimsy for arguing that, even where they might avoid it, firms will engage in technical rivalry which throws heavy expenses upon all but may well leave their relative position in the market unchanged. In the industries where heavy capital investments are needed, firms will normally prefer to introduce gradually changes in processes and products, so that equipment may be scrapped and replaced piecemeal. Of course, if one firm sets the pace, others may be compelled to follow. But the firms which would appear to have the liveliest interest in disturbing the *status quo*, in creating fluid conditions for their own purposes, fall into two groups. First, newer and smaller firms, striving to break into a market or interested in expansion by offering something different, and, second, firms which find themselves slipping back and which may introduce some radically new idea in a desperate effort to restore their position. But the larger, well established and successful firms might understandably hold back and allow others to take these risks ('pioneering don't pay', as a great industrialist once put it); they are strategically well placed in the sense that their size and strength enable them to acquire inventions and develop them rapidly when their potentialities have been thoroughly established.

A relevant case is that of the possible use of the gas turbine in the motor car. A number of large motor-car firms are ostensibly interested in the possibility of using gas turbines. But it is doubtful whether research in this direction is being pushed very energetically by many firms. And for very obvious reasons: it is far from certain that the gas turbine will ever prove suitable for the motor car, its introduction would presumably call for very great changes in car design, the cost of its manufacture might be high and it is not evident that the total sales of cars would increase in consequence. All these risks might be taken by some new firm which could find a way into the market only through an innovation or by a firm hoping to stage a recovery. But for established firms finding no difficulty in disposing of their output the judicious attitude might well be, not themselves to force the pace, but to keep in touch with what is going on and never to fall so far behind that, if needs be, they could not quickly bring themselves abreast of important changes.

IV

The industrial laboratory does not appear to be a particularly favourable environment for inducing invention. The organisation and administration of research is under any circumstances always difficult and, beyond a certain rudimentary stage, becomes impossible. When this research is being conducted by a firm which is, or should be, primarily concerned with profits, then the atmosphere may become even less propitious, for there will be a natural tendency for the firm to direct the research from above, to limit the objectives and to ignore the dangers of assuming that the best way of organising manufacture is also the best fashion of organising research. A laboratory designed to increase the flow of new ideas represents an attempt to *organise* research for three obvious purposes: first, to gather together more of the resources incidental to research, to provide the research worker with the best aids, devices and working conditions; second, to encourage co-operation between different minds, and third, to try to give some guidance about the kind of inventions which would be most useful to the firm. The question immediately arises as to whether these purposes can be achieved without destroying the atmosphere most favourable to innovation. Is the organisation of research an attempt to organise that which cannot be organised, to plan the unplannable? Or is it a case where co-operation can be brought about in such a way that a number of minds working in concert can produce better results than could be achieved by those minds working independently?

One of the three objectives mentioned is clearly feasible. The facilities for research can be planned; men can be given the tools for the job and assistance in the routine and secondary tasks provided. Judicious restraint is, indeed, needed even at this stage of organisation. Over-lavish equipment and attempts to release the research worker from the use of his hands and eyes can easily consume or blunt those very mental powers upon which innovation depends.[1] But it seems conceivable that by trial and error the milieu can deliberately be created in which nothing will impede the research worker except his own natural limitations.

The difficulties arise with the remaining two objectives. A firm might strive to reach these ends not by direct controls and orders but by the initial choice of its workers. What, it may be thought, could be simpler, once the proper environment has been provided, than for a firm to determine the general line of effort by appointing gifted individuals of a co-operative temperament who are known to be interested in the branches of thought most likely to be fruitful for the firm? Here, however, the fundamental difficulties of organised research begin to crop up. Who is to decide what lines of research will, in fact, prove most fruitful for the firm? Who is to know that the interests of a research worker may not shoot off at a tangent at any time? What compromises must be accepted if the men best qualified by achievement and intellectual power happen also to be individualistic and unco-operative? How is the probable performance of an amenable team of relatively mediocre men to be compared with that of a number of awkward sole performers?

This is the outstanding dilemma of organised research. Men with the gift of innovation are often in the grip of inner compulsions which lead them to assume the right of deciding how their special powers should be employed and how best a task should be approached, to resent interference and to be thrown out of balance by it.[2] It will only be by chance that the almost irresistible lure of intellectual and inventive interest will coincide with the

[1] 'The trend towards more and more complex apparatus should be carefully watched and controlled; otherwise the scientists themselves gradually become specialist machine minders, and there is a tendency, for example, for an analytical problem to be passed from the micro-analytical laboratory to the infra-red laboratory and from there to the mass spectographic laboratory, whereas all the time all that was needed was a microscope and a keen observer.' (R. M. Lodge, *Economic Factors in Planning of Research*, Nov. 1954.)

Many great scientists have shown a preference for simple apparatus and have considered it essential that a scientist should carry out the elementary and routine tasks associated with his work. J. R. Baker, *The Scientific Life*, p. 29, and P. Freedman, *The Principles of Scientific Research*, p. 135.

J. B. S. Haldane, *Science Advances*, p. 35, tells the story of C. V. Boys who 'was once asked why he did not employ a skilled mechanic to help him in constructing apparatus. He replied that his ideas only got into shape as the constructional work proceeded and that this work helped the thinking process and he would not get on quicker by having it done for him.'

[2] W. I. B. Beveridge, *The Art of Scientific Investigation*, pp. 47–8, has a most penetrating comment on the reason for this. 'There is an interesting saying that no one believes an hypothesis except its originator but everyone believes an experiment except the experimenter. Most people are ready to believe something based on experiment, but the experimenter knows the many little things that could have gone wrong in the experiment. For this reason the discoverer of a new fact seldom feels quite so confident of it as others do. On the other hand, other people are usually critical of an hypothesis, whereas the originator identifies himself with it and is liable to become devoted to it. . . . A corollary . . . is that a scientist works much better when pursuing his own hypothesis than that of someone else.'

best judgments of what will most profit their firm.[1] But if the firm, by reference to its own criteria of profit making, seeks to impose control from outside it runs the danger of checking the exercise of those very powers which it set out to foster for it is disturbing the 'creative, originating people who cannot be hurried but can so easily be delayed'.

There is traditionally a conflict between the men who think, contrive and observe and the men who must organise and act. This makes itself evident in many different ways; in the clash between those studying fundamental problems and the practical men,[2] between the research and production sides of a business, between the administrators and the research workers in a research organisation.[3] It is the source of interminable discussions as to whether a research organisation should be 'in the works' in close contact with the everyday life of the factory or in some suitably remote location where the thinkers will be shielded from the minor distractions of the moment, and whether research and development departments should be together or not. It is responsible for the attempts to create in the industrial research laboratory the conditions and atmosphere of freedom of the best university laboratories. It accounts for the efforts made in some laboratories to leave the research worker perfectly free to pursue his own interests for a part of his working time, even though, in the rest of it, he is expected to concern himself with planned projects. It leads to the search for some middle course (on the assumption that this is really to be found) between the 'pretentiousness and fussy unimaginativeness and premature claims of success' of an organisation dominated by administrators and 'the shamble of disorganisation and lack of fixed purpose' of the laboratory in charge of the scientific genius.[4]

There can be no final resolution of this problem, for the inventor and

[1] K. Ziegler, the discoverer of the low-pressure method of producing polyethylene, has vividly described (*The Indivisibility of Research*, Glückauf 91 (1955) Issue 47/48) how the pioneer may be driven on 'by the pleasure of finding somewhere and somehow something new' and not 'by the wish to provide new values apart from those of increased knowledge'. This essay should also be read as convincing proof of how the formulation of 'set' problems is destructive of discovery.

[2] 'For the study of fundamental problems, it is often necessary to shut the practical man out for a considerable time. The far-sighted advisers of the Empire Cotton Growing Corporation saw this; in 1926 a cotton research station was established in Trinidad to study the genetics and physiology of cotton. I had the good fortune to be the first geneticist in that research station. It was clear at the outset that the great advantage of Trinidad was that there were no cotton growers there. Our terms of reference were merely to get to know as much about the genetics and physiology of cotton as possible, regardless of whether the knowledge acquired appeared to be of use or not.' (S. C. Harland, 'Recent Progress in the Breeding of Cotton for Quality', *Journal of the Textile Institute*, Conference Issue, Feb. 1955.)

[3] 'In the course of our investigations we heard administrators and administrative functions described in terms which cannot be printed here.' (R. N. Anthony, *Management Controls in Industrial Research Organisations*, pp. 57–8.)

[4] These problems are very well described in Sinclair Lewis's novel *Martin Arrowsmith.*

the scientists will go on expecting to enjoy the benefits of organisation without its restrictions, whilst the administrator of research will constantly be striving to infuse predetermination into results essentially unforeseeable. A group of workers devoted to innovation cannot, in fact, be organised by the methods normally subsumed under the idea of administration.

The normal purpose of organisation is to place one directing head, of more than ordinary capacities, in such a position that he can be provided with more relevant information than anyone else and can thereby co-ordinate that knowledge and reach decisions which are mandatory for the organisation as a whole. Information flows up to him, he transmutes this into appropriate action and the action is then enforced throughout the hierarchy. If discretion is left to the lower echelons, the area of this discretion must be defined and agreed by the supreme authority. As seen in the organisation of a business, or even more markedly in that of a military force, the ultimate purpose is to make the greatest possible use of the special abilities of the directing individual, to spread the benefits of his qualities as widely and effectively as possible throughout the organisation. In such an organisation there must be one single, indisputable and overriding plan. Conflicting opinions must be resolved by decision. Co-operation must be complete. Of course, in practice no organisation ever reaches this standard; administration is never perfect. But in so far as it moves towards this standard it is moving towards a desired end.

In the administration of a group of persons devoted to innovation, such an arrangement would be meaningless. In such a group the administrative head will always know less than each member of the rank and file; these members are chosen because they know more than anyone else in a given subject and are expected by their work progressively to widen this disparity. There can never be any single and unitary purpose in the group, since specific directions can never be given to those in search of the unknown. Conflicting opinions and ideas need not be resolved by dictat, indeed it would be dangerous if they were so resolved, for, starting from a state of comparative ignorance, the imposition of authority from above might be ruling out the only possible line of success. The degree of co-operation within the group is best left to the participants, who will form their own coteries as interests, temperaments and scientific predilections determine. The only rational purpose in grouping together persons engaged in innovation is to make the greatest possible use, not of some directing head, but of the special qualities of each of the potential innovators. This, of course, explains the marked lack of sympathy sometimes found where a production organisation is bound up with a research organisation: their laws of life are different.[1] The attempt to impose upon a research organisation the principles

[1] The story has been told by C. F. Kettering of the head of a company who enquired from his director of research how long a certain piece of research would take to complete. The director estimated that he would need six men for two years. The head of the firm then instructed him to take on twelve men and complete the task in one year.

of successful organisation in production would be just as disastrous as the opposite policy.

Acute observers have, of course, long perceived all this. Research cannot be directed in the normal sense. Only by the use of administrative devices peculiar to the needs can a group of research workers be held together in a way which will bring greater success than could be achieved by the individual efforts of the group. Thus O. E. Buckley, then President of the Bell Telephone Laboratories:

> One sure way to defeat the scientific spirit is to attempt to direct enquiry from above. All successful industrial research directors know this and have learnt by experience that one thing a director of research must never do is to direct research, nor can he permit direction of research by any supervising board.

And C. E. K. Mees, of Eastman Kodak:

> The best person to decide what research work shall be done is the man who is doing the research. The next best is the head of the department. After that you leave the field of best persons and meet increasingly worse groups. The first of these is the research director, who is probably wrong more than half the time. Then comes a committee, which is wrong most of the time. Finally there is a committee of company vice-presidents which is wrong all the time.[1]

Alexander Fleming, the discoverer of penicillin:

> There are different ways of conducting research. One method . . . is to collect a lot of money and say 'now we will do research'. They collect somebody as the master and a lot of people as the lesser ones. They tell the master 'we want some research on some particular thing' and he has to do it. He tells the other ones what they have to do. It is labelled research. . . . They pool their ideas. . . . What are you going to get? Are you going to get the sum total of these or the average? I venture to suggest you are going to get the average. You get some work but you don't initiate anything. . . . A team is fine when you have something to go on, but when you have nothing to go on – well, I should think a team is the worst possible way of starting and it is impossible to start out to find something brand new with a team. I know that certain industrial places . . . put up a certain amount of money for research and hire a team. They often direct them on the particular problems they are going to work out. This is a very good way of employing a certain number of people, paying salaries and not getting very much in return.[2]

But to argue that there should be *no* direction from above in a research organisation does not really resolve the dilemma; it raises it in another

[1] Quoted from R. M. Lodge, *Economic Factors in the Planning of Research*, Nov. 1954.
[2] Quoted from L. J. Ludovici, *Fleming, Discoverer of Penicillin*, pp. 127–8.

form. Let it be supposed that a firm sets up a research organisation simply as a home for able men of inventive disposition. The firm might indicate roughly what kind of work it hoped each member of the group would engage in. It might choose persons who were likely, from past experience, to follow work allied to the interests of the firm. The director might be a scientist himself or a pure administrator, but in any case his functions would essentially be those of a servant, defending the interests of the group, shielding them from disturbing intrusions and making certain that amenities were provided most conducive to unbroken effort. Each worker, with a group of assistants for the carrying out of routine tasks, would otherwise be free to follow his own interests. This, of course, would not guarantee success in innovation but it might well maximise the chances of it.

But can a business, in fact, tolerate an organisation like this within itself? Some of the larger firms have indeed gone a long way in this direction, at least for periods.[1] But the chances are weighed against it for the following reasons:

1. The primary function of a firm is to make profits, not to extend the range of pure knowledge. This does not mean that it will seek to maximise its profits in the short period; it may well look forward over the long period and design its affairs accordingly. But in the life of all firms there will come periods of stress, if not of actual crisis, when the immediate future must be the paramount consideration, when survival must be the dominant thought.[2] At such times the firm must let the distant future look after itself, switch the energy and the resources of the business to the near future and cut down expenditure likely at best to bring uncertain returns over the indefinitely long period. It would be inexpedient for the firm to assume that there is one part of its activities, the research organisation, which must at all costs remain undisturbed. The scientists may grumble that 'the first thing which is cut down is research'. But in such cases it is probable that research should be cut down first. The disturbances inseparable from a profit-making system involve a departure from the conditions ideal for research and innovation.

[1] D. Masters, *Miracle Drug*, p. 40, quotes the case of H. Raistrick, the famous biochemist, who accepted an offer from Nobels 'to pursue any researches he liked into any moulds which he cared to select . . . they were prepared to give his researches all the requisite financial backing and leave him to follow his own lines of investigation without retaining any power to veto anything he might care to do. . . . Naturally he undertook, in return,to relinquish his claims in any discovery he might make during the term of his engagement.' Raistrick, however, later returned to academic life. In the General Electric Laboratories, Coolidge, Langmuir and Steinmetz worked under similar conditions.

[2] It was once said by a business man of Ferdinand Porsche, a genius in engine design: 'He is a very amiable man but let me give you this advice. You must shut him up in a cage with seven locks and let him design his engine inside it. Let him hand you the blueprints through the bars. But for heaven's sake don't ever let him see the drawing or the engine again. Otherwise he'll ruin you.' (E. Heinkel, *He 1000*, p. 63.)

2. The efficiency of business administration in recent years has been greatly increased by the growing reliance upon cost-accounting whereby the relations between cost and return are measured for shorter periods and for smaller parts of the whole of the firm's activities. To be 'cost conscious' is recognised as one way of weeding out inefficient operations and unremunerative practices. But the essence of research is that it cannot be costed in any systematic way: firms move towards or away from it almost as an act of faith. It is not unlikely, therefore, that research will come to be looked on as a 'luxury' and the research organisation as one which is privileged because it is not and cannot be subject to the tests to which other parts of the business are increasingly exposed. Time and again in the course of this enquiry cases have been noted where a research group has regarded itself as being 'starved of resources' while the board of directors have wondered what returns they were really getting from their research expenditure. This happens because costs can be counted exactly, but the two sides think of the returns to that expenditure in a different way.
3. Whatever their faith in the long-period returns from research, it is extremely difficult for firms to create an atmosphere in their research organisations which parallels that found in the best kind of university society (no one would, of course, deny that a university department of science or technology may, and sometimes is, organised in a manner which is destructive of its real purposes) or in the surroundings of freedom which an individual inventor may create for himself. It is often said that the research laboratory of a large firm can provide everything which a university can offer and then something in addition in the way of higher salaries. But that cannot be generally true; many of the finest innovating minds do not naturally gravitate to industry; they feel that 'it is not the same'. However wholehearted the efforts of a firm to create an environment in which there is no pressure for quick results and in which freedom for the worker is adequately safeguarded, the research worker himself finds it difficult to avoid the feeling that he *ought* 'to be doing something useful'.

All this is true even where a far-sighted and determined effort is made to create a favourable milieu for innovation. The chances of success are further reduced where the research group is organised in hierarchical fashion, with ideas and instruction flowing downwards and not upwards, and is held to be at the beck and call of the production and selling side of the firm; where the direction of research is so closely defined that it gains a momentum rendering it impossible for intriguing sidelines and odd phenomena to be followed up; where the allocation of functions is determined in such a way that voluntary and ephemeral groupings among the research workers are impossible or are frowned upon; where men are asked to report at regular intervals upon ideas around which their minds are still anxiously groping;

where achievements are constantly being recorded and assessed; where spurious co-operation is enforced by time-wasting committees and paper work and where painstaking efforts to 'avoid over-lapping' frequently quench originality. The awkward, lonely, enquiring, critical men – the men of 'wide-ranging, sniffing, snuffling, undignified, unselfdramatising curiosity' as Sinclair Lewis once described them – may well be a positive nuisance in such surroundings.

V

One symptom of the difficulties of grafting research activities upon a business which must be guided by profit calculations is that firms seem not to know *how much* should be spent on research and their attempts to explain the grounds of their decisions usually seem to involve circular reasoning or to be inconsistent with known facts. Firms explain that they spend upon research some fixed proportion of their turn-over, without explaining how the proportion itself is determined. Or they describe the methods employed to calculate the cost of one research worker without accounting for the number of such workers they deem it best to employ. Or they state that the limit to their expenditure is determined by the shortage of scientific personnel, implying that if there were more properly trained workers they would spend more – which still leaves the enigma of why they would spend more. Or they aim at expenditures roughly equivalent to those of other firms although this leaves obscure how the other firms decide what to spend. A somewhat unusual rationalisation is that 'a logical top limit of legitimate expenditure for a firm would be that sum which produced on an average as many successful inventions as the firm was able to exploit'.[1] Such a policy of self-sufficiency, however, would appear to be extremely difficult to implement. For first, it would involve an estimate of the probable return, in the way of invention, for a given expenditure on research; second, an estimate of what would be the probable development cost of inventions not yet made; third, it would imply that the firm would ignore in its calculations ideas arising from outside which might be better than those originating from within. Other methods of 'project evaluation' have been suggested but most of them are in fact more directly concerned with the past returns or the

[1] 'The Rate of Expenditure on and Evaluation of Results of Industrial Research', W. Akers (quoted from *The Organisation of Applied Research*, O.E.E.C., vol. II, pp. 13 and 14). Sir Wallace Akers, after admitting that 'there was no sound method for assessing the value of the results of research', went on to suggest that 'even if an accurate method for assessing the profits resulting from research could be found, it would be very wrong to assume that this was the sole justification for research and that the latter ought to be reduced if the estimated profits seemed to be less than the total expenditure. A company which did no research would almost certainly go bankrupt: and unless research of high quality was being carried out in the company, it would find it very difficult, or impossible, to recruit its fair share of the best brains leaving the universities, with consequent ill effects in production departments as well as in the research department.'

probable future returns from development as distinct from research as here being examined. In so far as they are concerned with invention, they ask the questions but appear to provide no answers.

VI

Finally, the research organisation may itself constitute an obstacle to the rapid acceptance and assimilation of new ideas offered to the firm from outside sources. When a company is already spending large sums on its own research organisation the board of directors may well feel that this should reduce their dependence upon outside ideas, that they should not be paying twice for the same thing. When the research organisation is accepted as the authority to which ideas emanating from outside should be submitted for judgment other difficulties arise. The research group, however original may be the individual minds at work within it, may easily prove a block to advance. The normal feelings of professional pride are here engaged.

> The mass of men, simply because they are a mass of men, receive with much difficulty every new idea unless it lies in the track of their own knowledge; and this opposition, which every new idea must vanquish, becomes tenfold greater when the idea is promulgated from a source not in itself authoritative. ... Is not every new discovery a slur upon the sagacity of those who overlooked it?[1]

Much more is involved, however, than mere jealousy or lack of generosity towards the achievement of others. Men devoted to research, if they are to be worth their salt, must always push on with a momentum sufficient to override the uncertainties, the anxieties and the frustrations suffered in common by them all. Their work is an effort of will. Being subject to this concentration they may easily underestimate or ignore ideas coming from outside even when they lie in their own track of knowledge. Even the most open and original of minds can at times reveal an unexpected purblindness. Watt and Boulton looked upon Murdock's idea of the steam-locomotive as a mental disease, which had to be cured.[2] Watt was strongly opposed to the idea of the high-pressure steam engine.[3] Edison fought against the use of alternating current. Marconi could not be brought to see the significance of wireless telephony; his mind was moving along another path.[4] Baird saw no hope for the cathode ray tube.[5] The story is told that Lawrence, the inventor of the cyclotron, missed the significance of a suggestion submitted to him which would have greatly simplified the particle accelerator. W. J. Kroll, the discoverer of the first practicable method of making titanium by the magnesium process, nevertheless discounted the possibility of the sodium

[1] G. H. Lewes, *The Life and Works of Goethe*, pp. 337–8.
[2] J. Rowland, *George Stephenson*, p. 107. [3] See p. 43.
[4] Maclaurin, *Invention and Innovation in the Radio Industry*, p. 52.
[5] S. Moseley, *John Baird*, p. 210.

process, a method subsequently perfected by I.C.I.[1] If such things can happen with individual inventors it is even more likely that inventors in large research teams, backed by the prestige and status of their companies, will be disposed to brush off ideas from outside as distractions to their own work, and this, undoubtedly, helps to explain why such research teams sometimes remain indifferent to ideas, even in their own field, which subsequently prove to possess path-breaking qualities – a point to which further reference will be made later.

Firms will in various ways seek to circumvent the obstacles to the reception of new ideas. One is to channel the suggestions from outside through a separate group of scientists and technicians capable of taking a wider view of the interests of the firm. Another is to enlarge and make a matter of routine the contacts between the research organisation and the outside world, deliberately to encourage informal contacts with universities, with individual inventors, officials of industrial research organisations and government research institutions; for the less insulated the research organisation the less likely is it that it will be confronted by some bolt from the blue against which it will instinctively take up a defensive attitude. Or the board of directors may make itself responsible for passing judgment on new ideas coming in from outside and even employ some outside agency to advise it. But none of these palliatives is without administrative or commercial embarrassments.

Despite the obvious difficulties of fostering invention within firms (difficulties which in some cases have led them to abandon the effort) an increasing number of firms do engage in research. The extent to which they have done so and the circumstances under which they are most likely to do so with effect are discussed in the following chapter.

[1] R. B. Mooney and J. J. Gray, 'I.C.I.'s New Titanium Process', *I.C.I. Magazine*, Jan. 1956.

CHAPTER VII

RESEARCH IN THE INDUSTRIAL CORPORATION: II*

What things can come into being, what things cannot – in short, what is the principle by which each thing's potentialities are marked out, its boundary stone set deep down within itself. – LUCRETIUS, *On the Nature of Things*, Book I.

I

IT is not easy to obtain any very exact idea of the scale on which corporations engage in research for the purpose of increasing the stock of invention. Most quoted figures of expenditure are for 'research and development'; that is to say they group together the costs of invention, of development (including pilot plants) and of routine testing, production control and a wide variety of everyday tasks, often minor but sometimes very important, commonly known as 'trouble shooting'. The proportion of this total directly devoted to real innovation will, of course, vary from case to case but it tends to be low.[1] Most industrial laboratories are small – in the United States, for example, more than one-half of them employ less than fifteen scientific workers – and in many of these it is more than likely that the whole, or the overwhelming proportion, of such costs are incurred in development or routine work. The number of firms in the world which, year in and year out, devote more than one per cent of their total costs to activities likely to emerge as inventions must be very limited.

II

Nevertheless the statistics regarding research *and* development throw some light upon the matters under investigation here and establish

* [Later and more complete statistics of the kind discussed in this chapter are given in Chapter X. It may be said that the later figures add emphasis to the conclusions drawn from the earlier.]

[1] In the United States in 1953 4·1 per cent of the total industrial research and development costs could be attributed to 'Basic Research'; the proportion for the chemical industry was 10·5 per cent; for petroleum 7·6 per cent and for scientific instruments 6·8 per cent (National Science Foundation, *Science and Engineering in American Industry*, Table A.12.) In a few cases where individual manufacturers have tried to separate research costs and development costs, the proportion of research costs varied widely from nothing to 70 per cent of the total.

certain presumptions about the occurrence of industrial research. The broad conclusions to be drawn from them are, therefore, summarised here.[1]

The spectacular increases in expenditure on research and development since 1940, before which year statistics are almost non-existent, have been largely due to the direct financing of the research activities of corporations by governments for the purpose of defence. In the United States and Great Britain between one-third and one-half of corporation expenditures are met by the government. This, of course, explains why a high proportion of these costs are found in such industries as aircraft, guided missiles, atomic energy and electrical and electronic equipment. In the United States 84 per cent of corporation expenditure in Aircraft is met by the Federal Government; in Electrical Equipment, 54 per cent; in Scientific Instruments, 45 per cent; in Telecommunications and Broadcasting, 52 per cent. It may be argued that preparations for war on this scale must now be regarded as normal. But to do so takes the matter outside the range of economic calculation; decisions about expenditures for war purposes are military and political decisions. Any study of the occurrence of research which is to include government-assisted research would lead to very simple and obvious answers: the industries which conduct the greatest amount of research are those which get the largest government grants and those which get the largest government grants are those whose efforts are, or are supposed to be, most intimately bound up with national defence.

Excluding government grants, American manufacturing industry seems to be spending something over $2000 million on research and development (0.7 per cent of the national income). In Great Britain the corresponding figure is perhaps 0.5 per cent of the national income. Statistical evidence of research and development carried out in other countries is fragmentary. It seems likely, however, that the proportionate rate of expenditure on industrial research is highest in the United States and Great Britain,[2] and that there are other countries, notably Switzerland and Sweden, where although much less proportionately is spent, it is generally considered that the flow of invention is nevertheless strong and technical achievement great.[3]

It is only in very recent years that thorough statistical studies have been made of expenditure by manufacturing industry. But the series which

[1] The more important statistical sources, upon which these conclusions are based, are these: *Scientific Research and Development in American Industry*, 1953, Bulletin 1148, U.S. Dept. of Labor; *Science and Engineering in American Industry*, National Science Foundation, 1956; *Expenditure on Scientific Research and Technical Development in Britain and America*, E. Rudd, Paper read to Section F of the British Association, Sept. 1956; *Industry and Science*, Manchester Joint Research Council, 1954; *Scientific and Engineering Man Power in Great Britain*, Office of the Lord President of the Council and Ministry of Labour and National Service, 1956.

[2] *The Organisation of Applied Research*, O.E.E.C., vol. I, p. 21.

[3] R. S. Edwards, *Industrial Research in Switzerland*, p. 12.

appear to be most securely comparable suggest that in the United States such expenditure is now about the same proportion of the national income as it was in 1940.

The American and British evidence shows, and the few facts from other countries suggest, that the vast proportion of manufacturing firms do not possess research laboratories or carry out research or development in any systematic way. Even in the United States only about one-tenth of the manufacturing firms operate laboratories for research and development. Most firms are manufacturers pure and simple. The world of technology and science in which they live has been created for them and not by them. They depend almost wholly upon outside technical sources for the improvements in what they make and how they make it. Machinery manufacturers may offer them better machines. They may profit from improvements in the materials they buy from outside. They may benefit from the generalised advancement of knowledge that comes from individuals, universities, industrial research associations, public laboratories and government services. But they themselves are not prime movers in such matters.

III

WHY FIRMS CONDUCT RESEARCH

Some firms are prepared to take the long odds involved in the search for invention and some are deterred by them. The Bell Telephone Laboratories produced the transistor as a consequence of basic research on semi-conductors of which the final brilliant achievement could not have been predicted.[1] Polyethylene was the unexpected outcome of wide general interest which I.C.I. had shown in the effect of great pressures upon the physical and chemical properties of matter.[2] When du Pont first established the group which ultimately was to discover nylon there was no means of knowing what the result would be and there were periods when, on any rational calculation, the presumption could only be that nothing would emerge.[3] When General Motors set their research workers to the task of discovering a fuel to deal with knock in engines or a satisfactory refrigerant, they were looking for answers which might well not have existed.[4] And today there are firms devoting resources to the study of low temperatures when no scientist working in this field could claim that this new body of knowledge offers any clear prospects of future practical value. Sometimes in such cases as these the need has been obvious but ways of satisfying it unknown; sometimes there has not been even a definable need but simply a vague impression that some natural phenomena might have practical consequences: but the risk is not demonstrably less in the one instance than in the other.

Can anything be said about the reasons for these decisions; why do

[1] See pp. 317–19.
[2] See pp. 279–81.
[3] See pp. 275–8.
[4] See pp. 312–14.

the risks sometimes seem worth while taking and at other times not?[1]

The personal element undoubtedly plays a large part here. Two firms in much the same circumstances may act in different ways because the board of directors of the one is more sanguine in outlook, or more interested generally in scientific matters, or more susceptible to the popular mood in favour of research, or more confident of being able to pick good men and fruitful objects than the board of the other. Firms which originally established themselves by successfully exploiting some invention of the founder may long retain, almost as a matter of tradition, a special interest in science and research. But these personal influences, pervasive as they are, hardly seem susceptible to systematic analysis and we turn, therefore, to three more general clues: the character of the industry, the size of the firm and the consequences of the presence or absence of monopoly.

IV

THE CHARACTER OF THE INDUSTRY

The greater part of the effort devoted to research and development falls within a few industries. In the United States, for example, more than one-half of the expenditure is incurred in the aircraft, electrical machinery and chemical industries. In non-subsidised research and development about one-half of the expenditure is to be found in chemicals, electrical and other machinery and petroleum refining. The proportions are much the same in Great Britain.

Some industries are technically more *advanced* than others; they make great use of existing scientific and technical knowledge: 'science is their business'. Such industries may be contrasted with others, perhaps in the majority, where empirical methods are widely used and constitute the only path to success. Firms in the technically advanced industries, even if they were concerned only with utilising existing systematic technical knowledge and not with extending it, will employ considerable numbers of scientists

[1] Two minor reasons are sometimes advanced. It is said that high rates of taxation, particularly taxes on profits, will increase the scale of research since research expenditure is normally deductible as a cost. On the other hand, while high rates of taxation may make research 'cheaper' to the firm, they also reduce the benefits accruing to the firm from any increase in profits derived from new inventions. There is probably more in the second point that research laboratories may be set up largely for prestige and with no clear conception of what other purposes they might serve. 'We saw show places of research which gave an impression of having been created because research is good advertising. We heard of still other examples where the board of directors had been caught in a research boom and had established research laboratories without any clearly defined ideas of what these laboratories could perform' (*The Organisation of Applied Research*, O.E.E.C., vol. I, p. 29). The suggestion that a high reputation for scientific achievement is regarded as a good 'selling point' by businesses is confirmed by their disposition to claim on occasions credit for innovations for which they were not really responsible, to minimise or conceal the help they have received from outside, and to stress the importance of their development of an invention as contrasted with that of the invention itself.

and technologists and will possess laboratory facilities simply to operate their plant, to maintain systems of testing and control and to conduct their commercial activities. In these firms, therefore, the additional costs of setting up an establishment for research and invention may not be considerable, and the presence of many officials, not merely in executive positions but at most levels of the hierarchy up to the top administrators, with a natural interest in technical and scientific matters increases the likelihood that the additional costs will be thought worth while incurring.

Independently of the actual stage reached in their technical advancement, some industries will be advancing technically more rapidly than others. A technically *advancing* industry is one which has access to a stock of scientific knowledge still largely unutilised and with fairly direct technical potentialities. Where fundamental scientific knowledge, clearly useful to the industrialist, has been pouring out of universities and other scientific institutions, the industries affected may enjoy almost an embarrassment of riches. They can be contrasted with other industries where, because the scientific knowledge available has already been thoroughly absorbed into practice, there appears to be very little chance of outstanding technical innovations.

In a technically advancing industry there will be strong inducements for a firm to engage in research. The new scientific knowledge is in the process of being turned rapidly and often empirically into useful technical procedures. For a time the factory methods may be rough and ready – the first ways that can be thought of to exploit the new possibilities. This explains the apparent paradox that, in industries which are technically advancing, production methods for the time being are often empirical and even primitive; there has not been sufficient time to think of anything better and the second or third best must suffice.[1] The so-called 'scientific' industries are then relying heavily upon 'know-how'. Important consequences follow: first, a firm in such an industry will require a large scientific staff partly to understand what is happening in the scientific world; partly to turn scientific possibilities into empirical procedures; partly to watch, guide and control the early, rudimentary technical methods and to try to improve them. Second, as a matter of urgency, the firm will want to know why the empirical methods work, for that is the first stage in trying to improve them. Third, a firm operating under such dynamic conditions will be more likely to expect a return from research because for the time being it will seem easy to make discoveries and produce inventions.

The propensity to research is likely to be strongest in firms in those industries which are both technically advanced and technically advancing. There the chances of success seem greatest, the collecting together of

[1] Good illustrations of this are the early methods of making penicillin and the first method of making polyethylene which involves the use of enormous pressures. Sometimes these primitive early methods, with their reliance upon 'know-how', persist for a long time; for example, in the manufacture of insulin it is still necessary to collect pancreatic glands from all over the world because no commercial method of synthesis has yet been found.

research facilities and personnel easiest and the personal interest of executives and administrators keenest.

The point may be illustrated by an analysis of the statistics relating to British manufacturing industry in 1956. In Table I of the Appendix to this chapter the various industries are first ranked in terms of the number of qualified scientists and engineers (as a proportion of the total number of persons engaged in the industry) employed by corporations on 'research and development' – this may be taken as a rough indication of their interest in research. In column 2 of the table is shown the number of scientists and engineers (again as a proportion) engaged in 'manufacture and other work', i.e. in operating the normal activities of the industry – this may be taken as a measure of how far industries are technically advanced. In general, the industries which are research-minded are also those which are technically advanced. The six industry groups showing outstanding activity in research and development are: Mineral Oil Refining; Aircraft; Chemicals and Allied Trades; Electrical Engineering; Rayon, Nylon, etc.; Precision Instruments. Of these Mineral Oil Refining, Chemicals and Allied Trades, Electrical Engineering, Rayon and Nylon are both technically advanced and technically advancing. Aircraft and Precision Instruments, while they do not rank by the measurements employed as technically advanced, would generally be regarded as among those industries most rapidly advancing technically.

Having said so much, it is important to reiterate that personal factors can and do upset the results which might be expected from general reasoning. It is conceivable that a businessman of more than special courage and foresight might decide to initiate research in his industry simply because little or nothing was being done, because past neglect or lethargy was in itself proof of opportunity. A striking case is that of the discovery of methods of producing crease-resisting fabrics by Tootal Broadhurst Lee, an outstanding inventive achievement of the cotton industry in the twentieth century. Sir Kenneth Lee, the Chairman of that company, in 1918 took the initiative in setting up a small research group in his firm because

> the cotton industry was founded largely on the work of a series of brilliant inventors: very little systematic work had been done either on the chemical and physical properties of cotton or on the physical basis of the machine processes to which it was subjected in the course of manufacture;

and in the choice of staff

> provided there was evidence of ability to conduct research, lack of experience was regarded even as an advantage, since such workers would not have got into ruts, and would be more likely to bring a new outlook.[1]

Other notable cases where one firm in an industry resolutely pushed forward to success, when it might have been expected that other firms in this world

[1] Kenneth Lee, *Industrial Research: A Business Man's View*, Royal Institution of Great Britain, Dec. 15th, 1933.

industry would have taken the lead, either because they were larger or had longer experience, are Pilkington Bros. with Float glass and the Beecham Group with semi-synthetic penicillins. Cases of a successful response to the challenge of adversity clearly cannot be left out of the picture.

V

THE SIZE OF THE FIRM

The greater part of industrial research is conducted by a few very large firms. About one-half of the industrial research and development workers in the United States are found in the 70 or 80 largest firms.[1] The same seems to be roughly true of British manufacturing industry. These frequently quoted facts have led to a number of over-hasty and perhaps incorrect deductions: that only large firms carry on research, or that very large firms are responsible for the greater part of technical innovation, or that research can only be conducted effectively on a very large scale. The proper picture, however, seems to be much less simple than that.

An examination of the statistics of industrial research and development in the United States brings out complexities which will surprise no one who has sought, by first-hand enquiry, to probe the minds of business executives in their attitude towards innovation. The figures support one generalisation and four qualifications to it. The generalisation is that for manufacturing industry as a whole, and for the major industrial groups separately, research and development on *some* scale is more frequently found among larger firms than smaller. Firms employing more than 5000 workers, it is virtually true to say, all do research and development; only one in ten of the firms employing less than 500 workers do so; there is a steady gradation between these limits.[2]

The four qualifications are these:

1. Whatever size group of firms is chosen, there is a very wide range of expenditure on research and development as between different industries. Thus for all firms with more than 5000 employees, the annual expenditure per firm ranges from \$1 million in the Food industries to \$36 million in the Aircraft industry.[3] The type of industry is an important determinant.
2. It is not true that those industries in which the average size of firm is large engage in research and development on a larger scale than those industries where the average size of firm is smaller. There is no simple connection between large-scale operation and research and development.[4]

[1] *Scientific Research and Development in American Industry*, 1953, Bulletin 1148, U.S. Dept. of Labor, p. 6.

[2] See Appendix to this chapter, Table II.

[3] See Appendix to this chapter, Table III.

[4] See Appendix to this chapter, Table IV.

3. Although small firms are less likely to spend money on research and development than large, when they do in fact engage in these activities, firms in the low-size groups appear to spend, on the average, in proportion to their size, as much as firms in the large-size groups.[1]
4. Although the evidence is not as complete as might be wished, it seems to be true that even in the same industry firms of about the same size engage in research and development to very varying degrees.[2]

Correlations between the size of firms and their other characteristics are notoriously tricky exercises and no definite conclusions are drawn from the foregoing figures except that the statistical case for a positive connection between the size of firms and their disposition to engage in research and development is not proven. And, even at the cost of repetition, the long-suffering reader may be reminded that this negative conclusion is two steps removed from the real question that is being asked at this stage. For what we want to know is whether the size of a firm will affect its inclination to seek for innovations and, more specifically, whether the size of the firm will determine its power, not merely to spend money for this purpose, but to get results.

Can qualitative reasoning throw more light on this matter? Obviously the same absolute sum spent on research by a large and a small firm will constitute less of a risk for the former than the latter. But the point at issue is whether the same *proportionate* expenditure by two firms of varying size will be a greater risk for the smaller firm. If the advantages lie with the larger firm there should be evidence

[1] The evidence is as follows:

In the United States, with all industries taken together and most industries taken separately, in those firms conducting research, the number of research workers per 100 employees is highest for the firms with less than 500 employees and lowest for the firms with more than 5000 employees. (*Scientific Research and Development in American Industry* 1953, Bulletin 1148, U.S. Dept. of Labor.)

From another (sample) enquiry it appears that companies with sales of less than $1 million a year spend a greater proportion of their sales dollars on research than companies with sales above this figure. (*Trends in Industrial Research and Patent Practices*, National Association of Manufacturers, pp. 3, 4.)

A sample enquiry carried out in British industry in 1951 showed that for the whole of the sample and for the chemical industry separately, there appeared to be no correlation between the size of the firm and the proportion of research expenditure to turnover. (See Appendix to this chapter, Table V.)

See also National Science Foundation, *Science and Engineering in American Industry*, Table A.30.)

[2] See Tables V, VI and VII in the Appendix to this chapter. In a sample enquiry conducted in British manufacturing industry in 1955, about one-half of the 100 largest manufacturing firms (all with assets greater than £13 million) provided information of the scale of their research and development activities. For each firm the proportion was worked out between annual research and development expenditure and total assets. For Chemicals the range was between 2·41 and 0·39 per cent; Paper, 1·45 and 0·22 per cent; Drink, 0·45 per cent and nil; Iron and Steel, 1·04 and 0·12 per cent; Textiles, 2·05 and 0·08 per cent; Food, 0·90 to 0·23 per cent. Within each industry, there seemed to be no consistent connection between size of firm and this proportion.

(*a*) that the minimum effective scale of research is large, that there is a critical dividing line above which results can be expected and below which results are impossible, and

(*b*) that, above the dividing line, there are progressively richer results as the scale of research is expanded.

What a firm can 'afford' to spend in the way of research tends to become indeterminate because, so long as research proves fruitful, there is no natural limit to the proportion of its resources which a firm might properly devote to it. There is, indeed, a continuous gradation among firms in this respect, an infinite number of mixtures of manufacturing and of research effort running from the firm which is wholly manufacturing to the firm which is little more than a research laboratory. But, of course, if the minimum scale of research were *very* large, this might completely outstrip the total means of a small firm.

Enquiries about the minimum or the optimum scale for conducting research tend to evoke colourless answers, such as that 'everything depends upon the director' or that 'everything depends upon the quality of the personnel' and certainly instances are easy to find where laboratories, without changing their size, have run through cycles of failure and success clearly attributable to changes in men and especially changes in leadership. Factual evidence about the isolated influence of size is not easy to come by. Successful inventors in industrial research laboratories are disposed to stress the virtues of small groups. The comments of Midgley, the famous inventor at General Motors, is perhaps typical of this group:

> A minimum size of laboratory is indicated for retaining efficiency. It would be dogmatic to define such a minimum since many variable factors are involved in making any particular determination. . . . It is fair to ask the question 'is there a maximum size about which efficiency declines?' I have an instinctive feeling there is. . . . The unit size seems to me to have rather definite limitations. . . . The obvious controlling factor is the capacity of the research director to maintain an efficient understanding of the various problems for which he is responsible. . . .[1]

Some research directors, while not prepared to give a definite figure for the optimum size of a laboratory, rather ruefully admit to their problems in retaining effective grip upon an organisation as large as their own. It seems relevant that some of the largest firms follow a policy of operating a number of smaller laboratories rather than one large one and even claim that there are virtues in 'competition' between laboratories, although they invariably also claim that there is 'cross-fertilisation' between their different research units.

All this is highly inconclusive. There are, however, some facts which are a little more solid. First, in the United States the average operating cost of

[1] T. Midgley, *The Future of Industrial Research*, p. 17.

a laboratory per research engineer or scientist is about $25,000 (the figure varying from industry to industry) so that a group of ten such workers, with their ancillary personnel and aids, would cost about $250,000 per annum – a sum which would not appear to be beyond the resources of a moderately sized firm. In Great Britain the cost per scientist or engineer seems to be about £8000. Second, a significant proportion of patents are held by firms of moderate size. In the United States in 1953, of all patents owned by firms conducting research and development, 51 per cent of the patents were owned by firms employing more than 5000 employees, 30 per cent by firms employing 1000–4999 employees and 19 per cent by firms employing less than 1000 employees.[1] In Great Britain big manufacturing firms (with a total stock of more than £3 million in 1948) accounted both in 1938 and in 1948 for about 10 per cent of the total patents sealed and, of the 160 firms in the list, the five companies[2] individually taking out the most patents accounted for nearly two-fifths of the total patents of the group. Third, small firms, or large firms known to have relatively restricted research facilities, have gained some notable successes in the field of invention in this century; for example, Terylene, the continuous hot strip rolling of steel, crease-resisting fabrics, 'Cellophane' tape and air conditioning.* And fourth, there are industries, not usually considered as large-scale industries, where research appears to be carried on widely.[3]

It is customary nowadays to argue that there are no limits to the advantages of size in a research laboratory because of the necessity for 'team work', the character of which is often misunderstood and the virtues of which are exaggerated. Of course, as knowledge grows and forces more specialisation upon scientists and technologists, systems of communication between the specialists must be progressively strengthened. But there is nothing new in this idea: Lavoisier[4] was expounding it in 1793 and Francis Bacon outlined schemes for the same purpose in Salomon's House in *New Atlantis*. And it is true that in some directions in recent years small teams are tending to replace the individual worker,[5] although this is often

[1] National Science Foundation, *Science and Engineering in American Industry*, Table A.35.

[2] I.C.I.; British Thomson-Houston; Marconi; General Electrical and Metropolitan Vickers.

* [Other recent cases will be found in Chap. X, pp. 207–8.]

[3] See pp. 123–4.

[4] 'Most of the work still to be done in science and the useful arts is precisely that which needs the collaboration and co-operation of many scientists. . . . This is why it is necessary for scientists and technologists to meet periodically in common assemblies and that these meetings should cover even branches of knowledge that seem to have the least relation and connection with one another.' (Quoted from *Antoine Lavoisier*, Douglas McKie, p. 259.)

[5] This is suggested by the growing practice of joint-authorship in scientific papers. Thus in the first 100 Papers in the British *Journal of the Chemical Society* for 1954, only 13 were by one author, 48 were by two joint authors, 23 by three authors, 12 by four authors and 4 by five authors; whereas in 1905 47 were by one author and 53 by two authors. (See R. M. Lodge, *Economic Factors in the Planning of Research*, Nov. 1954.) In the United States the proportion of Papers in the *Journal of the American Chemical Society* written by a single author fell from 45 per cent in 1918 to 14 per cent in 1950. It must, however,

because the man of original powers is given more assistants for routine tasks.

It is, however, a far cry from the useful, voluntary collaboration of a few like-minded people to the popular conception of serried ranks of Ph.D.s moving forward into the scientific unknown as an army guided by some common purpose. The working groups even in a large industrial research laboratory are normally small. The real moving spirits are few and the rest pedestrian, although of course useful, supporters. Quantity cannot make up for quality[1] and little purpose is served in lamenting the absence of what are in fact unattainable levels of intellectual co-ordination when there are always too few minds of the highest calibre and there is always a limit to the help that can be afforded them in their original thinking.

The reasons for the limitations of team-work are obvious. Team-work is always a second best. There is no kind of organised, or even voluntary, co-ordination which approaches in effectiveness the synthesising which goes on in one mind. Because of the vast growth of scientific and technical knowledge no one expert can be a master in all fields, and specialisation is inevitable. But, since it limits the angle of sight of each expert, specialisation is self-stultifying unless some device for integration can concurrently be created. Team-work is the only answer, but it carries with it a countervailing loss of power inevitable when several minds are groping towards mutual understanding and assistance. And the loss becomes the greater the larger the team and the less voluntary it is in character.

Many of the most inventive spirits have confessed a constitutional aversion to co-operation. 'I am a horse for single harness', wrote Einstein 'and not cut out for landau or team work.' Nor must it be overlooked that the members of a team must always go the same way; that the strength of a team may be determined by its weakest link; that friction even in small groups of men with original powers of mind is not uncommon; that all

---

be borne in mind that team work is perhaps commoner in organic chemistry than in other branches of knowledge. In the *Journal of Chemical Physics*, for example, the proportion of single authors remained at 40 per cent between 1940 and 1950. (See J. P. Phillips, 'The Individual in Chemical Research', *Science*, Feb. 25th, 1955.) Moreover statistics of this kind may simply reflect changes of habits – senior research workers, for example, may become more generous in attributing credit to their junior colleagues. It is noticeable also that when larger funds become available for research, assistants are often appointed for tasks which might easily have continued to be performed by seniors.

[1] Fermi is quoted as saying, 'There is much to be said for the small group. It can work quite efficiently. Efficiency does not increase proportionately with numbers. A large group creates complicated administrative problems, and much effort is spent in organisation.' *Atoms in the Family*, Laura Fermi, p. 185.

'I am not myself a blithe optimist about the future of the large Experimental Station for the solution of our kind of problem. To go into some of them is extremely depressing. You see crowds of people milling around with an air of fictitious activity, behind a façade of massive mediocrity. There is a kind of Malthusian Law acting on research institutes. Just as a population will breed up to the available food supply, research institutes will enlarge themselves as long as the money holds out.' S. C. Harland, 'Recent Progress in the Breeding of Cotton for Quality', *Journal of the Textile Institute*, Conference Issue, Feb. 1955.)

co-operation consumes time; and that a large team is essentially a committee and thereby suffers from the habit, common to all committees but specially harmful where research is concerned, of brushing aside hunches and intuitions in favour of ideas that can be more systematically articulated.[1]

Turning now to the factors more definitely in favour of the larger firm, there is substance in the argument that, given the strong element of chance in invention, the larger firm can better afford to carry the cost of the numerous, inevitable failures with the proceeds of a few sporadic successes. The fewer and larger the prizes and the more widely they are spread in time, the more valid this point would become. If one firm, with a given research effort, expects to make a winning hit once every five years and another firm, with one-half that effort, believes there is a reasonable chance of scoring once every ten years, the position of the two firms might appear to be the same. In practice, given all the other uncertainties in the business outlook, it is likely that the more distant prizes will be heavily discounted. Ten years in the calculations of the businessman will appear more than twice five years.

Again, as is often suggested, the smaller firm may be deterred from research because even if it fell upon some invention of real value it might not be able to afford the cost, or face the many other complexities, of development and marketing. This is a subject to which we shall return later. But the argument is strong only if it is true that a firm must always develop its own inventions, that there is no ready market for ideas themselves, or that, if there is one, the value of inventions which still remain to be developed will tend to be heavily discounted. The arrangements for the purchase and sale of inventive ideas are certainly not perfect; if they were, there would be fewer stories of inventions, ultimately proving of great value, being hawked about for long periods without finding a taker. But there is, after all, some kind of market. There are numerous cases where a smaller firm has sold its ideas to a larger better able to afford, or more competent in, development. Terylene provides an outstanding illustration. And it is far from unusual for a larger firm to purchase a smaller, together with the inventive minds employed in it, recognising in so doing that they were purchasing important assets in ideas and know-how.

Finally, there appear to be some *types* of invention which favour the larger firm. In an earlier chapter references have been made[2] to an important distinction between systematic and empirical invention, between instances where the new idea has a certain completeness and unity from the outset and cases where the new idea is in the nature of a discovery arising from a search, more or less informed, among many possibilities. This contrast is at its sharpest as between invention in some branches of mechanical engineering

[1] R. S. Edwards, *Industrial Research in Switzerland*, p. 49, points out that 'we found Swiss industrialists very lukewarm towards co-operation and teamwork. Many think, rightly or wrongly, that it results in too much discussion and too little research.'

[2] See Chapter III, pp. 63–4.

and that in some branches of chemistry. Many mechanical inventions are conceived of as a whole system before a model has been made or an experiment conducted. Harwood, with his self-winding watch; Farnsworth, with his system of television; Whittle, with his jet engine saw the end complete before the means had been perfected. Many modern chemical inventions, on the contrary, have arisen from the vague intuition that a chemical compound with a given structure might have certain desirable properties, followed by the experimental verification of the hypothesis. Whinfield, in his discovery of Terylene, was led by rough analogies to surmise that a substance of a certain molecular architecture would possess certain practical properties, but only after laboratory testing could he establish the soundness of his presuppositions and discover that the compound he had chosen had other, largely unsuspected, useful qualities. In many recent chemical and medical discoveries, such as that of penicillin, the element of chance has been even more marked, and there the results have flowed from exceptionally acute observation of phenomena which did not arise as a consequence of a deliberately pursued experiment.

The distinction between systematic and empirical invention cannot be pushed too far, for often the two ways of thought and research are closely intertwined, especially in such fields as chemical engineering or biochemistry. There are recorded cases of mechanical inventions arising unpredictably in the course of experiment and of chemical inventions arising as unitary conceptions independently of experiment. The reality of the distinction is, however, confirmed by the truth of the corollaries it carries with it. A mechanical invention involves something new, the device has never existed before; the chemical invention may consist of discovering something that has always been there. Accidental inventions are common, one might say almost the rule, in chemistry but rare in mechanical engineering. It is much more likely that a chemist would fortuitously turn up some new and surprising property in a known compound than that an engineer would group together pieces of metal with one idea in mind and discover that he had stumbled upon a device of a kind that he had not been seeking.

For the present purpose, however, the most important corollary is that when the main approach to invention lies in search and observation and not in systematic conception, then organisation may become more possible, since the ground to be covered can sometimes be methodically divided between different workers, and team-work may be of greater value since the accumulation of negative results is one method of finally identifying the correct line of attack. If a needle is known to be in a haystack then numbers will count in determining the speed with which it is actually discovered; but if it is not yet known how likely it is that needles will be found in haystacks it may be better for those research workers who are most likely to be able to establish the probabilities to devote themselves to a little quiet and isolated thinking rather than to spend their time organising groups of workers in extensive ransacking of haystacks.

To sum up: it is difficult to see any simple or consistent relation between the size of the firm and its inclination to engage in research in the hope of producing innovations. Just as the richer gambler who backs more horses in the race is, other things being equal, more likely to pick the winner, so the bigger firm with the larger investment in research is likely to draw more prizes. Whether its proportionate return is likely to be greater is another matter. The fact that large firms, by use of such well-known procedures as mass production, standardisation and a hierarchical organisation, have made a success of manufacturing, or that by throwing large teams into the task of developing an invention they have achieved speedy and great results, is by no means a guarantee that shock tactics will pay the highest dividends in innovation where the determinants of success are very different.

VI

RESEARCH, MONOPOLY AND COMPETITION

We pass now to the contention that the possession of monopoly powers will be an active stimulant to research.[1] Domination of the market may arise through the presence of one large firm (monopoly) or a few large firms (oligopoly); or it may arise where a significant number of firms fix common prices, control output, etc. (the 'ring'). Strict cases of monopoly are much rarer than those of oligopoly; these two can conveniently be considered together. The ring is a different type of case and must be considered separately.

Monopoly and Oligopoly

It is plausible that where an industry consists of only one firm (monopoly) it may be strongly disposed to engage in research because it feels confident that it will, at least for some time, be able to maintain a corner in any new ideas it produces. At first sight it seems odd that the same argument should be used in respect of a firm which is one among a few large firms in an industry (oligopoly). Nevertheless, it is contended that in such an industry the firms will shrink from price competition but that each will be inclined to allow the others to reap the full rewards of their innovations (these assumptions are examined in more detail below). Such, then, seem to be the grounds for supposing that the main stream of fruitful research will come from industries of one or a few firms.

Before accepting this supposition, it is reasonable to demand supporting evidence, either of a statistical or a qualitative kind. Some of the more obvious questions which call for examination are these:

1. Is it in fact the case that in the most highly concentrated industries technical progress is most marked? And, if so, have the outstanding

[1] 'There must be some element of monopoly in an industry if it is to be progressive.' (J. K. Galbraith, *American Capitalism*, p. 93.)

new ideas arisen in one or other of the big firms or have they come from the smaller firms or from outside the industry?

2. In the concentrated industries, do all the large firms show an equal interest in research or do their policies vary greatly?
3. Is it true that in those industries which have been concentrated longest research is most generously supported and technical progress outstanding?
4. In those cases where an industry has undergone fairly rapid concentration has this brought about important changes in the attitude towards research?
5. In those industries which can be broken down into fairly distinct sections, varying in their degree of concentration, can differences in the attitude towards research be discerned?

On all these matters it has to be confessed that the relevant facts are scrappy and otherwise unsatisfactory. But they are sufficient to indicate the contradictory nature of the evidence and to throw doubts upon the existence of a general rule.

There are certain industries with a high, although indeed not the highest, degree of concentration which devote large resources to research and enjoy rapid technical advance: chemicals, petroleum, aviation, electrical engineering and glass are perhaps the outstanding cases. There are, however, instances of the opposite kind.

The aluminium industry in the United States provided for a considerable period before the Second World War one of the nearest approaches to complete monopoly to be found in manufacturing industry, and although more recently other firms have made their appearance, the degree of concentration is still extremely high. Yet the author of the standard study of this industry has pointed out that of the three outstanding discoveries up to 1937 – the use of heat treatment and ageing for giving strength to certain alloys, the introduction of the silicon alloys and the perfection of methods of coating strong alloys with pure aluminium – the first two were the work of men outside the aluminium industry.[1]

In most countries the iron and steel industry is highly concentrated and it is generally regarded as the standard case of oligopoly, dating as it does from the beginning of the century. Yet it is an industry in which significant departures from prevailing practices have not been very common. The fundamental methods of steel-making are what they were at the beginning of the century, progress has been made largely in terms of size, the refinement of existing techniques and the increased use of instrumentation.[2] Of the two spectacular inventions in the subsequent processes – continuous hot strip rolling and continuous casting – the first was the result of the ideas of a new-comer to the industry and was developed by one of the smaller

[1] D. H. Wallace, *Market Control in the Aluminium Industry*, 1937, p. 59.
[2] F. Mortimer, *Model Research*, Iron and Steel, Feb. 1951.

American steel firms,[1] and the second largely the work of a German individual inventor.[2] Much of the pioneering work in connection with the use of taconite ores was done by an individual outside the industry.[3] And, among the American steel companies, the largest, the United States Steel Corporation, has frequently been described as a follower and not a leader.[4]*

In the last thirty years the motor-car industry nearly everywhere has fallen progressively into the hands of a few large firms, and it is indisputable that in this period the motor car has been improved by the embodiment of many devices of high inventive quality. In this advance the larger firms in the United States (although not nearly to the same degree in Great Britain) have themselves played a direct and important rôle. But this is an industry which has also benefited enormously from ideas which have come to it from outside, either from other industries or from individual inventors. This is especially true of automatic transmissions[5] and of power steering,[6] cases described in other pages. Beyond that, much has been contributed, as for example in new systems of suspension, by smaller motor-car firms on the continent of Europe and by small manufacturers of accessories and equipment everywhere.

Other instances can be picked out where oligopoly and rapid technical progress have not gone together. Although the British Oxygen Co. Ltd. has for long had a wholly dominating position in the British market for oxygen, it was not the originator of the important innovations in methods of production and distribution.[7] A recent study of the consequences of the monopoly possessed by the United Shoe Machinery Corporation reveals that important innovations arose outside that firm and suggests that competitive conditions might have produced the same rate of technical progress.[8] The linoleum industry is, both in the United States and Great Britain, one of the most highly concentrated, but the fundamental principles of manufacture are still those invented by Frederick Walton in 1860. In the agricultural machinery industry at least two important advances, the cotton picker and the use of the hydraulic lift, originated with individual inventors.

The general impression left by qualitative enquiries of this kind is that for the very slightly concentrated industries (say those where the three largest firms account for 20 per cent or less of the total output) interest in research has been slight and technical advance slow, but that in industries

[1] See pp. 241–3. [2] See pp. 239–41. [3] See p. 167.

[4] See *Fortune*, Mar. 1936; Bowman, W. S., 'Towards Less Monopoly', *University of Pennsylvania Law Review*, Mar. 1953; Testimony of G. W. Stocking, *Hearings before Subcommittee on Study of Monopoly Power*, Committee of the Judiciary, House of Representatives, 81st Congress Second Session, Serial 14, Part 4A (1950), p. 967.

* [The discovery of oxygen steel-making came from the smaller steel organisations. See pp. 338–41.]

[5] See pp. 231–3. [6] See pp. 281–3.

[7] Monopolies Commission, 'Report on the Supply of Certain Industrial and Medical Gases', p. 90.

[8] C. Kaysen, 'United States *v.* United Shoe Machinery Corporation', pp. 145–208.

with a higher degree of concentration than this the conditions vary greatly.

The impression is strengthened by other points which, taken separately, are not decisive but which, taken together, are difficult to ignore. In some oligopolistic industries, even in petroleum and chemicals, the large firms appear to carry on research on very varying scales.[1] The industries which, as a result of the great period of combination at the beginning of the century in the United States and Great Britain, first took on an oligopolistic structure are not those which would be considered the technical initiators par excellence.[2] Nor is it always true that where an industry has been drastically concentrated it has suddenly become research-minded. The British coal-mining industry since the end of the First World War has passed through three phases: that in which there was fierce competition between a large number of relatively small units; that after 1930 in which the industry was operated as a group of cartels; and that after 1946 when as a nationalised industry the coal-mines were run as one unit. But at no time has there been any considerable amount of research carried out in the industry. Indeed, the general claim made at the time that the nationalisation of certain British industries would lead to a great burst of research cannot be substantiated. In 1955 the total expenditure on research and development by the British nationalised industries appears to have been only about £4 million.[3] And finally, it cannot be universally true that where an industry can be broken down into sections, revealing very different degrees of industrial concentration, the sections show an interest in research corresponding to their degree of concentration. For in the various parts of the British cotton industry the percentages of total output attributable, before the last war, to the three largest firms were: Bleaching (61); Printing (56); Dyeing (34); Spinning (22); Weaving (4). But it would be difficult to find informed support for the view that, ranked in this way, they are also ranked according to the technical change which has occurred in them.

In view of these inconclusive results it may well be asked how the idea of a close connection between oligopoly and innovation has arisen. Perhaps the most forthright assertions on the subject have been made by Professor Galbraith.[4] He seeks to clinch his argument by comparing the oil industry – oligopolistic and progressive – with the bituminous coal-mining industry – competitive and backward. But the illustration is an unfortunate one for his case. For in the oil industry many of the outstanding ideas have come from outside the industry or from smaller firms.[5] As for the bituminous

[1] See Tables V, VI, VII of the Appendix to this chapter.

[2] In the United States these were iron and steel, agricultural machinery, copper smelting and refining, corn products, petroleum, tobacco products, sugar refining and distilling. In Great Britain, iron and steel and textile finishing.

[3] 'Expenditure on Scientific Research and Technical Development in Britain and America', Ernest Rudd, Paper read to Section F of the British Association, Sept. 1956.

[4] *American Capitalism*, pp. 96–8.

[5] See pp. 235–7 for a case history of the catalytic cracking of petroleum. P. H. Frankel, *Essentials of Petroleum*, 1946, p. 148, has summed up as follows: 'Looking back

coal-mining industry, it may be technically backward in the United States under competitive conditions, but it is probably even more backward in Great Britain under monopolistic conditions.[1]

The empirical evidence is, therefore, inconclusive. Can any more secure conclusions be reached by examining the motives of firms operating within different industrial structures? Here some very odd paradoxes and rival views present themselves. On the one hand it is held that, in the absence of monopoly power, innovation will be cramped, that firms in price competition will not find it worth while to undertake research. On the other, that competition will stimulate the search for novelty and monopoly inhibit it.[2]

It is a matter of definition that a monopolist producer is relatively free from the competitive pressure to introduce new products and to innovate; if it is assumed that no firm is likely to search for, develop and introduce innovations unless it enjoys, at least for a time, the benefits of being the only producer in the field, the odd conclusion is reached that, although a firm will not be interested in innovation unless it expects to acquire a monopoly, if it has a monopoly then it will have less incentive to innovate.

Again it is often said that if one firm in an industry is conducting research successfully and introducing inventions then other firms will be compelled to follow suit, otherwise they will lose their share of the market. But if in fact they do follow suit quickly and, in consequence, they do *not* lose their share of the market, then the first firm to introduce innovations will not gain from its efforts, except in so far as the general market is enlarged. This produces the equally odd conclusion that if one firm is interested in research,

---

dispassionately we may find that they (major oil companies) mainly took up and developed ideas which were brought to them by men who did not, in the first instance, belong to their own team'.

[1] A. Phillips, 'Concentration, Scale and Technological Change in Selected Manufacturing Industries, 1899–1939', *Journal of Industrial Economics*, June 1956, has made an interesting study of the statistical relation between indices of 'concentration' in a number of industries and the increase in output per head or horse-power per wage-earner 1899–1939 in these industries. He finds that the 'concentrated' industries show a rather more rapid technical advance than the others and, negatively, that 'the alternative hypothesis that industries with large numbers of small firms tend to be technologically more progressive, while not disproved, received no support from the data'. But the limitations of his statistical material, which he scrupulously catalogues, make it doubtful whether his conclusions throw any light on the sources of innovations. For instance, if a 'concentrated' industry, consisting of a few large firms and a number of smaller ones, shows rapid technical progress this may not be due to the concentration; the innovations may have arisen in the smaller firms and been taken up by the larger firms – a type of case of which there are some notable illustrations.

[2] *Report of the Attorney-General's National Committee to Study the Anti-Trust Laws, 1955*, pp. 317 and 318. 'Generally speaking, economists support competition . . . because the goad of competition provides powerful and pervasive incentives for product development . . . monopoly power implies . . . relative freedom from pressure . . . to develop new products, or otherwise to innovate. . . . It is an unsafe power to lodge in private hands, making the monopolist a judge in his own case.'

all must be interested; but if all are so interested then there will be a much reduced chance that any one of them will derive an advantage from it.

It is widely agreed that one form of monopoly must be present if the search for innovations is to be worth while – that which resides in patent rights. That point is not relevant to the present discussion. It can be assumed that all firms – monopolistic, oligopolistic and competitive – can benefit in the same degree from this kind of privilege. If not, it would simply constitute a case for some reform of the patent system. What is being examined here is the claim that industries with a monopolistic or oligopolistic *structure* will be the more likely to look for innovations.

Of three firms, one of which is a monopolist, the second of which is one of a few large firms in an industry and the third of which is one of many firms in a competitive industry, which will be most strongly disposed to search for innovations? In reaching its decision each firm would be compelled to ask itself three questions:

(*a*) Whether, if it does not engage in research, some other firm will do so and prove successful? This is the fear of being supplanted. Presumably the stronger this fear, the stronger the disposition on the part of any firm to engage in research.

(*b*) Whether, if it does engage in research and is successful, the firm will be able to preserve its new ideas for its own benefit? This is the fear of being dispossessed. The weaker the fear of being dispossessed the stronger will be the incentive to research.

(*c*) Whether any new idea is likely to enable it to extend its market? This is the hope of expansion. The greater the apparent opportunity of enlarging its market and increasing its profits the greater the incentive to research.

With the monopolist the fear of being supplanted, the fear of being dispossessed and the hope of expansion are all relatively weak. Since there is only one manufacturer, then not merely is it relatively unlikely that others will search for new ideas in that field but the monopolist, quite apart from the defence offered to him by patents, will possess a long lead because potential competitors would be confronted with the task of breaking into what would be for them a new area of manufacture. On the other hand, the prospect to the monopolist of expansion arising out of an innovation is relatively slight. By definition he already possesses the whole of the market; the innovation, of course, might enable him to lower his costs and either make larger profits on the same sales or expand his market by lowering prices. But he cannot expand at the expense of existing competitors. It may, therefore, be concluded that the incentive for the monopolist to engage in research is relatively weak on ground (*a*), relatively strong on ground (*b*) and relatively weak on ground (*c*).

As contrasted with this case, the firm in the competitive industry might greatly fear the risk of being supplanted and of being dispossessed but might be strongly tempted by the prospects of expansion. With a large number of existing competitors, any one of these might do what our firm fails to do. Similarly, in a competitive situation, a firm which introduces a new idea runs serious risks that others will take up, improve and employ in non-infringing fashion the original innovation or that the challenge of the new idea will be taken up by competitors who, thereby stimulated to effort, will produce different and superior innovations. On the other hand, in a competitive industry, a firm which can obtain and hold a lead for a time has high prospects of growth; for it can expand its market by eating into those of its competitors. To sum up: the competitive firm will have a strong inducement to engage in research on ground (*a*), a weak one on ground (*b*), a strong one on ground (*c*).

When the oligopolistic firm is considered, matters become much more obscure and differing judgments most common. The fear of being supplanted will presumably be great; any one of the other large firms in the industry may be actively pursuing new ideas. On ground (*a*) the incentive to research would appear to be strong. On the other hand, the chance of expanding its market would be relatively small; it would be reasonable to suppose that the other few large firms in the industry would not take lying down any cut in the proportion of the market for which they were responsible. The innovation, that is to say, would stir up competition to an unpredictable degree and with unpredictable results. The incentive to research on ground (*c*) would seem to be weak.

But what is to be said concerning ground (*b*)? It is a common presumption that whereas the oligopolist will be strongly disposed *not* to compete as to price with the other firms in the industry, he will be prepared to compete in the search for innovations. The argument here is somewhat obscure but it appears to be assumed that while price competition will be anathema because a price cut on the part of one firm will immediately bring a price cut on the part of others, leaving the relative positions unchanged, a firm which gains an advantage through innovation will not be challenged; it will be left in peaceful possession of the advantage it derives from its inventions. To that line of reasoning several objections can be presented. First, it is doubtful whether price competition is, in fact, as rare among oligopolists as is claimed. Second, it appears from experience that an innovation on the part of one oligopolist may well release challenging competitive energy by others who will seek to circumvent the original innovation or better it. Third, the supposed antithesis between price competition and innovation competition is false: they are different forms of the same competitive process. Innovation is competition. Thus a firm which introduces a new synthetic textile fibre immediately comes into price competition with cotton, wool and silk. There is, therefore, no more reason why an oligopolist should look upon the risks associated with innovation in any way different from the risks associated with price competition or take up different attitudes to the two. There is, in consequence, no strong reason for supposing that his fear of being dispossessed is any less than that of a firm in a competitive industry.

This simplified picture of motives necessarily leaves out some crucial factors. For example, it has been implied that firms invariably produce all their own inventions and never buy them from outside, which is palpably unrealistic – and it might reasonably be argued that industries of the monopolistic or oligopolistic type tend to dry up the flow of invention into industry from outside because they reduce the number of potential purchasers of the work of individual inventors. But sufficient has perhaps been said to indicate that the incentives to research involve both the fear of loss and the hope of gain and that a competitive situation may well be the one which will favour innovation.

The Ring

What evidence is there that, in those industries where monopoly is organised through the fixing of common prices and output quotas, innovation has thereby been fostered? If firms are engaged in vigorous price competition will they be much less disposed to engage in research and be unprepared to share new ideas with other firms?[1]

This argument is perhaps most frequently and vigorously advanced by British manufacturers.[2] That, in itself, raises doubts as to its validity. For in some other countries, notably the United States which certainly does not lag behind Britain technically, price and output rings are forbidden. Further, Britain is eminently the home of the Industrial Research Association – a type of institution to which the firms in an industry jointly contribute and from which all firms can take new technical knowledge on equal terms – where no question of common price fixing arises. Even in Britain it is a relatively new argument, for, despite the rapid proliferation of price and output associations between the wars, little was heard of the advantages of monopoly for technical progress until the Monopolies Commission began its work in 1948.

It is, of course, true that if firms are not competing as to price or output there is little or no reason why they should not share any new ideas they may fall upon. That, however, is not the point at issue. The real question is whether the absence of price competition will mean that there are more new ideas to share.

It is difficult to pin down, check and draw deductions from the claims made as to the virtues of common price fixing in inducing technical progress. For it is always a question of what would have happened if conditions

[1] 'Without such stability through price regulation we will not get, and could not expect, the technical advance which flows from co-operation and the exchange of techniques.' (Monopolies Commission Report, *Semi-Manufactures of Copper and Copper Based Alloys*, 1955, pp. 84–5.)

[2] See in particular the following Reports of the Monopolies Commission: *The Supply of Electric Lamps*, p. 83; *Semi-Manufactures of Copper and Copper Based Alloys*, pp. 84–5; *The Supply of Insulin*, p. 31; *Supply and Exports of Pneumatic Tyres*, p. 83.

had been otherwise. But the matter has been examined on occasions in the proceedings before the Monopolies Commission. Thus, in the case of insulin, the Monopolies Commission concluded that 'in view of the close technical collaboration between the manufacturers, the understanding between them that one manufacturer should not alter his insulin prices without first informing the others does not appear to us to be unreasonable'.[1] But it is worth while recalling that, since the discovery of insulin by Banting, protamine insulin was introduced by Dr. Hagedorn of the Nordisk Insulin-laboratorium before 1936; protamine zinc insulin came from the University of Toronto in 1937; globulin insulin came from the Burroughs Wellcome American subsidiary and N.P.H. insulin also came from the Nordisk-laboratorium. The outstanding innovations did not come from British firms either before or after the formation of their price association.

In the case of electric lamps, the Monopolies Commission reached the conclusion: 'We see considerable advantage in the exchange of technical knowledge within the industry and much force in the argument that it could not continue if there were price competition between the companies concerned'.[2] The Commission point out that the leading British manufacturers 'were not themselves responsible for any major development between 1878 and 1935'. But the definite claim is made that British manufacturers, members of the Electric Lamp Manufacturers' Association, invented the modern fluorescent lamp about 1935–6.[3] This claim is not, however, supported by other authorities.[4]

One of the most interesting cases brought before the Monopolies Commission was that regarding rubber tyres. In 1929 the industry, largely through the initiative of Dunlop, set up the Tyre Manufacturers' Conference. The manufacturers, indeed, denied that this resulted in agreements to fix common prices, but the Commission concluded that the members of the T.M.C. 'were under the impression that what they were doing was not materially different from agreeing on common prices'.[5] Nor did the T.M.C. claim in this case that the establishment of common prices was a condition precedent to the most rapid technical progress. But Dunlop claimed that

> within the United Kingdom they act as the 'corpus of knowledge' on tyre and rubber technique, and that this knowledge is eventually passed on to the other manufacturers. They regard this as a necessary part of their leadership. . . .[6]

The absence of price competition and the sharing in technical advance clearly went together after 1929.

Dunlop pointed out that their Company had played an important part in the technical progress of the industry: that it was first in the field with aero-

[1] *Report on the Supply of Insulin*, p. 31.
[2] *Report on the Supply of Electric Lamps*, p. 92.
[3] *Ibid.* p. 10. [4] See pp. 252–4.
[5] *Report on the Supply and Export of Pneumatic Tyres*, p. 112. [6] *Ibid.* p. 83.

plane tyres (1910), bullet-proof tyres (1917), giant pneumatic tyres (1921) and tubeless tyres of natural rubber (1953); and that it had been the first to use carbon black in the tyre tread (1904) and to have experimented in the use of rayon for tyre casings (1921).

Actually, as with the history of so many inventions, some of these advances are claimed by others.[1] Whatever is the correct story among these technical complexities, it is significant that many of the claims by Dunlop are for the period preceding 1929 when price competition existed and was at times extremely vigorous. It is indisputable that the B.F. Goodrich Company announced the tubeless tyre in 1947, put it on to the market in 1948 and produced two million tyres in the first five years. It is also worthy of remark that the retreading of tyres, which has now become an important part of the British tyre industry, was introduced and developed by small independent specialist firms. The tyre manufacturers attached no great importance to re-treading until 1932.[2] It is true that the various improvements in re-treading would hardly rank as great inventions, but in any case they were not the products of the tyre manufacturing firms, which subsequently bought up some of the largest retreaders.[3]

In the manufacture of metal windows it was a firm which was not a member of the price association which first installed the modern system of rust-proofing. Members of the association first sought to belittle and then to impede methods which subsequently came into general use.[4]

Clearly much more empirical evidence would be needed in order to reach conclusions along this route. In the meantime the argument that the practice of fixing common prices is a stimulant to technical progress must remain a perplexing one. For it is odd that stability, through the suppression of price competition, should be sought for as a means of speeding up innovation, which in itself is a form of instability. And it is difficult to see why competition between firms in the matter of research should be less productive of new ideas than an arrangement whereby each firm had a prescriptive right to share in the research results of other firms, so that any firm which succumbed to the temptation to leave the costs and risks of research to others would thereby lose nothing.

[1] Thus many authorities give the credit for the discovery of the value of carbon black to S. C. Mote, working in 1904 for the India Rubber, Gutta Percha and Telegraph Works at Silvertown. Giant tyres are said to have been put on the market in the United States in 1918 by the Goodyear Company. And some writers claim that the effective use of rayon in tyres came only during the Second World War after it had been discovered that the Germans were employing it in the form of high-tenacity fibre of 1100 denier and not with the ordinary filament of lower denier experimented with up to that time.

[2] Monopolies Commission, *Report on the Supply and Export of Pneumatic Tyres*, p. 78.

[3] *Ibid.* pp. 90–2.

[4] Monopolies Commission, *Report on the Supply of Standard Metal Windows and Doors*, pp. 12–13.

VII

THE LINK BETWEEN INVENTION AND DEVELOPMENT

Provided the other conditions are propitious, firms often appear to establish a research laboratory in an effort to solve the extremely awkward administrative task of linking invention and development closely together. The relations between the inventor on the one side and the developer and commercial exploiter on the other, for reasons already touched upon, are traditionally knotty and probably will always remain so. The inventor is only too likely to be blind to the tasks of bringing his ideas to an operating practicability. The developer and the manufacturer are more likely to be concerned with the early crudities of the invention, the ever present danger of insurmountable snags in development, the disturbances in methods of production and in markets often carried along in the wake of the new ideas.

In the past, many ways have been tried of smoothing out these incompatibilities. Inventors have turned businessmen and exploited their own ideas. Or they have found backers who have provided money and, what was often just as important, encouragement. Or they have employed a firm to carry out, as agent, the development of the invention. Or they have licensed or sold their ideas to a firm with which they have retained varying degrees of contact in the development stage. All these different ways of collaboration have their drawbacks. Certainly, most firms dealing with outside inventors are disposed to speak of them as 'awkward fellows'. It is, therefore, not unnatural that some firms should see a solution in having the inventors inside the firm, subject to the same disciplines and sharing the same loyalties as its other members.

It may well be that the research laboratory of the firm justifies itself both to its owners in the shape of increased profits and to the community in the way of more rapid technical progress because it provides a new form of accommodation for at least some inventors who might otherwise have knocked up against the rough corners of the world and, losing heart, failed to give what was in them. But it is clearly not a complete or universal answer as to how society should handle inventors. The inventor who becomes part of an organisation cannot expect to be wholly immune to the traditional frustrations. As one great research director put it, 'there are probably just as many frustrated inventors within industrial research laboratories as outside them'. Those who choose the security and the amenities of the large organisation must expect to find themselves surrounded by more cautious minds; their failures will be remembered when their successes have perhaps been forgotten; they will submit their ideas to a group of judges against whom there is no appeal; they will frequently have to choose between insisting upon the merits of their ideas to the point at which they become regarded as cranks or submitting to what they regard as unmerited neglect of their work. On the other hand, if they choose independence, they will theoretically have a wider possible market for their ideas but it will, on the whole, probably be a more sceptical and it may even be a hostile market. They will be freer to follow their own lines of thought but they will be more frequently forced to improvise in equipment, to spend time in selling ideas to outsiders and to fight in the courts, if necessary, for the establishment of the priority of their invention.

There has, in fact, been in the Western countries a line of distinguished and highly successful inventors who have carried on their work in large industrial research laboratories – Coolidge, Langmuir, Carothers, Steinmetz, Shoenberg, Midgley, Kettering, Wallis, Müller, Bertsch, Rochow, Shockley – to name only some of the outstanding personalities. Despite the confining influences of an organised institution – of which some of them have spoken pungently – they must have reached the decision that on the whole that was the best environment for them. But it was not the decision that was made, or perhaps could ever have been made, by other equally distinguished inventors – Edison, Whittle, Armstrong, Farnsworth, de Forest, Fessenden, Diesal, Land, Carlson, Cockerell, Moulton and many others.

VIII

CONCLUSIONS

A study of the occurrence of research in industrial corporations suggests that it is essentially *sporadic*. There are some large-scale operation industries, such as chemicals, where expenditures are high; there are others, such as iron and steel, where the expenditures are relatively low. There are some industries where the average size of firm is small, such as scientific instruments, which spend a great deal relatively; others, such as the food industries, which spend little. There are industries in which firms of the same size follow much the same policy, other industries in which firms of the same size follow different policies. So long as it is accepted, by and large, that business usages are rational, this variety of practices signifies that there is no one simple golden rule for the efficient operation of a firm, no unique behaviour without which nothing else avails. The combination of research and development with manufacture, like other institutional devices in the firm, such as vertical integration or the holding company firm, may well justify itself on occasions without necessarily constituting a universal prerequisite for profit-making.

If, from this and the preceding chapter, one conclusion can be drawn more confidently than any other, it is that the large industrial research organisation cannot be considered, either actually or potentially, the sole and sufficient source of inventions. This new type of institution is undoubtedly a valuable supplement to, but it cannot be a complete substitute for, what already existed. Three facts point in this direction:

1. The large research organisations of industrial corporations have not been responsible in the past fifty years for the greater part of the significant inventions.

2. These organisations continue to rely heavily upon other sources of original thinking.
3. These organisations may themselves be centres of resistance to change.

It has been shown in earlier pages that, although the tally of inventive successes by large corporations spending considerable sums on research has been by no means insignificant, yet much important invention has emerged from other quarters.[1]

Large corporations can hardly ever be self-sufficing in the matter of new ideas. They may get help from the outside independent inventor. In a sample enquiry among American firms, practically all the firms declared themselves open to receive new ideas from outside and about one-third of them suggested that the independent inventor was a potential factor in their business.[2] Firms such as General Motors and General Electric of America receive annually thousands of new ideas and suggestions and, although the proportion which really interests them is always small, the records show that of the few accepted some are of great importance.[3] The Borg Warner Corporation is an outstanding case of a firm which actively hunts for foreign and domestic patents, and among the most successful products brought to it in this way are the automatic over-drive and the Schneider torque converter. In 1954 the Company paid out nearly $2 million in royalties on patents.[4] Other cases are referred to in Chapter IV.

Again, large firms may buy up smaller firms in order to acquire ideas or men of promise even in the research fields in which they themselves are engaged. The General Motors research laboratories were originally founded by acquiring the Dayton laboratories and thereby bringing in Kettering and Midgley. When General Motors became interested in the diesel-electric locomotive they purchased two small firms, the Winton Company and the Electromotive Company, which had already made much progress in work on the engine and possessed men with talent and experience in the field. General Electric of America, by purchase of the Hotpoint Electric Heating Co., bought themselves into the field of electric ranges and electric irons, and by purchase of Warren Telecron Co. and the Walker Dishwasher Corporation acquired the earlier inventions and 'know-how' in electric

[1] See Chapter IV. It is significant that few of the Nobel Prize winners in Physics, Chemistry, Physiology or Medicine since 1900 have come from industrial research laboratories.

[2] *Trends in Industrial Research and Patent Practices*, National Association of Manufacturers, 1948.

[3] During the last war the U.S. National Inventors' Council examined 200,000 inventions and found 6000 of them worth passing on to various Army and Navy Bureaux. By the end of the war only 106 of these were actually put into production. But one of them was the land-mine detector which saved innumerable lives.

[4] 'Borg-Warner's Profit Factory', *Fortune*, May 1955.

clocks and dishwashers.[1] The General Foods Corporation purchased rights from Birdseye, the inventor of some of the methods of deep freezing.[2]

Within this category can perhaps be included those cases where a large firm conducting research on a considerable scale becomes indebted for important discoveries to another firm, mainly interested in a different field, which happens to be conducting research on a small scale outside its main activity. The discovery of Terylene is such an instance. Another extremely interesting case[3] is perhaps worth recounting in more detail.

In their early days the laboratories of General Motors produced two extremely important discoveries in their chemical department. Midgley was first responsible for the discovery of tetraethyl lead and its effective anti-knock qualities and he later went on to the discovery of the important refrigerant Freon. These chemical discoveries were made in a motor engineering firm and not by any one of the big chemical companies which afterwards manufactured these materials. In 1927 the same General Motors research group became interested in the possibilities of synthetic rubber. At that point the President of du Pont wrote to General Motors questioning this new line of research:

> I wonder whether it is a wise expenditure of money on the part of General Motors to go into the synthetic rubber investigation. A great deal of work has been done on this subject by very competent people and well organised research groups. I understand that the General Motors chemical department is neither well organised for this purpose nor is the personnel such as would likely prove successful. The same line of thought has so far kept the du Pont Co. from a general investigation of the subject, yet we think the du Pont Co. is better equipped for this purpose than General Motors.

Mr. Sloan was at that time President of General Motors and in reply to this letter he said:

> You say . . . that . . . General Motors chemical department is neither well organised nor is its personnel such as would likely prove successful. This statement may be right or it may be wrong. Frankly I do not know which it is, but I think that if I had told you six or seven years ago that General Motors research was working on some chemical scheme whereby we could inject some quantities of some unknown material into gasoline to enable us to increase the compression and produce a fuel of anti-knock qualities, you would have said exactly the same thing and would have thought, as I did as a matter of fact a good many times whilst we were

[1] T. K. Quinn, *Giant Business*, pp. 113–16. Also *Hearings before the Sub-Committee on the Study of Monopoly Power*, Committee on the Judiciary House of Representatives, 81st Congress, 2nd Session (1950). Evidence by T. W. Mitchell.

[2] T. K. Quinn, *Giant Business*, p. 111.

[3] U.S.A. *v.* E. I. du Pont de Nemours & Company – Plaintiff's Pre-Trial Brief.

> spending the money, that we were very foolish in mixing up with something that was purely of a chemical or fuel character and had nothing to do with the primary manufacture of motor cars. Then after we had discovered the material . . . you would very likely have questioned the ability of our research department to develop methods of making the material itself in a practical way, yet that was accomplished also.

The larger firms also make considerable use of consultants and of the facilities of non-profit-seeking and profit-seeking independent research organisations. The reasons they give for doing so are themselves significant. They may be short of trained people. Or they may be confronted with a task of a non-continuing nature which they prefer to hand out to others rather than set up an organisation of their own for the purpose. Or they may be confronted with a type of technical problem new to them which they feel they cannot handle at all. Or, having been continually defeated by some technical problem, they may hand out the task to others who will come to it with fresh minds and no preconceptions. These are all admissions that the specialised research agency has, on occasions, important merits.

There are some firms, although not always the largest, which have an almost unbroken record of sustained alertness in innovation. But, despite their watchfulness, established firms have frequently missed or overlooked important new departures or remained unconvinced of the merits of an invention which, it might have been thought, would have appealed strongly to them. Maclaurin has told the story of the apathy of the telephone, cable and electrical manufacturing companies towards the early progress in wireless telegraphy[1] and of how the Radio Corporation of America resisted Armstrong's ideas of frequency modulation.[2] When Ford, at the end of the last century, was casting about for a new form of transport, the head of the firm in which he was then employed, the Edison Illuminating Company, was continually urging on him the future of the electric motor 'but gas – no'.[3] No aero-engine firm either in Great Britain or Germany showed anything more than a perfunctory interest in the jet engine in the early days.[4] The Sulzer loom was offered to, and rejected by, the established European textile machinery manufacturers.[5] The retractable under-carriage was not thought to have a future by established aircraft companies.[6] The Swiss watch manufacturers were slow to see the importance of the British invention of the self-winding wrist-watch.[7] The British and American chemical firms took no part in the crucial discoveries connected with penicillin, and it was difficult enough to persuade them to take part at the pilot plant stage even when the chemotherapeutic virtues of penicillin had been placed beyond

[1] W. R. Maclaurin, *Invention and Innovation in the Radio Industry*, chap. II.
[2] *Ibid.* pp. 189–90.
[3] Allan Nevins, *Ford*, p. 175.
[4] See pp. 262–6.
[5] See pp. 303–4.
[6] N. Shute, *Slide Rule*, p. 182.
[7] See pp. 293–5.

doubt.[1] The newer methods of producing polyethylene were not found in the laboratories of the large chemical company which had originally discovered it.[2] The Kroll system of producing ductile titanium was strangely neglected by the metallurgical and chemical corporations.[3] When Baird called on the Marconi Company in 1925, he was told that they had no interest whatever in television.[4] When the Ford Motor Company in the United States wished to introduce automation in their factories, it was to the small specialist firms in the machine-tool industry to which they turned for assistance, 'the small uninhibited firms with no preconceived notions'. The manufacturers of navigational equipment played no part in the invention of the gyro-compass.[5] One of the most striking instances of a pioneering firm which ultimately proved conservative was the Ford Motor Company itself. Henry Ford resisted the introduction of the thermostat, hydraulic brakes and other devices into the Model T car.[6]

It would be foolish to decry the contributions to innovation made in large firms. But with the knowledge of how inventions arise, of the cases where an individual makes one invention in a lifetime, of the numerous inventors who would be quite unsuitable for employment within a research laboratory or by temperament would be averse to such employment, it seems equally wide of the mark to suppose that such firms could, of themselves, guarantee the maximum flow of innovation.

[1] E. B. Chain, 'The Impact of Biochemistry on Industry', *Financial Times 'Annual Review of British Industry'*, 1955.

[2] See pp. 279–81.

[3] See pp. 314–17.

[4] Moseley, S., *John Baird*, pp. 71–2.

[5] See pp. 254–6.

[6] Richards, W. C., *The Last Billionaire: Henry Ford*, pp. 155, 156.

APPENDIX

TABLE I*

GREAT BRITAIN, 1956

NUMBER OF QUALIFIED SCIENTISTS AND ENGINEERS EMPLOYED IN (a) RESEARCH AND DEVELOPMENT AND (b) IN MANUFACTURE AND OTHER WORK AS PROPORTION OF TOTAL EMPLOYMENT

Industry Group	Number of Qualified Scientists and Engineers, as Proportion of Total Employment, in:	
	Research and Development	Manufacture and Other Work
Mineral Oil Refining	1·84	3·40
Aircraft	1·49	0·40
Chemicals and Allied Trades	1·27	1·40
Electrical Engineering	1·18	0·86
Rayon, Nylon, etc.	0·61	0·48
Precision Instruments, etc.	0·56	0·30
Non-Ferrous Metals	0·32	0·76
Other Plant and Machinery	0·29	0·68
Other Manufacturing	0·24	0·32
Constructional Engineering	0·17	1·30
Motor Vehicles, etc.	0·16	0·26
Iron and Steel	0·12	0·38
Bricks, China, Glass, Cement	0·11	0·22
Other Metal Goods	0·09	0·41
Food, Drink, Tobacco	0·08	0·23
Wood, Cork, Paper, Printing	0·07	0·10
Other Textiles and Leather	0·07	0·25
Shipbuilding and Marine Engineering	0·03	0·33
Agricultural Machinery	0·03	0·34
Cotton	0·03	0·05
Railway Equipment	0·02	0·41
Wool Textiles	0·02	0·06
Clothing	0·003	0·012

* Based on Table I, Appendix III, *Scientific and Engineering Man Power in Great Britain*, Office of the Lord President of the Council and Ministry of Labour and National Service, 1956.

TABLE II

UNITED STATES

COMPANIES CONDUCTING RESEARCH AND DEVELOPMENT AS PER CENT OF ALL COMPANIES – BY SIZE OF COMPANY AND INDUSTRY – 1953*

Industry	All Sizes of Companies	Companies with Total Employment of: 0–99	100–499	500–999	1000–4999	5000 or more
	Per Cent conducting Research and Development					
Food and Kindred Products	5·9	4·1	12·0	39·3	50·0	89·5
Chemical and Allied Products	37·9	35·1	43·0	68·4	98·5	100·0
Petroleum	6·9	5·5	9·5	29·4	69·2	90·0
Rubber	28·5	23·6	29·3	40·6	70·8	100·0
Primary Metal	12·1	5·1	23·6	32·5	69·6	96·7
Fabricated Metal	17·5	12·6	39·6	40·7	69·4	100·0
Machinery	31·6	26·7	45·7	71·8	89·4	100·0
Electrical Equipment	46·6	40·0	53·0	72·8	91·7	100·0
Aircraft and Parts	57·2	62·8	25·9	87·5	89·5	100·0
Prof. and Scientific Instruments	32·2	27·9	39·7	65·4	97·0	100·0
Other Manufacturing	5·4	3·2	14·1	32·3	38·5	89·3

* National Science Foundation, *Science and Engineering in American Industry*, Table A.3.

TABLE III

UNITED STATES, 1953

EXPENDITURE ON RESEARCH AND DEVELOPMENT PER COMPANY FOR ALL COMPANIES CONDUCTING RESEARCH AND DEVELOPMENT – BY SIZE GROUPS OF COMPANIES AND INDUSTRIES*

($ million)

Industry	Expenditure on Research and Development per Company for Companies:	
	With Employment over 5000	With Employment 500–999
Food and Kindred Products	1·0	0·02
Chemical and Allied Products	8·6	0·44
Petroleum Products and Extraction	6·1	0·12
Rubber Products	5·2	0·09
Primary Metal	1·5	0·02
Fabricated Metal	1·7	0·15
Machinery	3·6	0·16
Electrical Equipment	28·7	0·42
Aircraft and Parts	35·8	1·95
Prof. and Scientific Instruments	11·4	0·70
Other Manufacturing	5·9	0·03

* Based on National Science Foundation, *Science and Engineering in American Industry*, Tables A.3 and A.9.

TABLE IV

UNITED STATES, 1951

RANKING (1–14) OF MAJOR INDUSTRIAL GROUPS AS TO RESEARCH AND DEVELOPMENT EXPENDITURE AND SIZE OF FIRMS

Industry	Ranking of 14 Industrial Groups According to:		
	Expenditure on Research and Development (non-Govt. financed), as % of Sales (1)	Average Size of Firm, No. of Employees (2)	% of Total Employment in Upper 1% of Firms (3)
Electrical Machinery	1	5	7
Professional and Scientific Instruments	2	9	4
Chemical and Allied Products	3	8	2
Stone, Clay and Glass Products	4	12	9
Transportation Equipment	5	1	1
Machinery (except Electrical)	6	10	8
Textile Mill Products	7	6	13
Rubber Products	8	3	5
Petroleum Refinery	9	2	3
Fabricated Metal Products	10	11	11
Other Manufacturing	11	14	12
Paper and Allied Products	12	7	14
Primary Metal Industries	13	4	6
Food and Kindred Products	14	13	10

Col. (1) based on Tables in *Scientific Research and Development in American Industry*, 1953, Bulletin 1148, U.S. Dept. of Labor.

Cols. (2) and (3) based on *Size Characteristics of the Business Population*, Survey of Current Business, May 1954.

TABLE V

CHEMICAL INDUSTRY – GREAT BRITAIN, 1951

Size Group (Largest First)	Average Research and Development Expenditure as % of Turnover	Highest Research and Development Expenditure as % of Turnover	Lowest Research and Development Expenditure as % of Turnover
1 (13 firms)	3·2	10·0	0·5
2 (13 firms)	1·8	5·4	0·3
3 (13 firms)	1·5	7·8	0·3
4 (13 firms)	1·4	3·8	0·3
5 (12 firms)	2·6	3·7	0·5

TABLE VI

CHEMICAL AND ALLIED PRODUCTS – UNITED STATES, 1951*
PERCENTAGE OF COST OF RESEARCH AND DEVELOPMENT
TO VALUE OF SALES

Size Group of Company	Average Percentage	Median Percentage	Lower Quartile Percentage	Upper Quartile Percentage
Less than 500 employees (162 companies)	2·5	2·6	1·3	5·8
500–4999 employees (58 companies)	2·4	1·7	1·0	2·7
More than 5000 employees (22 companies)	2·6	2·4	1·7	3·9

* *Scientific Research and Development in American Industry*, 1951, U.S. Dept. of Labor, Table C.20.

TABLE VII

FIFTEEN LARGEST OIL COMPANIES IN THE UNITED STATES, 1948

Size Group Total Assets $m.	Average Size of Research Staff	Smallest Research Staff	Largest Research Staff
Over 1500	811	455	1123
1000–1500	944	469	1129
500–1000	691	378	1314
230–500	184	95	347

CHAPTER VIII

THE DEVELOPMENT OF INVENTIONS*

> The writer may derive a more delicate satisfaction from the free confession of his ignorance, and from his prudence in avoiding that error, into which so many have fallen, of imposing their conjectures and hypothesis on the world for the most certain principles. – DAVID HUME.

I

THE ISSUES

ALTHOUGH when this work was started it was not intended to say anything in detail about the development of inventions, it subsequently became increasingly apparent that some comment on it was unavoidable. For even those who are prepared to accept the description and analysis of invention as given in the foregoing pages might well protest that this is, after all, the less important part of the story of technical progress and that the real determinants of the rate of advance will be the scale and the speed of the efforts made to perfect new commodities and devices and to contrive ways of producing them cheaply and in quantity.[1]

This, in effect, is a variation, much more convincing than the original, of the doctrine outlined in Chapter II. It can perhaps fairly be summarised as follows. The costs of carrying inventions forward to final use are high and are steadily increasing; development is very much a new function, only recently has it become a separable and important task for business. These costs are now so large that only a big firm can afford them. Indeed, the advantages of size may persist to the point at which a few firms, or even only one, within an industry can operate to the best effect. The risks are now so great that speedy and effective development is only likely where it is nourished by monopoly profits. Patent rights are one form of protection but, over and beyond these, a firm which finds little rivalry in its special field of manufacturing, and therefore has less to fear from technical developments closely parallel to its own, will carry out development most confidently and therefore with the greatest effect.

These contentions are obviously germane to policy; they imply that the

* [A discussion of the later studies on Development, together with additional material we have ourselves been able to collect, will be found in Chapter X, pp. 212–19.]

[1] Thus *The Economist*, May 5th, 1956, speaking of titanium: 'Even though Dr. Kroll, back in the thirties, may certainly fit the stereotype of the individual inventor, it is hard to fit the whole titanium exploit, including the last chapter of it that forms I.C.I.'s titanium venture, into the traditional folklore of invention in the free capitalist economy. But it may well epitomise the ruling pattern of innovation in our time.'

future will lie with the bigger firms and that monopoly has its merits. The whole complicated subject is, indeed, worthy of thorough examination. It has not been possible here to undertake such a study nor to bring to bear any information beyond that acquired incidentally in the present narrower enquiry. What follows, therefore, is an attempt, not to reach settled conclusions, but to pose questions and clarify issues.

The most important specific questions seem to be these:

(*a*) Are the costs of developing inventions becoming progressively greater? If so, why?

(*b*) Can the outstanding twentieth-century successes in development usually be attributed to large firms?

(*c*) An invention once made, are there reasons for assuming that the development will more readily be entered upon and more wholeheartedly pursued where one or a very limited number of firms can enter the field and share in the profits? Is monopoly the most favourable environment for development?

II

DEFINITION

One reason why systematic knowledge of development seems so scrappy, a few generalisations passing on unchanged and unchallenged from author to author, lies in the awkward problem of definition. We speak of the development of some revolutionary idea, such as the jet engine or the Wankel rotary piston engine, meaning the building of one machine which establishes the fact that results can be obtained by the new principle. At the other extreme we speak of the development of a new aircraft design, which of itself embodies no radical innovation, meaning the systematic testing and modifying in numerous minor ways of the design to satisfy standards of performance and safety for the special conditions which the aircraft will encounter. And, between the two extremes, of course, there are to be found innumerable intermediate conditions.

Even if study is confined to development in the first sense, to the work following closely upon and directly connected with the original innovation, useful lines of demarcation are not easily drawn. Sometimes development consists of finding ways of producing on a large scale the same thing – or broadly the same thing – as has been produced already on a small scale; a new man-made fibre, for instance. But the development of the jet engine, in the course of which power and reliability were increased, also involved changes in the physical form of the product.

Assuming this hurdle is surmounted, how are the costs of development to be ascertained? If they are wholly incurred by one firm which brings the development to a successful conclusion the task is easier than where a number of firms have pursued the same development, one or more successfully and others perhaps fruitlessly. Some of the heaviest development costs have probably been spent in wholly abortive efforts, such as the attempts to store electricity in large quantities or to make perpetual motion machines. In assessing the costs of the final success in transmuting metals, would it be necessary to include all the expenses of all the alchemists of the past fifteen hundred years?

For practical purposes, the important thing is to fix upon the 'period' of development. In one sense this period has no end – at least until the process, commodity or machine has completely disappeared from use. Improvements were still being made to the sailing ship right up to the end of the nineteenth century; indeed the competition of the steamship stimulated numerous improvements in the sailing ship. In this chapter, the discussion is confined to that period, intimately associated with an invention, which ends when commercial utilisation appears at least to some people to be feasible.[1]

III

DEVELOPMENT IN THE NINETEENTH CENTURY

The contrast between the relative cheapness of inventing and the heavy costs of perfecting and developing was frequently remarked upon in the nineteenth century. The United States Commissioners of Patents frequently refer to it in their reports and it is one point on which the witnesses before the House of Lords Committee on the Patent Law Amendment Bills of 1851, the Royal Commission on Patent Law of 1865, and the Commons Committee on Letters Patent of 1871, were almost unanimous. The views of Sir William Armstrong are typical. He told the Royal Commission in 1865 that 'mere conception of primary ideas in invention is not a matter involving much labour, and it is not . . . a thing demanding a large reward; it is rather the subsequent labour which a man bestows in perfecting the invention, a thing which the patent laws at present scarcely recognize'.

In the middle of the century, Goodyear is known to have spent some $130,000 on the vulcanisation of rubber, mainly in evolving a satisfactory process for commercial use. Richard Roberts, the inventor of the self-acting mule, said that he had spent thousands of pounds on perfecting it; there were many inventions on which very large sums had to be spent, 'for there are many things which cannot be seen through all their parts without experiment'. The Bessemer process was an invention which required the solution of certain development tasks before it could be used, especially

[1] Thus the first stage of development of Terylene or nylon might be thought to end with the pilot plant stage; that of the jet engine when the Germans and British put fighters into operation; that of high-altitude rockets when the V2s started to fall with some regularity on London; that of the steam-turbine around 1900; that of viscose rayon around 1910. But when it was announced in 1955 that the first British diesel-electric locomotive had been 'developed' at a cost of £500,000, when nearly all the main-line locomotives in the U.S. had become diesel electrics, this clearly is a reference to a later phase of development.

those arising out of the presence of phosphoric pig-iron in the converter. Bessemer told the 1871 Committee that he had spent £16,000 on solving this problem. He argued that no man would go to the expense of introducing a new process without patent protection; they would let another have that expense. Sir William Siemens, another witness, referred to the great expense and delay involved in developing a process; without patent protection men would not go to this expense. E. K. Muspratt, the alkali manufacturer, said that in the chemical industry an idea had to be elaborated by practical men for several years, often with a very large expenditure, before it became practical.

Towards the end of the century, it is known that Parsons spent considerable sums, perhaps in the neighbourhood of £100,000, on the development of the steam turbine. The development of viscose rayon from the first patent in 1892 to the first commercial marketing in 1910 in Great Britain by Courtauld's seems to have cost, including the first manufacturing plant, not less than £250,000. The patents were also licensed to German and French firms, which are known to have incurred heavy expenditure for the same purpose.

IV

THE GROWTH OF DEVELOPMENT COSTS

Development costs were, therefore, sometimes heavy in the nineteenth century, and it is unfortunate that more detailed comparisons are not available, for example, of the costs of development of the steam turbine and the gas turbine, of the first aniline dyes and the later ones, of the Solvay process for producing soda ash and polyethylene; of the Bessemer steel-making process and the Kroll titanium process; of the methods of making aluminium and those for producing magnesium.

Even in the absence of such studies, however, it seems difficult to escape the conclusion that there is nothing in the nineteenth century to match some of the spectacular cases of costly development to be found in the twentieth. Many of the figures quoted are, indeed, unduly swollen by the inclusion of the cost of the final manufacturing plant. But it appears that about $1 million was spent on the research and development of nylon; some £4 million was spent on the development of Terylene (although this figure apparently includes the cost of pilot plants, some part of the product of which was sold commercially). The Radio Corporation of America, one of a number of American firms engaged in the task, spent $2½ million on research and advanced development of television; in Great Britain, E.M.I. is said to have spent £550,000 for the same purpose.[1] The cost of development of penicillin ran into millions of dollars; that of the hot cathode fluorescent lamp is given as $170,000; of the long-playing record, $250,000; of the Houdry catalytic cracking process, $11 million.

[1] Lord Brabazon of Tara, *The Brabazon Story*, pp. 151–2.

Is there, then, some general law arising out of the nature of modern science, the character of modern technology or the pattern of modern markets, indicating that a new idea of a given inventive content will increasingly cost more to develop? If so, this would in one sense be surprising, for it might be supposed that scientific and technical advance would make for economy in the effort of development; that costly empirical testing could be replaced by cheaper and quicker scientific calculation; that cul-de-sacs could be perceived before they had been pursued too far; that scientific market research would render less hazardous the commercial introduction of a new product. Why should technical progress make everything cheaper except the process of development?

Nevertheless, the grounds for assuming that development steadily becomes a more formidable, intricate and costly task are strong. They will be discussed, first in terms of technical forces and then in respect of market factors.

The cost of perfecting an invention will depend partly upon how far a small-scale experiment or model will suffice to provide adequate knowledge for conducting operations on a larger scale. Whether the one will reveal reliable data for the other and how far this can be known beforehand are matters on which there has long been dispute.[1] Although it is only too easy to find exceptions to any generalisation, it appears that there are significant differences, for instance, between engineering at its various levels, where quite small models will often provide as much useful working knowledge as full-scale machines and where the cost of even a full-scale machine may be relatively modest, and the chemical industries, where relatively large pilot plants are often needed for the empirical accumulation of 'know-how'. In so far, therefore, as the growing industries, such as chemicals, are those where very large scale experiments are a condition precedent to commercial operation, development costs will tend to increase.

Development consists of applying existing technical knowledge to exploit some new idea. It follows that the greater the stock of technical knowledge, the wider the range of effort which may be brought to bear upon any development task. There are now more different possible routes for reaching a predetermined target; more ways of spending money which offer some chance of success. This, of course, gives heightened value to a new type of judgment, the choice of the methods combining in the best way the highest chances of success and the lowest cost. (Conversely, it is probably easier to waste money in industrial development than it has ever been.) But so long as technical knowledge accumulates, it seems reasonable to suppose that potentially worth-while avenues of expenditure will increase in number.

Much development, especially in the field of chemistry and chemotherapeutics, has in this century consisted of empirical search and observation in wide new fields suddenly opened up by one crucial discovery or

[1] Thus Leonardo da Vinci: 'Vitruvius says that small models are of no avail for ascertaining the effects of larger ones: and I here propose to prove that this conclusion is false'.

invention. When the success of penicillin drew attention to the possible virtues of moulds, then the area of search for new strains of penicillin or other moulds with similar qualities became virtually unlimited. Of course, the opening up of one new field of search might simultaneously close down another – doubtless the success of penicillin brought to an end much more traditional work on antiseptics. But it seems that frequently the new fields of effort are additional to, and not a complete substitute for, the old.

The accumulation of technical knowledge is a two-edged weapon; it helps to determine what is possible but it may also define what is impossible. With the area of probable successful search narrowed down, firms may be more ready to risk development expenditures which, though great, appear tolerable. Suppose, for example, it were known that a satisfactory method of synthesising insulin existed as one among 10,000,000 possibilities. The risk involved in a systematic search, which might not reach success until the 10-millionth experiment had been performed, might be too great for any firm to undertake. Under these circumstances, of course, a firm might be prepared to try to find the answer without a systematic search: it might be prepared to devote a limited sum to a search for the answer, relying upon intuition or the possibility of a lucky shot. In that case the firm would have to be prepared to spend with the chance of a nil return. But if, through increased technical knowledge, it became known that the correct answer must be found as one within 100,000 possibilities, the firm might be prepared to spend a large sum knowing that a systematic search must bring success, even if the chance of getting the winner at an early stage in the search be completely discounted.

Finally, on the technical side, the increased costs of modern development may be attributable to the greater caution of manufacturers in not taking their problems out of the laboratory or moving beyond a pilot plant stage until they are convinced that full-scale manufacture is wholly feasible. The absence of such discretion in the nineteenth century sometimes led to mistakes and back-tracking, and consequent loss of time and money.[1]

Market factors, it is argued, will also tend to increase the cost of development. The more discriminating becomes the taste of the consumer the more important is it that a new commodity should not be put on to the market until its quality and reliability can be guaranteed. The producer now searches in the laboratory for weaknesses which might in earlier times have been discovered by putting the commodity on the market and allowing the consumer to discover the defects through use. New commodities must force their way into markets against the resistance of competing traditional types of goods. If they are launched prematurely while still imperfect, their reputation may be finally and fatally damaged. Perfecting in the laboratory cannot always be made a substitute for testing through widespread use but it will probably become increasingly so.

[1] J. M. Cohen, *The Life of Ludwig Mond*, p. 218, points to this as the cause of some of the vicissitudes in the early business career of Ludwig Mond.

It seems, too, with many modern processes and commodities, that the handling of the materials can be extremely dangerous, calling for special security devices and most scrupulously standardised procedures. This helps to explain the high development costs inseparable from work in atomic energy or in the manufacture and mass application of drugs such as polio vaccine.

On occasions, a new commodity may have to be put on the market on a minimum scale if it is to make an appropriate impact.[1] An advertising campaign to break down the initial resistance of the consumer is an overhead cost which it is desirable to spread as widely as possible. A car manufacturer might be reluctant to introduce a new form of transmission unless it could be embodied in at least one full assembly line. The older commodity, presumably, is already being manufactured on the scale which brings lowest costs; the introduction of a new competing commodity will meet additional obstacles unless it can also be manufactured from the outset on an economical scale. The more extensive the scale on which the new product must initially be manufactured, the more formidable may be the task of development.

But the most important reason for these heavy expenditures on development seems to be one in which technical and market factors are both involved: the high competitiveness of an economic system where technical progress is general and rapid and where, in consequence, one invention may quickly supplant another. The point can best be established by considering the puzzles which confront a business trying to decide whether or not to develop an invention and, if the development be decided upon, how much should be spent and at what rate.

Such a firm is faced by a group of questions, the answer to each being mutually dependent upon the answers to all the others. The discussion can, therefore, begin at any point. The firm might first ask: is the development technically feasible? But the feasibility of development will, at least in part, turn on the resources made available for it. The second question would be: the development once successfully completed, is it likely that a profitable market will be available for the final product? The answer here will of course depend upon a whole series of guesses: the probable cost of production, the probable volume of sales (which will help to determine cost per unit) and so on. But the answer to this second question pushes the firm back to the answer to the first question: for the probable cost of production of the finished product will depend upon how successful the developers are in devising cheap methods of production, and this, in turn, may be a function of how much the firm is prepared to spend on development.

The firm, therefore, finds itself arguing in a circle. If it knew how cheaply a commercial product would finally be put on the market it would know whether development was worth while. But whether development is likely to be successful will partly turn on how much is spent on development.

[1] In the case of both nylon and Terylene in Great Britain an output of about 10 million pounds per year was considered the minimum commercially desirable.

And the costs of development may help to determine the price at which the final product can be sold.

Practically every important decision on the development of a new product represents a heavy plunge by the firm concerned; there can be no precision or certainty in what is done. Let it be supposed, however, that the firm has actually decided that development is technically feasible and that, with expenditures of an acceptable magnitude, it seems that the final article can be put on the market at a price which will give a volume of sales providing a margin over cost (cost including the cost of development) yielding a return on investment probably larger than that obtainable by investment in any other possible direction.

The firm decides to spend up to a given sum on development, always of course hoping that good fortune will be on its side in that the cost of development will prove to be less than expected, or the final product will have unexpected merits. Another question then arises: how quickly should this money be spent? Rapid expenditure may be wasteful. Pressure for swift development may bring quicker results but cost more finally. To gather together quickly a group of technologists of second-best quality when more deliberate methods would have made possible the appointment of more able men; to plunge a large number of technologists into a task before the real problems have been clearly perceived (which may partly be a function of time); to pursue simultaneously likely and unlikely avenues when it would have been more economical to try out the more likely first;[1] to attempt to make progress at the same time in the different stages of development which could more conveniently be carried out successively; to have research groups standing by in case they may be needed: these are all forcing tactics, justifiable in emergency, which can rapidly produce decreasing returns to cost. Speedy development, therefore, is likely to be expensive.

On the other hand, speed may have its attractions whatever its cost. Development consummated over five years rather than ten brings profitable sales all the sooner: the firm is standing out of its money for only half the time. There may be other good though less tangible reasons for haste: the excitement of an all-out effort may itself be a stimulus and may contribute to results; the shorter the period of development the less hazardous will be the guesses about the potential future market.

The main reason for haste, however, seems to be the fear of competition. The most favourable circumstances in which the firm could find itself would be where it had fallen upon an invention so novel in character that it was unlikely to be challenged or superseded for a long time; where the firm held

[1] For a very interesting account of how, in the efforts to get quick results, a good deal of wasted effort has necessarily to be tolerated, see the account of the development of the I.C.I. sodium process for the making of titanium. R. B. Mooney and J. J. Gray, 'I.C.I.'s New Titanium Process', *I.C.I. Magazine*, June 1956. Of course the outstanding example of prodigal expenditure in the interests of speed is that of the development of the first atom bomb.

what seemed to be an unbreakable master patent; where it felt confident that by secrecy or subsequent patenting it could keep to itself the fruits of development, and where it possessed skills and experience in the relevant manufacturing processes enjoyed by few or no other manufacturing firms.

Although instances are to be found of firms which have in fact enjoyed such good fortune, they are likely to be few; in even fewer cases will the firm believe, before the event, that its luck will hold to this degree.[1] Even a firm with good master patents, power to defend them and to retain exclusive use of its know-how will face the future with some trepidation. Those who have taken out a patent have given a hostage to fortune; for the secret, thereby publicly revealed, will set other minds working along related, and perhaps non-infringing lines, to ends which no one can foresee. Beyond that, there is always the chance that an invention, along entirely different routes, will bring onto the market a competitive final product. Where the strategical and tactical position of the firm is less favourable than the ideal pictured above (which surely must very often happen), the natural anxieties of the firm lest it be supplanted will be all the greater.

Firms which once set their hand to development will usually be anxious to produce results quickly, even if this means a higher total development cost. Whether they feel simply that their own innovations will breed competition against them or whether they fear generally that the constant forward pressure of technical knowledge on a wide front will in some unpredictable way bring rivalry, time will appear to them to be of the essence for successful exploitation. If it is true that in these days there is a greater chance than formerly of one invention being superseded by another moving in from an entirely different direction, this would help to explain the upward trend in development costs.

To reduce its risks, therefore, the firm will increase the speed of development, and this will enlarge the annual, and perhaps the total, expenditures, in respect of any one invention. The greater the risks the stronger the justification for higher development investment. But it is part of the tantalising antithesis of this subject that if the risks are *too* great, if the chances of being supplanted *too* palpable, the development may not be attempted at all. A small wind may fan the flames but a strong wind extinguish them completely.

V

SOME CONTRARY DATA

The solid value of heavy development expenditure in many instances cannot be denied; further supporting testimony is provided by a number of cases

[1] Note that anyone who has solved a puzzle by methods which after the event seem simple, finds it difficult to believe that others will not also discover the critical clues. That was the terrifying anxiety in the minds of most atomic scientists on the Allied side during the war.

where development was suddenly speeded up by a heavy increase in the funds devoted to it, as with Terylene when taken over by I.C.I. and du Pont, the jet engine when taken up in Britain by Rolls-Royce and the diesel-electric locomotive in the United States when developed by General Motors.

On the other hand, particularly in the earlier phases of development, much progress has often been made by the use of very moderate means. Even with the jet engine:

> The total cost of Power Jets' work from its beginning in 1936 to the middle of 1939, when it had definitely shown that the turbo-jet was not a dream but a practical new type of propulsive system, was only some $100,000. The total cost of Heinkel's development over the same period of time was little if any greater than this.[1]

Of the seventy cases studied in detail in this volume, successful development appears to have been carried out by individuals or smaller firms without enormous cost in air conditioning, automatic transmissions, bakelite, 'Cellophane' tape, electric precipitation, magnetic recording, power steering, quick freezing, shell moulding and the synthetic light polariser. The discovery of the crease-resisting process for textile fabrics merits special attention.[2] This was a path-breaking discovery which has been universally employed and has long remained unchallenged, where the invention and development was carried through by a firm of medium size with expenditures very much smaller than those normally associated with twentieth-century chemical inventions.

There are also a number of striking instances outside the seventy cases where development has been carried out successfully by small firms, particularly in the electronics industry.

Again, instances are not unknown where firms of different size have pursued parallel tasks of development and the victory has not always gone to the large battalions. In the evolution of the cotton picker one large combine and two smaller firms reached results at about the same time.[3] Television provides an extremely interesting illustration.[4] By 1932 it had been clearly established that mechanical systems offered little prospect of further advance and that the future lay with electronic systems. Zworykin, working in the laboratories of the Radio Corporation of America had already produced his 'iconoscope', perhaps the crucial invention in the emergence of modern television. From then on, the search for the perfection of a practical electronic television system went forward in several quarters. Farnsworth and his associates, with more limited resources than R.C.A., contributed much to the final results. And in Great Britain in 1936 the first regular system of high-definition television broadcasting in the world was established as the

[1] R. Schlaifer and S. D. Heron, *Development of Aircraft Engines and Fuels*, p. 90.
[2] See pp. 245–8. [3] See pp. 243–5. [4] See pp. 307–9.

result of the work of a small team in Electrical and Musical Industries Ltd. under the direction of I. Shoenberg. Comparisons of this kind can, of course, never be exact or conclusive. In this instance, it would have to be pointed out that the system introduced by R.C.A. in the United States was one of 525 lines as compared with that of 405 of E.M.I.; that in the United States the task of producing a satisfactory system was complicated by the great distances involved; that E.M.I. held licences under the Zworykin patents. Even so, it should counter any facile assumption that the size of a firm is the sole determinant of its commercial imagination or its power in development.

VI

MONOPOLY AND DEVELOPMENT

The grounds for believing that inventions will be most actively developed under conditions of monopoly are various: they may amount to little more than a plea for the merits of the patent system; they may constitute simply a case for the economies of large-scale development for, if the bigger the better, the best will be where an industry consists of only one firm; monopoly then appears as an incident to size.

But a third and more sweeping suggestion is now to be examined; that the more sheltered and exclusive the position of the firm the more likely it is to embark upon, and succeed in, development. To put it concretely, that the ideal industrial organisation for snapping up and transmuting inventions into serviceable products is that where the firm holds a master patent on an important invention, is confident it can rope off for itself the fruits of development, and is dominant in the relevant field of manufacture.

At this point scepticism appears pardonable. For if, as suggested earlier, the large development costs of modern times arise partly out of the threat of competition, it seems inconsistent to assert at the same time that only under the security of monopoly will these large sums be spent.

The case for competition, even in the process of development, would run as follows. Competition is in itself a stimulus. The knowledge on the part of one firm that other firms are on the same track forces them all to move more rapidly, the prize lies in what can be gained from priority and, transient as the lead may be, it still remains a powerful motive. Development under such conditions, to be sure, may mean overlapping of effort, it may result in much waste, but it brings about a larger total effort and, carried out by different firms pursuing different lines, may not only make success swifter but sometimes make the difference between success and complete failure. Why, it may be asked, should the knowledge on the part of one firm that success will bring to them the whole of a long-continuing reward, be a sharper spur to effort than the recognition on the part of a number of firms that opportunity beckons but only to a relatively short-lived reward, the size of which depends on their own forcefulness and skill?

Whether a radical invention is more likely to be developed swiftly and effectively if the industry most directly concerned consists of one very large firm holding a secure patent than if the industry consists of a substantial number of small firms and the patent is in the public domain or is freely under licence, will depend partly upon factors other than the degree of monopoly: whether businessmen are temperamentally disposed to competition; the character of the invention; the apparent feasibility of the development; whether a market seems to be there for the taking or would have to be built up against the solid resistance of existing commodities; whether it is possible for firms to retain the secrets of accumulated know-how; whether firms of the same size are roughly equal in the facilities they possess for development; and how each firm may rate its chances against others in a struggle for technical and commercial priority. General reasoning, that is to say, cannot carry us to a final conclusion. Undoubtedly the *capacity* of the monopolist to carry through development is greater than that of any one of the firms in the second set of conditions, for any one or all of these firms might find it impossible to take the risks involved. But it is relevant to consider not merely the *power* to do something, but also the *will* to do it.

At this point there seems to be no choice but to have recourse to actual experience. This provides some support for three conclusions:

(*a*) that certain examples, frequently put forward as proof of the virtue of monopoly, can bear interpretations different from those normally placed upon them and, in any case, are hardly typical;
(*b*) that there are a large number of instances of development occurring under keenly competitive conditions;
(*c*) that the fewness of the producers in an industry where an important invention arises may in itself constitute a block to development.

These conclusions will be illustrated in turn.

The Case of Nylon

The case of nylon is perhaps the most commonly quoted as proof of the success of a one-firm development. There, a very large firm with a dominating position in the chemical industry supported for several years in its research laboratories a brilliant scientist whose chances of success must at times have appeared depressingly slight. In the event, a remarkable discovery was made, the first synthetic fibre was produced upon which du Pont obtained patents and to the development of which it devoted very large sums. Within a short period the product was available commercially in quantity.

It is not surprising that this story should excite the imagination and colour the popular view of how progress can most expeditiously be effected. In fact, however, it is reasonable to assume that potential competition played some part in determining the policy of du Pont. Their patents were strong but it was known at the time that at least one German firm had made progress in parallel work on synthetic fibres. In the event, the progress of the German technologists proved slower than might have been expected. The war interrupted their efforts and they encountered unexpected obstacles in development. The German fibre, Perlon, however, came on to the market during the war and has since enjoyed large sales in Europe.

Beyond that, the nylon case, it might be argued, is untypical. There can hardly ever have been another example in industrial history where, the discovery once made, so large and obvious a peace-time market presented itself if only the invention could be successfully developed. In the case of nylon the market was that for women's stockings. Fashion had long decreed that women in the Western world should wear thin stockings of light shades; cotton and wool fibres were unsuitable for this purpose; silk itself was expensive, artificial silks of various types were suitable but had a very short life and stood up badly to frequent washing. Stockings of the new fibre were an answer to a multitude of prayers;[1] for nylon could be drawn very fine, was relatively hard-wearing and was easy to wash. When it is further added that the fibre by good fortune proved very suitable for use on the knitting machine, and that, once the first fibre had been produced, the tasks of development, while formidable, did not appear insuperable, there can have been little doubt that here was a windfall of great magnitude.[2]

Further, and without in any way seeking to belittle the great achievements of du Pont, is it unreasonable to suggest that, if Carothers had made his discovery in a University instead of in the du Pont laboratories, and if non-exclusive licences had been available or the patents had been in the public domain, many chemical or other firms would have been prepared, in competition, to devote large sums to its development? And, if this had happened, might not successful development have resulted just as quickly as it actually did?

Development under Competition

From the very large number of instances which might be quoted where firms have been prepared to take the risks of competition in development a few of the most interesting may be referred to here.

The first practicable method of producing *Titanium* was evolved by an individual inventor, who found the chemical and metallurgical firms

[1] During and immediately after the war when nylon stockings were not produced in Europe, the scale of smuggling in them and their commerce on the black market were both enormous.

[2] The scale of the commercial scoop can be judged by the fact that the nylon stocking, where available, rapidly replaced all others. In the United States in 1954, 97 per cent of the women's stockings produced were of nylon. From 1939 until 1955 du Pont was the only maker of nylon-type yarns in the United States. Nylon 66, the type du Pont first chose, has remained unchallenged among the different types available. Nylon has proved to be by far the greatest of du Pont's products by sales and almost certainly by profits; in 1954 it may have accounted for one-third of du Pont's total profits.

extremely lukewarm about the potentialities of his discovery.[1] The early stages of development were carried through during the Second World War by a United States Government research agency. The problem of the companies at first was to find a market for this remarkable metal, but, from 1950 onwards, this was partly solved for them by the demands of military agencies for titanium for use in aircraft. In Great Britain I.C.I. adopted its own sodium process after experimenting with and discarding the original Kroll magnesium reduction process. In the United States, a number of firms, some of them government-financed and some acting independently, took substantial risks in the production of the metal and in research directed both towards the further development of the Kroll process and to improvements upon it.

Shell Moulding was a brilliant invention whereby castings are produced with smoother surfaces and more accurate dimensions, thus enormously reducing the need for machining. It was invented by a German during the Second World War; after the war it was found by the Allied technical teams and was published as belonging to the public domain. Its potentialities were so great and so obvious that dozens of firms, large and small, old and new, vigorously took up the perfection of the idea.[2]

Polyethylene was discovered in the research laboratories of Imperial Chemical Industries, which took out patents on it before the Second World War. Their method of manufacture involved the use of very high pressures. After the war two new processes, which seem to have some similarities, were discovered in which high pressures were no longer required. Of these, one was discovered by Professor Ziegler of the Max Planck Institute. The Ziegler process was eagerly seized upon for development by a number of firms in the United States, Germany and Great Britain. In the United States licences were taken out by du Pont, Union Carbide, Koppers, Monsanto and other chemical companies. In Great Britain an exclusive licence was taken out by Petro-Chemicals Ltd., a firm afterwards absorbed by Shell.[3]

The development of the *Jet Engine*[4] is perhaps not a clear-cut case, for the engine was badly needed for war purposes and the firms concerned undoubtedly took a wider view of their national responsibilities than would have been determined by purely commercial considerations, but very vigorous competition occurred and, of course, still continues.

Insulin in Great Britain is a highly relevant case.[5] Between 1922 and 1941, five British concerns held licences to develop and manufacture insulin; there was no technical collaboration between them. There was further competitive prodding from abroad in the form of improvements principally arising from the brilliant work of Dr. Hagedorn of the Nordisk Laboratorium in Copen-

[1] See pp. 314–17.
[2] See pp. 295–6.
[3] See pp. 279–81.
[4] See pp. 262–6.
[5] The authoritative story is told in the report of the Monopolies and Restrictive Practices Commission, *Report on the Supply of Insulin*, 1952.

hagen.[1] Under these conditions rapid progress was made. With improvements in the method of manufacture, the price of the standard insulin pack was brought down from 25s. in April 1923 to 1s. 6d. in July 1935. In 1941, however, these conditions changed, originally because of war-time needs. Since that year, with one short break, the manufacturers of insulin have engaged in the fullest exchange of technical information and have collaborated at every stage in research, development and production and have fixed common prices. But it would be misguided to exaggerate the improvements that accompanied co-operation after 1941 and ignore the more spectacular improvements that went along with competition in development before 1941.

There are other instances, too numerous to describe in detail here,[2] of successful development under competitive conditions: the cotton picker, air conditioning, the ball-point pen, television, automatic transmissions, the catalytic cracking of petroleum, the new forms of three-dimensional films, the electron microscope, the hardening of liquid fats, magnetic recording, acrylic fibres, power steering and synthetic detergents.*

Security as a Block to Development

The very *fewness* of the firms in an industry may (it would be going too far to say will) result in delay in exploiting new ideas or even in their permanent loss to society. Here we move again among the intangibles. A firm with a dominating position, conscious of its power to pounce if its position should suddenly be put into jeopardy, may be so confident of being able to deal with incipient competition as to become sluggish.[3] And the general knowledge that it can so pounce may deter potential rivals and thus leave the field empty.

In an industry with one or a few firms the chances are obviously greater that a new thing may be missed either simply because it is overlooked or, even where it is known about, because it happens not to excite the imagination. Illustrations are easy to come by. In the 1930's the large oil companies

[1] An import duty was imposed on insulin, but, as a result of public agitation, the Government was forced to remove it in 1934.

[2] But some detail is given in the cases recorded in Part II.

* [It is interesting that firms responsible for important innovations may license them to other firms with which they remain in competition both commercially and in the matter of research. Thus the Beecham Group, with semi-synthetic penicillins, has followed a policy of selective licensing in foreign markets but retained the right to sell its own brand names in the same territories.]

[3] A. A. Bright, *The Electric Lamp Industry*, p. 389, has pointed out that General Electric, content with its dominating position in artificial lighting, was sluggish in taking up the development of fluorescent lighting, but that, once it was aroused by the presence of other pioneers in the field, its bountiful resources and the assistance it derived from the innovating work of others, quickly enabled it to reassert its predominance. Similarly R. Schlaifer, *Development of Aircraft Engines*, pp. 100–3, suggests that the monopoly possessed between the wars by Stromberg in aircraft carburettors and by General Electric in superchargers delayed improvements.

moved slowly until Houdry, an outsider, with his catalytic cracking process forced the pace and finally galvanised the industry into action. In the United States the big firms were slow to take up and exploit the invention of power steering for passenger cars; in Great Britain they have been tardy in the development of automatic transmissions for the lighter motor cars. For a long time the large photographic firms ignored the possibilities of xerography. For a number of years the leading companies in the non-ferrous and electrical industries showed no interest in ideas for producing ductile titanium which have since been taken up eagerly.

The case of taconite is especially relevant. Taconite is a rock which forms the underlying bed of the Mesabi deposits of high-grade ore in the United States and constitutes an almost unlimited reserve of iron. Quite recently it has been increasingly mined by a number of American companies as an important supplement to the dwindling domestic supplies of iron ore. But much of the credit for the final achievement must be given to one man, E. W. Davis, a scientist in the University of Minnesota, who began in 1912 what proved to be a lifelong struggle to arouse interest in the possible use of taconite and to solve the highly intractable problems of extracting the iron from the rock. An experimental plant was set up in 1916 but closed down in 1922. Davis, however, persisted and made great progress both in the breaking down of the rock and the forming of the ore into pellets for use in furnaces. But most of his funds were obtained from public bodies, for the big eastern steel companies were unresponsive. It was not until 1943 that they started their own development on the basis of his work. Since 1948 the scale of development has enormously increased, although even now it is by no means clear that the pace is being set by the largest iron and steel companies.

The presence of a significant number of possibly interested firms can be expected to bring about the trial of a wider range of innovations than where fewness is the rule. Society may thereby avoid putting too many eggs in one basket. A dominating firm in an industry may well be confronted with an embarrassment of riches; such firms frequently assert that they have many more promising ideas than they can possibly afford to develop. If they are disposed, for reasons already enlarged upon, to choose a limited number of innovations and to push these on at great cost with the utmost speed, this may be to the detriment of the chances of other new methods and products ever seeing the light of day.

The dangers of fewness can, of course, be exaggerated. It must not be supposed that every alleged case of resistance to change is a mistake: it may represent a perfectly legitimate reluctance to embark on projects where the balance of economic advantage is against action. And there are always safety valves: an innovation may be taken up by a firm outside the industry where it might have been expected to be developed, and many new firms have come into an industry for the sole purpose of exploiting an idea which established interests had set on one side. But there are enough instances since the beginning of the nineteenth century of the neglect, sometimes prolonged, of potentially fruitful ideas, to indicate that there is a real risk of loss through development being pursued on too narrow a front.

VIII

Against the claim for the efficiency in development of the biggest and the most securely established industrial organisations, may be set, therefore, the advantages of the attack from many angles. The tasks of development are themselves of such diversity and of so varying a scale that it may be a great and a dangerous over-simplification to suppose that they can always be best handled by any single type of institution. It may be that the happiest situation for a community, the condition most effectively contributing to general liveliness, will be found where the variety in form and outlook in industrial institutions matches that of the problems with which they have to deal; firms of varying size, some disposed to pursue plans deliberately and with an eye on the distant future and others inclined to plunge heavily for quick results; some mainly concerned with holding an established status and others prepared to dare much to restore a lost position or to break into a new industry; some that regard their forte as lying in rapid innovation and others which feel that their strength is found in following up and improving. It may well be that there is no optimum size of firm but merely an optimum pattern for any industry, such a distribution of firms by size, character and outlook as to guarantee the most effective gathering together and commercially perfecting of the flow of new ideas.

CHAPTER IX

CONCLUSIONS AND SPECULATIONS

It is a fitting enquiry what are really the intellectual characteristics of this age: whether our mental light has not lost in intensity at least a part of what it has gained in diffusion; whether our 'march of intellect' be not rather a march towards doing without intellect, and supplying our deficiency of giants by the united efforts of a constantly increasing multitude of dwarfs. . . . Where, then, is the remedy? . . . It is in the distinct recognition that the end of education is not to teach, but to fit the mind for learning from its own consciousness and observation. . . . Let the feelings of society cease to stigmatize independent thinking. – JOHN STUART MILL.[1]

I

THE CONTINUITY OF THINGS

THERE is nothing in the history of technology in the past century and a half to suggest that infallible methods of invention have been discovered or are, in fact, discoverable. It may be true that in these days the search for new ideas and techniques is pursued with more system, greater energy and, although this is more doubtful, greater economy than formerly. Yet chance still remains an important factor in invention and the intuition, will and obstinacy of individuals spurred on by the desire for knowledge, renown or personal gain the great driving forces in technical progress. As with most other human activities, the monotony and sheer physical labour in research can be relieved by the use of expensive equipment and tasks can thereby be attempted which would otherwise be wholly impossible. But it does not appear that new mysteries will only be solved and new applications of natural forces made possible by ever increasing expenditure. In many fields of knowledge, discovery is still a matter of scouting about on the surface of things where imagination and acute observation, supported only by simple technical aids, are likely to bring rich rewards.

The theory that technical innovation arises directly out of, and only out of, advance in pure science does not provide a full and faithful story of modern invention. As in the past three centuries, there is still a to-and-fro stimulus between the two; each has a momentum and a potential of its own. The case for scientific enquiry is not a utilitarian one. It may be that the flow of inventions is just as likely to be increased by stimulating the fuller exploitation of the myriads of technical possibilities inherent in the existing stock of scientific knowledge as by increasing that stock.

The history of invention shows no sharp break in continuity. The inventive drive continues to be found in people who, because of their temperament

[1] 'On Genius', *Monthly Repository*, n.s., vol. vi, 1832, pp. 649–59.

and outlook, are not easy to organise. The sharp contrasts sometimes drawn between the present and the last century seem to be the product of distortions, in the one direction, of what was happening in the nineteenth century and distortions, in the opposite direction, of what has been happening in this. The impression thereby created of the sharp passage from one epoch to another can be reinforced by comparing extremes – for example, an important discovery in a large modern industrial research laboratory contrasted with the struggles, such as would have called forth the sympathy and interest of Samuel Smiles, of some impoverished and obscure individual inventor of a hundred years ago. Extremes today are undoubtedly wider apart than they were. But the broad band of middle cases provides little support for spectacular interpretations. In both periods there were inventors of scientific outlook, inventions of the intuitive and empirical type, men who worked in teams, men who worked essentially alone.

Of course, the immediate future may have something entirely different in store for us. It may be that we have now passed into a stage of very rapid transition (dating from about 1950 or perhaps the day when the first atom bomb blew off) which amounts to a violent break in the nature of technical change. But if this is to be the assumption made, and if public policy concerning innovation is to fly in the face of history, the onus for substantiating this assumption properly lies with those who make it.

II

THE PREDICTION OF INVENTION

There is nothing in the foregoing pages which lends support to the view that inventions can be predicted or that forecasts of their consequences can provide secure grounds for anticipatory social action. Peering into the future is a popular and agreeable pastime which, so long as it is not taken seriously, is comparatively innocuous. But the claim that there is, or can be, a 'science of prediction' might easily do more harm than good.[1]

[1] At the beginning of the century H. G. Wells advocated the encouragement of a science of prediction as a part of a system for conscious social planning: 'In the past our kind had been hustled along by change; now it was being given the power to make its own changes'. He himself made many predictions of highly varying accuracy. More recently a group of distinguished sociologists associated with the University of Chicago developed these ideas in some detail. Their views are summarised in the Report of the United States National Resources Committee, *Technological Trends and National Policy*, 1936. A great deal of the literature on the conservation of natural resources is linked up with the prediction of invention. For the attempts to determine how long certain natural resources will last involve a comparison of the probable future supply (which is affected by the possibility of inventions improving the methods of locating fresh supplies) and the probable future demand (which will be partly determined by inventions leading to the use of substitutes). It is interesting to note that, whereas those writers who wish to predict invention for the purpose of controlling its consequences are usually highly optimistic about the rate of invention, conservationists tend to be pessimistic about the speed with which invention is likely to relieve the pressure upon existing natural resources.

Experience suggests that most *specific* inventions were not foreseen: they had an element of the accidental in them, they represented the last, and therefore the crucial, step between the uncertain and the certain. And the more revolutionary they were the less foreseeable they were. No one, least of all the inventors concerned, predicted the discovery of penicillin, nylon, polyethylene, the transistor, insulin, radio, the cyclotron, the zip fastener, the first aniline dye, the vulcanisation of rubber or many other cases which could be quoted. More generally, in so far as specific inventions are empirical, they cannot be predicted.

But if the details of the future are hidden, are there reasons for believing that it is still possible to perceive in a broader way what is to come? Is there a valid parallel, for example, between plotting the broad surge of technology and drawing up (say) a table of the movements of the tides? An affirmative answer can be given here only if invention is accepted as a 'social process', as a movement which has a direction and a force independent of the influences of individuals.[1] The reasons put forward for supposing that inventions will emerge, in their due and proper season, as the inevitable incidents in a forward sweep of history which is humanly comprehensible and can be tabulated may now be examined.

First, it is said that each new scientific discovery will reveal a new range of technical possibilities. This, in effect, is to admit that scientific discoveries cannot be foreseen, but to claim that their technical aftermath can. One possible objection to this is that many inventions do not, in fact, arise from recent prior scientific discovery. Another is that new scientific findings do not necessarily point to the uses to which they may ultimately be put. Even those in the best position to know have often been completely misled.[2]

Second, there is the argument derived from the phenomenon of 'simultaneous invention'. In a considerable number of cases inventions have been made at roughly the same time by different investigators apparently unaware of each other's work. It has therefore been suggested that inventions

[1] It is interesting to watch H. G. Wells, in his efforts to establish the proposition that we can generalise about the future of humanity, trying to escape from the embarrassing fact that there are great men, whose appearance cannot be predicted, but who presumably have some influence on events. He concludes, 'the great man cannot set back the whole scheme of things; what he does in right and reason will remain, and what he does against the greater creative forces will perish'. ('The Discovery of the Future', *Proceedings of the Royal Institution*, vol. 17, 1902.) In this form the cult of prediction is really a kind of religion.

[2] Up to his death Rutherford believed that the use of nuclear energy on a large scale was unlikely (A. H. Compton, *Atomic Quest*, p. 279). Robert A. Millikan, 'There is no appreciable energy available to man through atomic disintegration' (*Science and the New Civilization*, 1930, p. 163). Hertz did not think that the wireless waves he had discovered would have any practical application. (W. R. Maclaurin, *Invention and Innovation in the Radio Industry*, p. 15.)

Even Sir Winston Churchill, whose prescience has been as outstanding as his distrust of long-range prediction, said of atomic energy in August 1939: 'It might be as good as our present-day explosives, but it is unlikely to produce anything very much more dangerous.' (*The Second World War*, vol. I, p. 301).

occur when the possibility of them is 'in the air', and many minds are being directed towards the same ends. Inventions are the product of deep-seated forces, of which the final inventor himself is but a creature; if these forces can be understood, the inventions arising from them can be forecast. This is a highly plausible line of reasoning but some of the links in it are weak. What is meant by simultaneous? In at least some quoted cases the simultaneous inventions were far enough apart in time to suggest the possibility of a transfusion of the idea from one point to another. What number of simultaneous inventions constitute a significant proportion of the whole? For every one simultaneous invention which could be named, certainly many more could be quoted as not falling into this category. Indeed a special group is often spoken about, 'inventions which come before their time', where one man has produced an idea which has aroused no general interest. Even, however, if the phenomenon of simultaneous invention were general why should it be supposed that simultaneous inventions are more predictable than isolated inventions? Their very nature suggests that they have broader origins but not necessarily origins more easily identifiable. Even if inventions of the simultaneous variety could be predicted in this way, what of the inventions not of this kind? And just as men may simultaneously stumble upon the truth, they may simultaneously fall into error:[1] those who predict are still left with the task of deciding whether a certain concatenation of social forces will lead men towards the right answer or the wrong.

The third argument is that there is always a considerable period between the 'conception date' of an invention and the time when it comes into general use and begins to exert its effect upon standards of living and that this interim period will provide a favourable opportunity for studying the probable consequences of the invention and preparing to meet them. It will be noticed that two stages of prediction have now been dropped – that regarding the course of science and that concerning the emergence of inventions. In so far as inventions come into use slowly then it would seem less necessary to plan consciously for adjustments to them; such adjustments are more likely to take place spontaneously. In so far as the inventions make their impact speedily then this third argument falls to the ground.

Fourthly, it is argued that technology is now so versatile, there are so many different technical routes to the solution of a given problem, that once a general need has made itself evident it can be confidently assumed that this need will be met, that in one way or another an answer in the form of an appropriate invention will emerge. Here an entirely different predictive technique is being suggested: all unsatisfied human needs would presumably be surveyed and some picked out as exercising greater force than others. This in itself would be a formidable task and it carries with it a serious drawback for policy-making; for if it is predicted that *some* kind of invention

[1] F. S. Taylor, *The Alchemists*, p. 68. 'In our present state of knowledge we must treat this early Chinese alchemy as showing remarkable parallelism to Western alchemy but not as being connected with it by any known contacts.'

will emerge to satisfy a given need but not exactly *what* kind, prognostications of its social consequences will be the more difficult to make. Another objection here is that it is simply not true that urgent needs automatically evoke a method of satisfying them. Inventions which have not been made, although there is an obvious need for them, are much more numerous and just as striking as those which have.

It seems quite as hazardous to try to anticipate the broad sweep of innovation as to spot future specific inventions. It is, however, at the second step – that of deducing the probable social consequences of any invention – that the proposals for a science of prediction seem most chimerical. For where policy-making is the aim, it would have to be decided how much the new products or processes will be utilised and this introduces all the uncertainties of the market: how rapidly and to what degree the new methods will supplant the old; how likely it is that the new methods may themselves have only a short life and be supplanted by something even newer. These hurdles having been surmounted, it would then be for the planners to assess the social consequences of the predicted technical and economic changes, to decide whether they should be welcomed and, if not, to design countermeasures and explain how to put them into operation. This is a nightmarish labyrinth which perhaps need not be followed any further. Except to note that it is often impossible to get agreement about the social consequences of an invention, as for example television, even after it has come into widespread use. And that, given the fashion in which, in democratic societies, important policy decisions on current and pressing national economic matters are arrived at, the suggestion that social planning for hypothetical conditions in the remote future would be seriously entertained or, if so, wisely and successfully conducted, appears unreal.

Against objections of this kind the defence is sometimes offered that, in fact, some outstanding successes in prediction have already been recorded and that, if such encouraging results have been achieved mainly by amateurs working in an embryonic science, much more could be hoped for from professionals with a more elaborately developed technique. It is, of course, true that many prophecies, often in the most unlikely quarters,[1] have been made of things which have come to pass. Considering the total volume of this prophecy it would have been surprising if the target had not been hit from time to time. The significance of such successes, however, depends upon the relation they bear to the number of failures, whether the proportion of successes is increasing, whether success can be repeated and the secret of it transmitted to other persons. At all these points there are grounds for extreme doubt.

A catalogue of errors in prediction, made even by unbiased and knowledgeable observers, would be an enormous document, limited only in

[1] For example, Marie Corelli seems to have predicted the atom bomb in 1911, and Rudyard Kipling described Atlantic air transport with all the paraphernalia of radio communication and landing priorities in 1904.

size by the patience and available time of its compiler. The errors fall into two groups: the foretelling of things which in fact have not come to pass[1] and the denial of the possibility of things which have, in fact, made their appearance.

Most of the outstanding technical features of modern life have crept upon us almost unaware. In 1906 the Engineering Editor of *The Times* was asserting that 'all attempts at artificial aviation . . . are not only dangerous to human life but foredoomed to failure from an engineering point of view', and in 1910 the British Secretary of State for War could argue that 'we do not consider that aeroplanes will be of any possible use for war purposes'.[2] The possibility of radio broadcasting was overlooked until it happened. Until 1944 it would have been impossible to say with any confidence that atomic energy for peace-time purposes would be available in this century, and even now it is problematical how far it will be usefully employed.[3] The use of electronic devices in industry was not talked about until after 1945. Up to 1930 it was generally believed that aeroplanes would never be suitable for the carriage of passengers over great sea areas and stubborn, but abortive, efforts were therefore made to develop the airship for this purpose.

Conversely, predictions made with confidence have often proved unsound. The idea that the diesel engine would be universally adopted for aircraft, especially large ones, was widely accepted in the 1920's. After the Second World War it was commonly assumed that the gas turbine would almost wholly replace other prime movers as a source of power, which has proved far too sweeping a conclusion. In 1945 it was generally supposed that the age of the private aeroplane was at hand; it was to replace the motor car just as the motor car had replaced the horse-drawn vehicle; nothing of the kind has happened. The views, strongly held but a few years ago, of the dominating rôle which would be taken by the chemical laboratory in the synthesising of materials have recently been modified in important ways: medicinal plants may be preferred because they produce effects at

[1] In 1924 J. B. S. Haldane predicted that within fifty years light would cost one-fiftieth of its then price and there would be no more night in our cities (*Daedalus*, p. 18); this ought to be checked in 1974.

[2] L. J. Ludovici, *The Challenging Sky*, pp. 40, 75. Even Lord Kelvin had little confidence in the future of dynamic flight (Lord Brabazon of Tara, *The Brabazon Story*, p. 48).

[3] Thus in June 1954 the United States Atomic Energy Commission, in its document *Probable Course of Industrial Development of Economic Nuclear Power*, made this extremely cautious statement: 'No one can say with any assurance now how long it will take and how much effort will actually be needed to convert the present technical feasibility of nuclear power into future economic reality. Nor can any one foresee now the shape of industrial development and the extent of benefits that may come from economic nuclear power. These are strictly speculative subjects about which there are certainly wide ranges of opinions. The results of our speculation are, of course, transient and replete with broad guesses, plausible presumptions and personal judgments. We hope no one, therefore, will take these comments as either predictions or considered judgments. They are neither. They are simply some working ideas presented to illustrate what may be in store if nuclear power becomes economically worth while.'

present impossible for the chemist to reproduce, or because they are cheaper, and it is now realised that biochemical processes are sometimes so subtle and economical that artificial synthesis is unlikely to rival them. More generally, between the wars, much informed opinion held that rapid technical progress was coming to an end; no one forecast the technical advance after 1945, which was due largely to the effort to overcome the shortage of labour, itself an unpredicted phenomenon.

The balance between predictive successes and failures is important for, even if it is acknowledged that the successes might have some social value, it has to be borne in mind that, conversely, errors may do harm. For both over-optimism or over-pessimism would bring trouble. If, for example, it is wrongly assumed that some general need will shortly be catered for by a new invention, this may inhibit the efforts to improve the existing, older ways of meeting the need. Thus improvements in the more traditional methods of producing insulin were held up by the widespread belief that a synthesised product would soon be found. And exaggerated claims regarding the part to be played by atomic energy in the generation of electricity led to the comparative neglect of ways of improving traditional methods of generation. On the other side, pessimism may lead communities to impoverish themselves unnecessarily. Thus most policies for conserving natural resources have in the past been based either upon over-gloomy views about the probable future supply of such resources or about the possibility of substitute materials or methods; this has led to reduced consumption in the present without necessarily providing a *quid pro quo* for the future.

Finally, it may be asked, what body of men could be entrusted with the duties of prediction and of planning to change the shape of things which would otherwise come? How would they put themselves into a position in which they could establish such an overwhelming prestige and command such general confidence, on matters on which most people tend to regard themselves as experts, that communities would in the present patiently accept rearrangements, some of them painful, in order to avoid hypothetical future troubles. Scientists and technologists would hardly be suitable to act as the assessors; they are frequently among the most hardened sceptics about the possibilities of foreseeing the future for, by reason of their daily work, they are brought up sharply against the barriers to the unknown and the uncertainties about ever finding a way through them. Historians are for the most part inclined to fight shy of extrapolating into the future. Businessmen, toughened by their experience of the uncertainties in developing inventions, might well be reluctant to speculate too far ahead, particularly in unfamiliar fields.[1]

[1] Thus Mr. Crawford H. Greenewalt, the President of du Pont, has said: 'I dislike making specific prophecies . . . in the first place they are much more likely to be wrong than right. Second, they are almost always more pessimistic than the actuality. And, finally, spectacular new developments in technology by themselves are unlikely to determine the material progress we will make over the next 25 years.' (*Fortune*, May 1955.)

It has been suggested that sociologists – standing back from the evidence, having no vested interests in these matters and no special reasons for optimism or pessimism – might carry out these duties, for, with their universal interest in the affairs of men, they could avoid the danger of not being able to see the wood for the trees. But their disabilities, too, are obvious. Not constituting a part of the active society of science and technology, they might fail to see even the trees. Up to now, their experiments in this field have not been free of serious mistakes.[1] In any case it is doubtful whether it would be wise to entrust activities so highly charged with the possibilities of error to those who would carry no immediate and personal responsibility for the consequences of their mistakes.

The correct attitude would seem to be to accept the changes brought about by invention and to deal with these changes, as and when they occur, by methods adjusted on each occasion to the character of a specific and perceived problem. It is well to remember that, in the past, the prophets have tended to press upon us hasty and hysterical action because their views about the future have been founded upon a narrow and unsubtle picture of possibilities and an inadequate grasp of the power of communities to resolve their problems step by step. It is easy to list the major alarms that they have raised: progressively extensive unemployment as machines came to be more widely used;[2] absolute shortages of raw materials because the use of natural resources was not being controlled; the social dangers of increased leisure as communities became richer; the spiritual and moral hazards of societies where more and more tasks were mechanised and less work left to the healthy use of hands and eyes; the prospects of science and technology being devoted wholly to the designs of individuals plotting to concentrate power in their own hands.

Some of these impending horrors have, in fact, proved to be nothing but shadows cast by the peculiar temperament of the prophet himself. It is by no means evident that those who have claimed the gifts of longest sight have proved to be the wisest counsellors. It may, of course, properly be argued that *if* men are determined to act now about events they believe to lie in the future, much better that they should so act with a sense of history and with as wide a knowledge as possible of all the different intertwined forces which are now creating their future. There is, however, a possible alternative. It might be better to recognise that all the knowledge of the

[1] For example, Mr. S. C. Gilfillan, who has probably made more advanced claims about the possibility of prediction than any other writer, in his book *Inventing the Ship*, 1935, speaks of the startling invention of the rotorship. He obviously imagined that there was a great future for this type of ship. But nothing has been heard of it since 1935.

[2] A prediction which was most confidently and most widely made in the 1930's was that the introduction of the cotton picker would create unemployment and the most serious social problems in the southern states of America. The cotton picker is now very extensively employed but none of the predicted consequences have arisen. The latest illustration of this is the scare about 'automation' which it is said threatens us with unemployment on an unprecedented scale. (N. Wiener, *The Human Use of Human Beings*, p. 162.)

past, even if it could be accumulated, provides no secure basis for seeing the future and that adjustment to observed change is an intricate enough task without piling upon it the confusions inseparable from the practice of soothsaying. For, as Francis Bacon put it:

> Men are wont to guess about new subjects from those they are already acquainted with, and the hasty and vitiated fancies they have thence formed: than which there cannot be a more fallacious mode of reasoning, because much of that which is derived from the sources of things does not flow in their usual channels.

III

THE PACE OF THE DEVELOPMENT OF INVENTIONS

If the fundamental sources of invention are much what they always have been, and if the future is concealed from us so that there is no more reason to suppose that it will produce a vast burst of technical progress than a period of comparative technical quiescence, why are science and technology looked upon increasingly with such admiration and awe? Why has it become so widely accepted that, since we have suddenly and unexpectedly been launched into what is variously described as the 'new technical age' or the 'new scientific civilisation', everything else must rapidly be brought into conformity – the methods and purpose of education, the structure of industry, even the forms of government? Why has it become almost an axiom that no country can ever spend too much on science and technology?

Some of the more obvious reasons for this growing emphasis upon the revolutionary features of scientific and technical change are non-economic. There is a neurotic preoccupation with the lethal powers of some modern discoveries: although in fact technical progress in ways by which people can kill each other has gone on steadily for a number of centuries. And there is the knowledge that the powers of offence have greatly outstripped the powers of defence: although to attach too great importance to this constitutes a paradoxical lack of confidence in the power of technology to redress a balance which it has for the time being disturbed.

From the narrower economic point of view, the basis of the belief in the novelty of our surroundings is essentially optimistic: that we have in recent years discovered ways of using existing and new technical ideas more swiftly and more effectively for the purpose of improving standards of living; that it is the enhanced powers to *develop* inventions to the point of commercial success which differentiate the present from the past. Here, again, there is a dichotomy of thought; for while it is assumed that if the new practices and devices are freely accepted and employed, almost unlimited benefits will accrue, it is also believed that to neglect or reject them will lead to disaster. The choice, apparently, is either to go quickly upwards or quickly downwards: we are not free to stand still.

Is it in fact true that nowadays inventive ideas are more quickly seized upon and exploited and their potentialities more swiftly diffused throughout the community in the form of higher standards of living than heretofore? We do not know because these things cannot be measured in their totality. Is it true, for example, that standards of living are rising more rapidly than ever? Because this cannot be measured, it is seductive to reach snap generalisations based on a few spectacular illustrations. On the other side, however, it may be asked whether there is much cause for self-congratulation in the fact that although a jet-propulsion plane flew in 1939, it was not until the end of 1958 that the jet-propelled airliner became established as a regular part of air transport? Or that colour television was introduced so slowly? Or that the work over 30 years of the scientists in connection with silicones only began to find commercial applications after 1945? Or that the diesel-electric locomotive was brought to the point of general use only in 1935 although the diesel engine was invented at the beginning of the century and systems of electric transmission had long been known? Or that no automatic transmission system was adopted on any scale for the light motor car until 1959? Or that it took a quarter of a century for industrialists to recognise the value of taconite? Or that the improvement in the electric battery has been so incredibly slow.

The widespread faith that we are much cleverer and more energetic in development than were our ancestors rests not upon any measured assessment of results but upon two critical assumptions. First, that because there are more persons equipped with technical knowledge employed in this task, results are bound to emerge. This seems to deny the possibilities of diminishing returns. The second assumption is that by employing technicians in larger, organised groups, the effectiveness of their work will be enhanced, that in a community with a given number of technicians it is better to have a few large groups than a larger number of small groups, or a mixture of groups of varying size. Here we are confronted with such a tangle of intricate economic and administrative factors that it may be sheer futility to try to find answers. The conclusion arrived at in an earlier chapter is that, since the tasks of development vary from case to case, it is likely that some corresponding variation in the size and form of the groups pursuing these tasks is called for.

IV

THE INSTITUTIONALISATION OF RESEARCH

Since the Statute of Monopolies, 1624, laid down for 'the first and true inventor' rewards in the form of grants of conditional and limited monopolies, industrial societies have shown themselves peculiarly reluctant to experiment with, or to do much serious thinking about, new methods of fostering the exercise of inventive powers and of exploiting these powers for economic progress. It is true that, ever since the eighteenth century, there have been vague exhortations to 'bring science and industry into closer contact' and associations have been formed to pursue these ends. In the nineteenth century there were long public wrangles, which came to very little, about the merits and demerits of the patent system; a few instances of public prizes offered for desirable inventions; some cases of *ex gratia* payments to inventors whose lot seemed to be exceptionally unfortunate. The emphasis usually was upon ways of increasing the incentive of the individual inventor or of making his position more secure.

In the twentieth century, apart from the special awards to war-time inventors, the emphasis of public endeavour has been upon institutions and not upon incentives. Governments have sought to encourage invention by setting up research organisations under their own control, or have subsidised research groups set up otherwise. Firms themselves have established industrial research laboratories and, at least in the United States, private enterprise has created specialised research organisations working for profit. The underlying principle, rarely formulated precisely but ever present, has been that originality can be organised; that, provided more people can be equipped with technical knowledge and brought together in larger groups, more new ideas must emerge; that mass production will produce originality just as it can produce sausages.

Under the influence of this doctrine the process of discovery and invention is becoming progressively institutionalised. The disposition of individuals to pursue their own ways with their own resources is weakened in many ways. High taxation (which in part is high because governments collect resources from the citizen for the purpose of stimulating public scientific and technical research) makes difficult the accumulation of private means. The lure of adequate equipment, congenial intellectual society and a secure livelihood provided by the institution is strong. In turn, institutions will naturally place emphasis upon the formal training and academic qualifications of those they employ: they will therefore become increasingly staffed by men who have been subject to common moulding influences. There is a possibility of in-breeding from which the more eccentric strains of native originality may be excluded.

Even as between the various types of research institutions the scales seem to be weighted in favour of those which, in the nature of things, will tend to rely upon advanced organisation and planning and which are, in consequence, less able to give rein to the autonomist bent of genius. Thus, measured by the scale of effort, though not necessarily by results, the universities and the independent non-profit-seeking research organisations lose ground to the government research organisations, the industrial research laboratories and the profit-seeking research institutions. Room for independence is gradually narrowed down. This, of course, is not to say that in what has been happening in recent years there has been any deliberate intention of stifling individual originality. The purpose presumably has been to supplement, to enlarge, to make fuller use of innovating ability. It may, therefore, come as a surprise to find how completely, at some points, the institution has ousted the individual and in what unlikely quarters, and in what extreme terms, the doctrine is now promulgated that industrial society can get along without individual independence and generate its necessary innovations by mass effort.

For example, contrary to conditions in many other branches of technology, there are few individual chemical inventors. It is widely regarded as unlikely that any important chemical discovery could arise outside a university or an industrial research organisation of one kind or another. The fact is clear: the reasons for it highly obscure. It may be, although there are some grounds for doubt, that chemical experimentation is especially costly, the background of necessary knowledge greater or the possible new discoveries less on the surface. It may be that the inherent amateur interest in invention, so widely found in other fields, is less prevalent in chemistry, that chemistry will never be a hobby in the way in which other branches of devising and discovering clearly are. It may be that various research institutions in the chemical field are prepared to devote such large resources to research and absorb so completely the original chemical thinkers that they have swept the boards clean. It may be that independent chemical inventors have been discouraged by the treatment they have received (for such people certainly used to exist). But, whatever the cause, the fact is indisputable and intriguing.

More generally, there are not wanting voices in industry claiming that it would be a fatal mistake to rely in these critical times upon the individual research worker; that this route towards discovery is outdated and that only the compounded effect of large numbers can match the needs of the moment.[1]

Even the autonomy of the universities may be in peril because of the stress placed upon the virtues of direction and co-operation. In the United States at the present time over four-fifths of the sums devoted to research and development in science and technology in universities are derived from government sources, mainly from the Department of Defence, for work the general aim and purpose of which is designated beforehand by the authorities providing the money. And university administrators have been known to argue that, under modern conditions, it is not sufficient for universities to look upon their functions as those of teaching and autonomous research; the universities must recognise that they have 'a public service function' and be prepared to undertake work of an organised kind having specific utilitarian ends, as conceived of by some body outside the university.[2] In Great Britain,

[1] Thus Daniel P. Barnard, Research Co-ordinator, Standard Oil Company of Indiana, *Proceedings of the Ninth Annual Conference on the Administration of Research*, Sept. 1955, Northwestern University, p. 26. 'We find the self-directed individual being largely replaced by highly organized team attack in which we employ many people who, if left entirely to their own devices, might not really be research-minded. In other words we *hire* people to be curious as a group . . . we are undertaking to *create* research capability by the sheer pressure of money. . . .'

[2] Clifford C. Furnas, Chancellor of the University of Buffalo. *Proceedings of the Ninth Annual Conference on the Administration of Research*, Sept. 1955, Northwestern University, p. 94, *et seq.*

where it is commonly supposed that the autonomy of the universities is perhaps more jealously guarded, scientists who call for larger government funds for science in the universities are prepared to employ arguments which, in effect, destroy the case for university autonomy. The approval given by some Western scientists[1] to the methods of research pursued in Russia and the somewhat strident exhortations that we would be wise to follow the Russian example is, at its worst, a claim that research is best conducted when it is wholly government controlled and, at best, a somewhat naïve assumption that the scale of organisation can be enormously magnified, with all the advantages of size, without sacrificing any of the virtues of small-scale operation and autonomy.

Although it is impossible to measure these things quantitatively, some countries have gone further than others in attempting to direct all research as a vast co-operative effort. Russia presumably has gone the whole way. In the United States there seems little doubt that the individual inventor continues to play an extremely important rôle (much more important than in Great Britain, for example) and the Americans, with their inclination to try everything, have a remarkably wide range of different kinds of institutions concerned with research, running from the institution wholly financed and operated by government, through organisations partly Government-financed, independent research bodies established by private munificence, research corporations of many different types and the research laboratories of manufacturing firms. There is no lack of experiments in organisation in the United States and, so far, these have not strangled, indeed in some directions they may have strengthened, the efforts of individual inventors. In Great Britain, although the individual inventor persists, the scales are apparently more heavily weighted against him. Beyond that, the British have been peculiarly unenterprising in trying out new ways of organising research. It is true that they have evolved a chain of industrial research associations to which there seems to be no counterpart elsewhere.[2] But there

[1] Thus the late Sir Francis Simon, *The Listener*, Jan. 19th, 1956, referred, with obvious approval, to the Russian Academy of Sciences, which is 'directly responsible for all fundamental scientific research and controls about 100 research institutes, employing over 10,000 scientists of university standard. In addition, it co-ordinates in a loose, but efficient, manner all the activities concerning science and technology.' It is only fair to point out that, of course, he had no sympathy with the Russian political system. In another place (*Lloyds Bank Review*, Apr. 1955) he pointed to the need for 'achieving a proper co-ordination without the unpleasant methods of the dictatorial countries'. But in this he was demanding mutually incompatible conditions.

[2] The British suffer from the delusion that they are, and always have been, a highly inventive people, whose ideas are frequently stolen by other nations. This certainly has not always been the case.

'The reputation of English industry stood high among European countries, but England was not generally regarded by her neighbours as an inventive nation. Defoe's remark that the English perfected other people's ideas was apparently a commonplace on the Continent. . . .' (A. P. Wadsworth and Julia Mann, *The Cotton Trade and Industrial Lancashire, 1600–1780*, p. 413.)

We can find no convincing evidence to support the present popular view that the British

is nothing in Great Britain which corresponds to the Mellon Institute or the Battelle Institutes in the United States, or to the Max Planck Institutes in Germany, and very little development so far of profit-seeking research organisations such as Arthur D. Little, Inc. In this enquiry, German methods of research have not been studied at first hand but, from the easily accessible published records, it would appear not only that the individual inventor is still important there but that organised research is conducted in a multiform and flexible fashion, giving much play to freedom of thought.

It is opportune to ask whether Western societies are fully conscious of what they have been doing to themselves recently; whether, even although there now may be more people with the knowledge and training that may be a prerequisite for invention, they are used in ways which make it less likely that they will invent; whether there is truth in the idea that, although the institution is a powerful force in accumulating, preserving, discriminating, rejecting, it will normally be weak in its power to originate and will, therefore, carry within itself the seeds of stagnation unless the powers and opportunities of individuals to compete with, resist, challenge, defy and, if necessary, overtopple the institutions[1] can be preserved.

To the majority of people the suggestion that technical progress is likely to be endangered by the very means now so commonly employed to stimulate it may appear nothing short of fantastic pessimism. For if present-day technical advance is so swift, how can it be supposed that brakes are being applied? One possible rejoinder is that perhaps our age is not unique, that the belief in our technical virtuosity is at least partly an hallucination born of the self-generating clamour of popular writing. Be that as it may, it is incontestible that if the current trend for institutionalising research is not bringing ill effects, then we are succeeding in a most difficult task: that of organising originality without diminishing its power, a task in which in the past there are many more failures to record than successes.

There is always a serious danger that institutions will tend with time to grow stiff in outline and rigid in outlook; their early successes may enhance the dangers, for their very confidence may prove their undoing. It is in the nature of an institution that it must be organised in some degree, organisation involves centralisation in some measure and centralisation is the process of reducing the number of points of autonomous thought and action. When

---

are 'good at' invention but 'bad at' development. Indeed, in other countries, the opposite is often believed. Thus American aero-engineers have suggested that the British are quick and good at development since their relatively small firms, because they are not constricted to ideas of mass production, can often knock up by empirical methods development models much more quickly than can the bigger American firms.

[1] In his last days, A. N. Whitehead continually came back to this fear: 'The vitality of thought is in adventure. . . . One of my anxieties . . . has been lest a rigid system be imposed on mankind and that fragile quality, his capacity for novel ideas, for novel aspects of old ideas, be frozen and he go on century after century . . . until he and his society reach the static level of the insects. . . . The vitality of man's mind is in adventure.' (L. Price, *Dialogues of Alfred North Whitehead*, p. 251.)

centralisation is complete, or approaches that point, then completely irrational actions may follow. It is known, for example, that the direct and overriding intervention of Hitler blocked technical innovation at certain crucial periods and may have lost him the war. The bizarre Lysenko episode in Russia is a similar case.

It is natural enough to look upon these as cases of the barbaric mishandling of knowledge arising from the over-concentration of political power and to deny their relevance in assessing the consequences of changes in the organisation of technology in the Western countries. On the other hand, it cannot be denied that in every hierarchical organisation, in proportion as it gradually accumulates traditions and respect for precedence and authority, originality begins to fight a losing battle with the forces of ossification. The evidence is overwhelming and need not be quoted in full. Military organisations have been peculiarly susceptible to resistance to ideas.[1] Social classes united to break one set of privileges have become defenders of another. Religious orders created to establish freedom of worship have become the agents of intolerance; universities have impeded the free play of thought;[2] associations founded for discovery have become centres of resistance to change. Society has to find ways both of recognising and fostering talent and of leaving itself open to the acceptance of genius. The encouragement of mere talent is a task which might be extremely well conducted in a highly institutionalised society: for all this goes on within the accepted rules. But genius breaks the rules and thereby confronts the institutions with challenges which may at first sight appear to constitute a danger to their very existence.

It is, therefore, not wholly perverse to pose the following question. *If* present trends continue for any length of time, it is not improbable that all technical research will be carried on by men with high university degrees in institutions where a measure of organisation is necessarily imposed on their work; that such institutions will be looked upon as the sole source of technical ideas; that emphasis will be laid on the need for, indeed, the duty of, research workers to submit themselves to team work; that the activities of the groups will be planned with a view to eliminating overlapping efforts and 'filling obvious gaps'. Is this the kind of system most conducive to innovation?

[1] Of the hundreds of examples which could be given, one must suffice: 'The reluctance of all naval chiefs in every Allied country to adopt convoy finds its counterpart only in the reluctance of the military chiefs of all the armies, Allied and enemy, to comprehend the significance of the tank. In both these cases, these means of salvation were forced upon them from outside and from below.' (W. S. Churchill, *Thoughts and Adventures*, p. 116.)

[2] 'When Rutherford was working at McGill on some of his early path-breaking work, on several occasions colleagues in other Departments of the University expressed the fear that the radical ideas about the spontaneous transmutation of matter might bring discredit on McGill University. . . . He was advised to delay publication and proceed more cautiously.' (A. S. Eve, *Rutherford*, p. 88.)

V

THE VIRTUES OF ECLECTICISM

Knowledge about innovation is so slender that it becomes almost an impertinence to speculate concerning the conditions and institutions which may foster or destroy it. But the evidence in this volume points to the conclusion that, in seeking to provide a social framework conducive to innovation, there are great virtues in eclecticism. The conditions under which inventions have arisen up to the present day are so diverse that safety would seem to lie in numbers and in variety of attack. If past experience is anything to judge by, crucial discoveries which add to the conveniences of life may spring up at practically any point and at any time. The only danger would seem to be in plumping for one method to the exclusion of others. In so far as society can usefully interfere – and there is much truth in the belief that 'the only thing men of power can do for men of genius is to leave them alone' – its task would be to try to maintain a balance between the different sources of inventions, to strive to prevent any one dominating to the exclusion of others.

The prospects in the Western countries for keeping open numerous channels are not altogether unpromising. There are some natural defences against the dangers of a vast, tidy, monolithic national organisation of technology and science. Autonomy of science and technology in the universities, while threatened by recent events, is still strongly believed in; it is much to be hoped that scientists and university administrators will recognise and resist the encroachments of outside bodies and the lure of tempting grants which turn the universities into tied houses. If science has a 'social' function it is to defend its self-government. It is all to the good that there are very many different types of research institutions, ranging from the university laboratory to the private enterprise research organisation, providing a wide range of experiment and competition in the different ways of handling research workers and, conversely, offering to each worker some choice in the conditions under which he operates. It is perhaps not altogether accidental that the two countries which are often claimed as possessing the greatest inventive energy, the United States and Germany, have the greatest diversity of types of research organisations. And in Great Britain it would be a mistake to suppose that the existence of an extensive chain of industrial research associations, which show very varying levels of achievement, render unnecessary or undesirable other technical research organised in different ways and working with different motives. Again, the very number and diversity of approach of the research laboratories of industrial corporations might create a bar to excessive uniformity of method and outlook. So long as firms are in competition it may be supposed that their rivalry will extend to the best manner of handling their research policy and that the better system will prevail. Thus there are not a few firms which are very conscious of the danger that prolonged university training may inhibit or destroy the inventive faculty, and which therefore recruit their research workers from local technical schools or even from among the promising youngsters at the benches. There are some firms which try, difficult though it may be, to provide for at least a few men of promise the kind of freedom which an individual inventor or the university worker would possess. There are some who feel that the best use they can make of whatever resources they can afford for research is to support university activities.

The range of experiment into the best possible ways of utilising the funds of industry for enlisting the assistance of the more inventive brains of the community would be narrowed if manufacturing became more monopolistically organised, either in the sense that one firm or a few large firms dominated an industry or if all the firms within one industry frustrated competition by fixing common prices and associated such cartel arrangements with a common research centre. Yet here we are confronted with a modern, and by now widely held, opinion that monopoly encourages, and may even be a condition precedent to, innovation. This doctrine has been examined from several angles in preceding chapters. It may, however, be worth while asking here, in more general terms, not whether the idea is correct but why it has recently commanded such general assent. For, at first glance, it appears odd that novelty should be thought to be bred by restriction and originality flourish most vigorously if the number of its sources are kept to a minimum.

Some of the reasons for assuming that monopoly and progress are bound together are simply shallow thinking. Thus it may be said: this is the age of the large monopolistic or oligopolistic firm, it is also the great age of technical invention, therefore the two must be cause and effect. Big firms go to great pains to bring to the public notice the part they have played in technical advance and they rarely say anything about the work of others which preceded their own development efforts: and the public is disposed to accept the expurgated stories at their face value. It is symptomatic that in these days no firm ever boasts about its high profits, which are the real test of its achievement, but will vaunt its innovations.

There are, however, reasons which go deeper. One of these is the spurious parallel often drawn between monopoly powers as embodied in the patent right and the monopoly possessed by a firm as the only producer in its field of manufacture. Most people regard the patent as, on the whole, a beneficial institution; why, then, it may be argued, boggle at the logical extension of it in the form of industrial monopoly? They are, in fact, not the same kind of thing. The patent right is conferred upon the inventor after the invention has been made, a reward for services actually rendered; it is limited in time and conditioned in other ways. The privilege of industrial monopolising is claimed for services to be rendered in the form of more rapid innovation and is not dependent upon proof of delivery; nor is it limited in time or area.

It may be that the attempt to correlate monopoly and innovation is the latest manifestation of that ingrained belief in the virtues of size which has run through so much economic thinking about industrial organisation since the beginning of this century. In the earlier years it took the form of a claim that very large factories or firms would inevitably be more efficient in *manufacturing* than smaller, that there was an inner 'logic' of this kind in the structure of industry. Less is now heard of this form of argument; it has largely been replaced by the conviction that research on a very large scale is the all-important advantage that the large firm or the large group of firms has over the smaller. Certain economists, who are themselves highly individualist by nature and would probably be horrified by the suggestion that they would do better work as members of a large team, confidently assert the merits of scale of operation in other branches of knowledge. But in earlier pages it has been suggested that the correlation between the size of research institutes and their inventive achievements is not clearly established.

It seems impossible to establish scientifically any final conclusion concerning the relation between monopoly and innovation. The arguments in favour of monopoly are not very good arguments, and they do not wholly fit the facts. On the other hand, the case made for competition is certainly not conclusive. In the last resort, those who have to frame public policy must judge whether competition, with its stimulus but its uncertainty, will be a more effective force than monopoly with its security but its absence of the driving force of rivalry.

VI

CAN THE INDIVIDUAL INVENTOR BE HELPED?

The multiplication of different institutional forms for the carrying on of research may do something to offset the dangers of rigidity. Are there any good grounds for going further and attempting to encourage and assist the individual inventor? Are there ways of doing so which will not do more harm than good? The case for trying to help him is that he has provided, and to an important degree still provides, a dynamic element in the industrial system which it would be difficult to replace. Economic progress does not occur wholly under the spur of exact economic calculation and technical advance is often attained by those who do not count the odds too carefully. The individual inventor may often work under a sense of illusion, which an institution with its accountants, market researchers and informed scientists and technologists would never do; but blind pushing has not infrequently brought returns under conditions where cataloguing the obstacles would have been tantamount to failure. Further, the disappearance of the individual inventor would constitute a dead loss because there are so many of his kind who, if they cannot operate independently, cannot operate at all. Formally organised research institutes, numerous as they may become, could no more absorb all the members of the community capable of inventive effort than could all the University Schools of Literature cater for all those capable of writing one good poem or novel in their lifetime. There are part-time inventors, inventors capable of one burst of innovation and nothing more, inventors unable to pass examinations, inventors who by temperament would not survive a week in an organised régime.

To put the matter concretely: the next time that someone invents, as Whittle invented, a new type of engine, is there any way in which the individual inventor can be spared the neglect, discouragement, or active obstruction which was Whittle's experience? The answer may be no. It may be that, clumsy and wasteful as the process seems, no more effective way can be found of separating wheat from chaff among individual inventors than to submit them to this kind of rough jostling. For, it may well be argued, if there were no method for enforcing standards and side-tracking the charlatans, we would be overrun by a jungle of cranky ideas. In science the task of sifting the spurious from the real is conducted by the profession itself; in commerce the test is that of the market; in art in most countries it is settled by a combination of market forces and public arbitrament. With invention, except in war-time, the final test is, and of course must always be, again the market; but under the patent system the State makes a preliminary and rudimentary distinction between those ideas which have inventive merit and those which do not, and then confers upon the originators temporary protection in the market. The question at issue is whether this is the right combination of obstacles and spurs for blocking abortive novelties but not suppressing seminal minds, whether we pay just about the right price in good ideas lost to society in order not to be pestered unduly by the charlatans. Could we in any new ways provide a sanctuary for the inventor which would not become a rest home for freaks and oddities?

One conclusion can be put forward immediately with confidence. So long as the survival of the individual inventor is not utterly despaired of (and the evidence of the preceding chapters suggests that there is no need to do this) and so long as nothing better can be suggested for the purpose, there is a very strong case for the retention of the patent system.

It is easy enough to perceive the weaknesses, even the absurdities, of the patent system and the reasons why conflicting opinions as to its value are to be found.[1] Its very principles are paradoxical. It is meant to encourage

[1] Contrast, for example, the conclusions of the Final Report of the U.S. Temporary National Economic Committee: 'In many important segments of our economy the privilege accorded by the patent monopoly has been shamefully abused. . . . It has been used as a device to control whole industries, to suppress competition, to restrict output, to enhance prices, to suppress inventions and to discourage inventiveness' with those of the British Departmental Committee on Patents and Designs Act, Second Interim Report: 'The history of industrial development seems on the whole to have justified the theory (of patents). . . . It is easy to over-estimate the part played by patents in creating and maintaining cartels. . . .'

over the long period the widest possible use of knowledge, but it starts out by conferring upon the inventor the power to restrict to himself the use of that knowledge. It grants statutory monopolies but it arose out of an Act to curb monopoly. It flourished most vigorously in the nineteenth century, the great period of economic competition, and even now it is most robustly defended and embodies the most extensive monopoly rights in those countries which most tenaciously adhere to the competitive system of private enterprise.[1] It is a crude and inconsistent system. It is based upon the assumption that the right and proper reward for the innovator is the monopoly profit he can extract in an arbitrarily fixed period. It offers the same reward to all inventors, irrespective of the intellectual merits of their inventions. It provides rewards for certain kinds of discoveries but usually confers no such rewards for other kinds of discovery, such as scientific principles; commercial devices and institutions; biological knowledge; the arts of agricultural and horticultural cultivation; systems of ciphering; methods of teaching. The standards of patentability, the patent period, the conditions attached to the patent have varied greatly from time to time in the same country and vary as between different countries.

The patent system lacks logic. It postulates something called 'invention' but in fact no satisfactory definition of 'invention' has ever appeared, and the Courts, in their search for guiding rules, have produced an almost incredible tangle of conflicting doctrines. This confusion has led to extensive and costly litigation. Its critics have described the patent right as merely 'something which has to be defended in the courts' and, because it may put the individual inventor at a disadvantage against the larger corporations, as 'a lottery in which it is hardly worth while taking out a ticket'.

The system, too, is wasteful. It gives protection for sixteen years (or thereabouts) whilst in fact over nine-tenths of the patents do not remain active for the whole of this period. It is dangerous in that the monopoly it confers can often be widened by its owner into fields and forms which it was never intended he should possess.

It is almost impossible to conceive of any existing social institution so faulty in so many ways. It survives only because there seems to be nothing better. And yet for the individual inventor or the small producer struggling to market a new idea, the patent right is crucially important. It is the only resource he possesses and, fragile and precarious as his rights may be, without them he would have nothing by which to establish a claim to a reward for his work. The sale of his ideas directly or the raising of capital for exploiting the ideas would be hopeless without the patent.

In the circumstances it is surprising that so little attention has been paid in recent years to ways of improving, or of providing substitutes for, the existing patent system. Professor Polanyi has, however, proposed radical

[1] The United States, for example, is one of the few countries where the patentee has the right to withhold his patent from use entirely without bringing into action public provisions designed to prevent the 'abuse of monopoly'.

reforms[1] which, although they have much wider implications, are relevant here. The dilemma he sees is that whereas 'it is in the nature of knowledge . . . that the more people use it at the same time, the more it tends to grow and to benefit each of its users', the patent system restricts the use of knowledge. The escape can be only through 'relieving inventors of the necessity of earning their rewards commercially, and must grant them instead the right to be rewarded from the public purse'. Every inventor, so long as he revealed his discovery, would have the right to claim a public award; all other persons could use the invention freely, so long as they submitted regularly to the awarding body the information requisite for determining the social value of the invention. Professor Polanyi's case is argued so thoroughly, and the possible objections to it faced so squarely, that it is regrettable that it has not received more public attention. The fundamental difficulty, of course, lies in the proper assessment of individual awards. For not only is equity between different individuals involved, but also the broader question of the general level of the awards. If they were pitched too high, too large a proportion of the effort of the community might be diverted to invention. If they were set too low, which might appear the greater danger in communities dominated by the idea of economic equality, the system might frustrate itself through lack of incentive. It is, of course, true that the present patent system can be challenged on exactly the same grounds.[2] But, in matters of this kind, ancient anomalies perhaps seem easier to tolerate than those newly created. For our present purpose, it is sufficient to note that Professor Polanyi's proposals would strengthen the position of the individual inventor in society.

There are other, less radical, changes which might tip the balance in the same direction. They fall into two groups, distinguishable although not of course wholly independent: measures which assist the inventor in the process of invention and measures designed to increase his incentives by enlarging his rewards or making them more secure.

Something will always turn on the degree of public respect which is accorded to the inventor. In certain countries, notably Great Britain, he is in these days too often regarded as an object of sympathy or of tolerant amusement. A man of ability who seeks to pursue his own lines when the organised research institutions offer such attractions is, naturally enough, considered as not knowing on which side his bread is buttered. This, combined with the habit of belittling the contributions made in the recent past by individual inventors and the superiority felt by highly trained members of organised research institutions towards the amateur or the lone worker (the attitude formerly of pure scientists towards industrial technologists),

[1] M. Polanyi, 'Patent Reform', *Review of Economic Studies*, vol. xi, p. 61. H. S. Hatfield, *The Inventor and His World*, chap. xiii, also has suggested public rewards for inventors.

[2] See in particular the classic article by Sir Arnold Plant, 'Economic Theory concerning Patents', *Economica*, Feb. 1934, the reading of which should constitute the departure point for any modern study of the patent system.

cannot help but be discouraging. Nothing much can be done here except to try to provide a more faithful story of the part that the individual inventor has played.

In days of war, when novel needs define themselves and the urgency of the times sharpens men's wits, the individual inventor comes into his own. For then he is looked on as the servant of the community and not as its butt, his wasteful mistakes are condoned because of his occasional sucesses, his singleness of purpose is considered a virtue and not a nuisance. It is therefore all to the good that the achievements of the 'backroom boys' are now being recorded in detail[1] and that the debt which is owed to them should have received permanent and grateful acknowledgement in the highest quarters.[2] No one, of course, would suggest that what is possible or desirable in war-time could or should set the standards for more normal periods. But even in peace-time the inventor deserves something of the prestige so generally accorded to the scientist and the artist.

There are at least some firms which are prepared to house and encourage individuals who have original but still embryonic ideas and who, whilst they value the contacts with more centrally and systematically organised groups, would hardly ever succeed as units in a team. It has sometimes been suggested that, in the field of chemistry, medicine and physiology, experimenters would be helped if they could draw upon a central supply of synthetic chemicals. The idea might be enlarged to cover, from private, philanthropic or public funds, the provision of equipment and materials for a wider range of inventors.

The merits and demerits of most schemes for assisting the inventor are immediately brought to a focus by examining the suggestion that a public institute might be set up in which inventors were allowed to work, free of charge and with the provision of facilities, whilst retaining the rights to their own discoveries. Such an institute would presumably choose its members by reference to the apparent novelty of their ideas and not to their technical qualifications, academic distinctions or past performance. But how to decide what sums should be spent for this purpose, how adequate the technical facilities should be, how to choose among the applicants for assistance, for how long support should be given to one individual, how to preserve the secrecy of each inventor's work, how to overcome the sense of anxiety by inventors that their ideas will be stolen: all these are questions difficult to answer. It seems certain that most of the enterprises would be a dead loss, that many of those who used the institute would be pursuing ideas which finally proved to be half-baked. Yet three things might be said in favour of an experiment of this kind. The first is that one occasional success, which otherwise might not have been achieved, would counterbalance many failures.

[1] As, for instance, G. Pawle, *The Secret War, 1939–1945*, or P. Brickhill, *The Dam Busters*.

[2] Sir Winston Churchill in his history of *The Second World War* has perhaps gone further than any other writer in paying tribute to inventors of all kinds.

The second, that many individual inventors of this century, known because they have ultimately struggled through to success, might have found their path less stony if they had had access to such an institute. And third, whilst undoubtedly such an institute would have its crop of impostors, it is not at all improbable that society loses much less from the activities of the charlatans than it does by the failure of many new ideas to see the light of day. Society subsidises many activities – ballet, music, the theatre, sport, research in government institutions, research by private firms – where results must largely be taken on trust and where there can be no guarantee that each and every outlay will yield returns. There seem to be no good reasons why the community should become more exacting and punctilious in judging of the long-period returns from help given to inventors.

An adequate discussion of how, apart from the measures already discussed, the proper rewards of inventors could be made more secure and even more generous would take us too far afield. But one or two general points may be made.

The market for new inventive ideas is imperfect. To some degree that is inevitable, for in the trading of ideas the sellers are fearful lest their property be stolen and the purchasers especially anxious not to acquire something worthless. But, all this conceded, there still remain deficiencies difficult to explain, unhappy in their consequences and perhaps capable of remedy. Firms looking for new products to purchase often hardly know where to begin their search. Inventors wishing to sell ideas often have not the faintest notion which firms may be interested in them. In the past few years, a new type of organisation, which helps to fill the gap, has been serving some American firms: it makes a study of its client's facilities, needs and potentialities; tries to find some inventor or invention which the client might take up and brings the two together. Generally speaking, however, it is the inventor who remains ignorant of commercial possibilities. Patent lawyers may negotiate licence agreements for inventors, but frequently they do not possess sufficient knowledge, interest or time to spare from other legal duties to fill the gap completely.[1] It seems odd that agencies have not sprung up to act as brokers in ideas, each perhaps specialising in one technical field, and thereby to relieve the individual inventor of the frustrating and often uncongenial task of hawking his intellectual property.

The high levels of taxation in many Western countries are often alleged to be inimical to risk-taking, and attempts to secure profits from inventions and their development are, above all, the risky activities of society. Are there, then, good reasons for supposing that high taxation inhibits technical progress?

So far as industrial corporations are concerned, it is sometimes suggested that high taxes on profits actually encourage research and development. For if expenses incurred for such purposes can be counted as costs, the higher the tax rate the smaller the net cost of the research to the corporation. An

[1] A study would be well worth while of the part that has been played in technical progress by the patent lawyer and of how far a widening of his functions would be desirable.

obvious rejoinder is that if high taxes make research cheaper, they also reduce the returns accruing to the corporation from any inventive success. On this subject there are, indeed, many pros and cons and, in reaching a conclusion, it would be necessary to bear in mind that the United States, where corporation taxes are very high, is also the heaviest spender on research and development.

If taxes on profits are high now, but it is expected that they will fall in the more distant future (and if the high rates of tax are a relatively recent phenomenon most people will be disposed to think in this way), then this would encourage a corporation to incur considerable expenditures now on research and development which would bring returns, if at all, in the long period. Again, if the tax rates on profits earned are high but the rates on capital gains are low or are non-existent, this would provide an inducement to pile up potentially valuable inventions and development knowledge, for the possession of such assets would tend to increase the market value of the company's shares. This may be one reason why companies are usually anxious that the public should not under-estimate the scale of their research effort.

Where there is a marked disparity between the rates of tax on distributed and undistributed profits this might also appear to be favourable to research. For in such cases firms will be encouraged to retain profits within the business and investment which takes the form of accumulating potentially valuable innovations may well appear as attractive a form of ploughing back as any other. As against this, however, the disposition on the part of existing firms to box up resources may mean that less risk capital finds its way on to the open market, so that new firms will find it less easy to acquire capital and outside innovations will find fewer backers.[1]

Whatever may be the truth as regards corporations it seems difficult to avoid the conclusion that high taxation works against the individual inventor. For his risks remain the same, his net returns become smaller. If the earnings of the inventor are taxed as income, then the problem arises of the permissible spread of income earned in one year over earlier periods of work, and tax authorities are inclined to be as niggardly here as they are in the case of authors. In countries where relatively low taxes on capital gains exist one escape route is provided, for the sale of an invention can be regarded as a sale of capital. It is also true that in some countries an inventor, by forming a company to take over his patents, may escape much of the taxation which would fall upon him if he sold his rights for income or capital to a second party, but many individual inventors may not be in a position to engage in manœuvres of this kind. And it could hardly be disputed that, other things being the same, the position of the individual inventor is a happier one in some countries than in others, for example in the United States than in Great Britain.

[1] The whole subject is ventilated by Sir Arnold Plant, *The Substance and the Shadow*, Third Fawley Foundation Lecture, 1956.

The more adverse consequences of high and progressive rates of taxation for the individual inventor appear to arise, however, in more general ways. Among the cases that have been examined in this study, where the inventions have arisen in America the instances are numerous in which private financial backers have supported the individual inventor at crucial periods. That is not nearly so true of Great Britain. The reason is well known: there are many more individuals in the United States with considerable fortunes who are in a position to back long-odds ideas. If, through high and progressive taxation, large fortunes tend to disappear, then one of the props of the individual inventor will have disappeared, and it appears doubtful whether it can be adequately replaced by public or private financial institutions designed for the same purpose.

CHAPTER X

THE PAST TEN YEARS: A RETROSPECT

Not far from the invention of fire we must rank the invention of doubt . . . for it is out of doubt of the old that the new springs; and it is doubt of the new that keeps invention within bounds.—T. H. HUXLEY.[1]

SINCE *The Sources of Invention* was first published in 1958 much has been written on the subject, especially by authors in the United States and those connected with various international organisations. Research and Development has itself become the subject of much research and development. The purpose of this chapter is to summarise the new knowledge and to indicate how far it calls for changes in our own original conclusions.

Definitions remain the bugbear of this subject. Between the discovery of a new scientific principle and the marketing of a new, commercially useful product, there is a broad band of activities merging imperceptibly into one another yet differing significantly in method, purpose and result. Different writers prefer their own dividing lines within what is really a continuous spectrum. We have thought it best to continue with our original threefold distinction between science, invention and development. This corresponds fairly closely to the division, now commonly adopted in the compilation of official statistics, between basic research, applied research and development.[2] These dividing lines are discussed and illustrated on

Abbreviations in the footnotes to Chapter X.

Hart Committee: *Economic Concentration*, Hearings before the Sub-Committee on Anti-Trust and Monopoly of the Committee on the Judiciary, U.S. Senate Eighty-Ninth Congress.

Enos: John L. Enos, *Petroleum Progress and Profits* (M.I.T. Press, 1962).

Miller and Sawers: Ronald Miller and David Sawers, *The Technical Development of Modern Aviation* (Routledge and Kegan Paul, 1968).

Schmookler: Jacob Schmookler, *Invention and Economic Growth* (Harvard University Press, 1966).

[1] Quoted from Cyril Bibby, *T. H. Huxley* (Watts, 1959), p. 257.

[2] The definitions given by the National Science Foundation run as follows:

Research and Development – Basic and Applied Research in the sciences and engineering, and the Design and Development of prototypes and processes. Excluded are routine product testing, market research, sales promotion, sales service, research in the social sciences or psychology and other non-technological activities or technical services.

Basic Research: Original investigations for the advancement of scientific knowledge that do not have specific commercial objectives, although such investigations may be in fields of present or potential interest to the reporting Company.

Applied Research: Investigations that are directed to the discovery of new scientific knowledge and that have specific commercial objectives with respect to products or

pp. 26–32. They should clarify what we have in mind and enable the reader to adjust the classifications adopted by other writers to our framework.

I

INVENTION

Ten years ago we suggested that the volume, character and sources of inventions had not been changing as rapidly as was often believed; that independent inventors[1] were still playing an important role; that inventions continued to flow from firms of very varying size; that the cost of invention may have been increasing but not universally or in spectacular degree; that most invention consisted of comparatively minor improvements; that while economic forces undoubtedly were the main determinant of the flow of inventions, elements of chance and non-economic motives were also important.

The new evidence does not seem to call for substantial modification of those views.

II

THE VOLUME OF INVENTION – PAST AND PRESENT

Despite occasional wild talk,[2] there is still no solid evidence that technical advance is now going on more rapidly than before or that the inventions of the present century are any more significant for mankind than those of the nineteenth. It is perhaps even truer now than it was formerly that an

---

processes. This definition of applied research differs from the definition of basic research chiefly in terms of the objectives of the reporting Company.

Development: Technical activities of a non-routine nature concerned with translating research findings or other scientific knowledge into products or processes. Development does not include routine technical services to customers or other activities excluded from the above definition of research and development.

[1] We now prefer to use the term 'independent inventor' instead of 'individual inventor' as used in the first edition. But a glance at the definitions on pp. 81–3 will show that the two are strictly synonymous.

[2] Dr. Edmund Leach, the Provost of King's College, Cambridge, in his 1967 Reith Lectures, has claimed that 'scientific knowledge was changing our lives with ever-accelerating speed' and that 'science offered us total mastery over our environment'. It is interesting that in 1904 Lytton Strachey, a leading member of an intellectual group in Cambridge, also closely associated with King's College, was writing: 'We are the mysterious priests of a new and amazing civilisation . . . what is hidden from us? We have mastered all.' But Strachey was registering these extraordinary claims not in the name of science but in that of philosophy. (Michael Holroyd, *Lytton Strachey*, vol. I (Heinemann, 1967), p. 198.)

An official view of the same kind is found in O.E.C.D., *Reviews of National Science Policy – United States* (1968): 'Under the influence of scientific and technical progress society is changing so profoundly that it has become possible to contemplate the deliberate application of the research and development effort to the achievement of new goals, widely different from the major strategic orientations hitherto prevailing.' (p. 289)

invention may satisfy one need but only at the expense of creating others for which technology appears to be unable to find an answer; as, for example, when the modern aeroplane speeds up travel but creates apparently insoluble problems of noise; when nuclear fission is employed to generate electricity but creates the task of disposing of radioactive waste; when road transport is improved but creates traffic congestion and air pollution; when synthetic detergents and insecticides pollute streams and lakes; or when new drugs reveal dangerous and widespread side-effects.

There are three rival doctrines about the probable future scale of invention. The first is that there will be an exponential growth in inventions since each one opens up the way to several others. The second is that the flow of inventions is carried forward by major advances around which a large number of smaller inventions appear, so that the flow over the long period will depend critically upon the frequency of major inventions. The third is that since all easy inventions have already been made, the future flow may well decline. In the absence of reliable measurement, all three continue to command their supporters.

Although much remains obscure, there are two directions in which, in recent years, knowledge has been increased. Much more is known about expenditure on research and development and much more detailed analyses have been made of the patent statistics. Even this fresh material, however, must be handled with more caution than is sometimes shown.

In the United States the cost of R and D increased from $150 million in 1921 to $14,000 million in 1960 and over $20,000 million in 1965. These are astronomical figures but, in order to keep them in their correct perspective, it must be remembered that

(i) of the total of $20,000 million, only about one-quarter was directed to what could be regarded as normal economic growth; nearly three-quarters was spent on research into atomic energy, space travel, defence and welfare services;[1]

(ii) of the total of $20,000 million, about two-thirds came from the funds of the Federal Government;[2]

(iii) of the total of $20,000 million, only about one-eighth was spent upon industrial 'applied research' (which is most closely linked with invention); the larger part went into industrial 'development';

(iv) of the total of $20,000 million, only about one-eighteenth was spent by industrial firms on 'applied research' *out of their own funds in industries other than aircraft, missiles, and electrical equipment and communication*;

[1] O.E.C.D., *The Overall Level and Structure of R and D Efforts* (1967), Table II. It is, of course, often claimed that there is much 'spin-off' even from military expenditure. Serious doubts, however, have been expressed as to whether this 'spin-off' is considerable. Indeed, outside the electronic and space industries, industrialists in the United States are inclined to argue that the major Federal R and D programmes diverted talented research workers from more useful work.

[2] See Table VIII.

TABLE VIII

UNITED STATES: RESEARCH AND DEVELOPMENT EXPENDITURE 1965

($ million)

	Total R and D Expend-iture	Industrial R and D Expenditure											
		All Industries				Aircraft and Missiles; Electrical Equipment and Communication				All Other Industries			
		Total	Basic Research	Applied Research	Develop-ment	Total	Basic Research	Applied Research	Develop-ment	Total	Basic Research	Applied Research	Develop-ment
Total Gross Expend-iture	20,470	14,197	607	2,673	10,918	8,287	215	1,176	6,897	5,910	392	1,497	4,121
Drawn from Federal Funds	13,070	7,759	191	1,052	6,516	6,478	94	775	5,609	1,281	97	277	907
Private Industrial Funds	6,530	6,438	416	1,620	4,402	1,809	121	401	1,287	4,629	295	1,219	3,115
Other	870												

Sources: Statistical Abstract of the United States 1967.
National Science Foundation: *Basic Research, Applied Research and Development in Industry.*

(v) these proportions are not very dissimilar from those for the United Kingdom.[1]

It is evident that in these two western countries industrial firms themselves, looking for a profitable return, are prepared to spend upon essentially inventive activity in normal peace-time industries only a fraction of the total sum spent nationally on research and development.

Figures of expenditure on research and development suffer from the drawback that they measure costs and not returns. Patent statistics have the merit that they measure results and not costs but they still have serious limitations as indices of inventive achievement. These difficulties are described on pp. 88–90 and no more need be said except that, as a result of the work of Schmookler[2] and others, most students of the subject are now prepared to recognise that patent statistics are more useful than had been supposed.

When we turn to the information about patents, another enigma presents itself. Over long periods in the present century, the annual number of patents issued has not been increasing or, where increasing, has not grown nearly so rapidly as expenditure on research and development or as the employment of R and D workers. For example, the annual number of patents issued in the United States was less in the middle of this century than it had been forty years earlier. And although the number of patents has been growing in recent years, the rate of increase bears no relation to expenditure or employment. So that, for each one thousand of the population or of scientific and technical workers or for each dollar expended on research and development, the number of patents has been falling sharply.

Does this mean, what at first sight it seems to mean, that technical progress has not been the unique phenomenon that some writers have suggested? Those who are reluctant to accept this conclusion claim that on the average each patent in recent years represents a more radical industrial advance than formerly; that firms, particularly large firms, patent fewer of their new ideas now than in earlier years; and that a higher proportion of patents are being made use of than was generally thought.[3] But even if these suggestions, which have not gone unchallenged, contain some truth, it still remains difficult to square the patent statistics with sweeping claims that we live in a period of progressively accelerating technical advance.

Even with a constant flow of invention, technical change might increase, for a time, if new ideas were more quickly taken up and developed for profitable use. But there is no evidence that the average period of development is shortening. It may be becoming longer, which would help to explain the apparent increase in the cost of development in recent years. The development of Float Glass occupied seven years; the successive major advances in the cracking of petroleum took five years, six years, six years,

[1] H.M.S.O., *Statistics of Science and Technology* (1967), Tables 5 and 6.
[2] Schmookler, chap. II.
[3] Schmookler, pp. 55–6.

twelve years, four years and nine years;[1] nylon six years; penicillin twelve years; the transistor five years; crease-resisting fibres six years; Corfam nine years; semi-synthetic penicillins thirteen years.

The hypothesis of an ever-accelerating rate of technical advance runs into another snag. If it were well-founded, a growing rate of increase in income, industrial production and productivity might have been expected. This has not occurred. In the United States and the United Kingdom, the two countries which relatively spend most on research, the rate of economic growth in the past two decades has not been higher than in earlier periods. Moreover, American and British growth in recent years has often been exceeded in other countries where the scale of expenditure on research has been much lower. The country with perhaps the most striking achievement in industrial rate of growth since 1950, Japan, still ranks among the lowest in research expenditure in relation to income or population, although in very recent years spending for this purpose has been increasing. In the U.S.A. 'even disregarding expenditure on military R and D, no correlation can be established between R and D expenditure and the growth of G.N.P.'[2]

It is not easy to fit together these awkward facts. It should be kept in mind that the greater part of industrial research expenditure is confined to a narrow segment of industry. In the United States the so-called research-intensive industries – aircraft, motor vehicles, electrical equipment, chemicals, machinery and instruments – account for over four-fifths of total expenditure but employ only about one-third of the total manufacturing population. In the United Kingdom the corresponding proportions are about nine-tenths and about two-fifths. And it may be that there has been a good deal of waste in the industrial research boom. Edwin H. Land, one of the successful inventors of this century, has spoken of 'billions of dollars on research and invention all over the globe for an output that I predict will in retrospect seem scandalously limited with relation to the size of the investment'.[3]

III

THE RELATIVE RATES OF SCIENTIFIC AND TECHNICAL PROGRESS

Exaggeration of the volume of invention may have been fostered by the belief that scientific knowledge is growing at an unprecedented rate and that

[1] Enos, p. 238.

[2] This is a direct quotation from O.E.C.D., *Reviews of Scientific Policy – United States* (1968), p. 255. It is worth noting as an isolated and somewhat reluctant admission in official quarters of an awkward fact, which has always been evident. When one of the authors (J. J.) in his Presidential Address to the Economic Section of the British Association in September 1960 – 'How Much Science' – ventured to point to this fact, he was much taken to task by British scientists for what was then considered to be heresy.

[3] Edwin H. Land, 'The Role of Patents and the Growth of New Companies', *Journal of the Patent Office Society, U.S.A.* (July 1959).

it is closely linked with technical progress. Since there appears to be no satisfactory method of measurement, we hazard no view on whether science is expanding more rapidly now than it did in the nineteenth century. But there are grounds for doubting whether science and technology always march together; whether science, that is to say, always carries with it an economic 'pay-off'. The subject is important enough to call for a digression.

Up to the middle of the nineteenth century science and technology flourished at different times or, where at the same time, in different regions. Britain was a leader in technical innovation at a time when it was considered backward scientifically while, in the first part of the nineteenth century, technically backward France was the leader in European science.[1] After the middle of the century, the link between the two became closer and industries sprang up which depended for their growth upon scientific knowledge, especially of electricity and chemistry. In the twentieth century this process has continued, particularly in connection with atomic energy and electronics.

These 'science-based' industries, however, even in the advanced industrial countries, still constitute a comparatively small segment of the whole. It may be that this segment will expand. But too much should not be taken for granted here. Schmookler, after studying petroleum refining, paper making, railroading and farming, concluded that, 'While many important inventions in the four fields depended on science, few if any were directly stimulated by specific scientific discoveries'.[2]

The history of the aeroplane is of special interest. Before 1930, aircraft manufacturers neglected many of the lessons that could have been derived from available theoretical knowledge. In recent years the aircraft industry has become more directly based on science because of the problems in supersonic flight. But Miller and Sawers have concluded: 'The degree of empiricism and so the margin of error remains a large one, for so elementary a problem as the inaccuracy of wind-tunnel tests continues to plague the industry some sixty years after it was first recognised'.[3]

So that, in the future as in the past, there may be periods and industries where empirical knowledge will outstrip science or where scientific knowledge will for long periods remain unexploited.

Attempts have been made, especially in Britain by some scientists,[4] to establish a close correlation between the growth of science and that of technology and thereby to justify increased expenditure on science by strictly economic criteria. These efforts are far from convincing. They claim credits for science but ignore the debits – it is true that the discovery of the atomic nucleus has made possible the nuclear power industry (if that has produced

[1] T. S. Kuhn, *The Rate and Direction of Inventive Activity* (Princeton University Press, 1962), pp. 450 *et seq.*

[2] Schmookler, pp. 63 *et seq.*

[3] Miller and Sawers, p. 249.

[4] See in particular *Second Report on Science Policy*, Council for Scientific Policy (1967), pp. 6–14.

an economic advance) but it is equally true that the appearance of the atom bomb, and the consequent world-wide proliferation of these fearsome armaments, has probably led to more economic loss than has any other new piece of knowledge since the beginning of time. Increased length of life and the elimination of some diseases are due in part to scientific discoveries but are more largely attributable to general rises in the standard of living.[1] Those who seek to prove that science will always pay its way tend to ignore the cases, such as Japan, where extraordinary economic progress has been made with only a narrow scientific base. And in attempting to judge the economic results flowing from fresh scientific knowledge, they have failed to ask the critical economic question: what would have happened if the money spent on science had been spent in other ways?

Whether in recent years science has, in fact, made a greater contribution to economic progress and human satisfaction than studies in other fields – the arts, law, economics or administration – and whether, even if it has done so, this was inevitable in the nature of things or was a fortuitous happening which might not be repeated, are questions worth examining in spite of their elusiveness. For example, was Fermi or Keynes the greater benefactor to mankind? When Hertz, with his scientific discoveries, opened up the way to the broadcasting of the music of Bach, should the scientist or the composer be given the greater credit for this benefaction?

We do not propose to enter into such wide questions. But, narrowing the issues, the sceptic may well express surprise that apparently no one has shown why the scientist, as a free agent in 'an imaginative adventure of the mind seeking truth in a world of mystery',[2] should always or frequently be guided by a kind of hidden hand to extend knowledge in the form in which it will contribute to the fuller satisfaction of the ordinary wants of the consumer. For, despite the doctrine of 'the two cultures', it seems indisputable that the scientist and the artist both seek for perfection in terms of elegance, symmetry and generality and not in terms of practical applications.[3]

It is sometimes argued that the concept of 'pure science' is unreal; that

[1] John and Sylvia Jewkes, *Value for Money in Medicine* (Blackwell, 1963), chap. V.

[2] To quote the late Sir Cyril Hinshelwood.

[3] It is interesting to find that the historian Froude, nearly a hundred years ago, was defending science and extolling the scientific life on grounds just the opposite of those now so commonly used:

> 'So far, perhaps, the finest result of scientific activity lies in the personal character which devotion of a life to science seems to produce. While almost every other occupation is pursued for the money which can be made out of it, and success is measured by the money result which has been realised – while even artists and men of letters, with here and there a brilliant exception, let the bankers' book become more and more the criterion of their being on the right road, the men of science alone seem to value knowledge for its own sake, and to be valued in return for the addition which they are able to make to it . . . they are happy in their own occupations, and ask no more; and that here, and here only, there is real and undeniable progress is a significant proof that the laws remain unchanged under which true excellence of any kind is attainable.'
>
> James Anthony Froude, *On Progress*, 1876.

the scientist will, in fact, be guided by his economic, social and political ideas and that, being normally a sensible person, his work will be directed towards what is likely to be more immediately useful.

No subject is more elusive than the compulsions which determine the work to which creative individuals will turn. Our own impression is that few scientists with a free choice do deliberately steer their research into directions which they think will be technically fruitful and thus enable consumers' demands to be met more fully. It is, indeed, well known that some of the scientists who have most forcefully stressed the 'social' purposes of science have, in their own scientific work, and to their great honour, engrossed themselves in subjects only remotely related to possible practical benefits to mankind.

Even if it were true that all scientists were 'socially-orientated', what grounds are there to assume that, guided by his extra-scientific motives, the scientist would necessarily accept complacently established opinions on public policy? He might have eccentric ideas. He might believe in the need for the destruction of capitalist civilisation. He might believe that the emphasis on greater material wealth through improved technology is a false social ideal and should be frustrated. Even if his social conscience naturally guides him towards the work which he thinks would increase the national economic wealth, why should he necessarily make the correct decisions to bring this about? His success would depend upon correct guesses to three most awkward questions: what will be the scientific outcome of his work? what technical possibilities will be opened up by his scientific results? what solid economic and commercial gains will follow from those technical possibilities? On the first question, he will probably be better qualified than anyone else to reach a correct answer. On questions two and three he would be out of his depth, more likely than not to get the wrong answers and even, in consequence, to do damage to the economy.[1]

Scientists have at times belittled empirical knowledge and discouraged its use because the underlying principles had not been identified. In the nineteenth century the Astronomer Royal advised the Government that Babbage's calculating machine was 'worthless'.[2] Electrical scientists stigmatised the claims of Edison as 'bombast' and doomed to failure.[3] Some scientists disbelieved in the possibility of powered flight until it had happened. Lord Rutherford did not think that atomic energy would ever

[1] When the economic ideas of scientists are made explicit they frequently amount to little more than vague, popular half truths about the economic system. Thus in 'Brain Drain', the Report of the Working Group on Migration, 1967, a Committee which consisted largely of eminent scientists, the statement was made repeatedly that 'the source of national wealth is in the creative productive services of this country'. The statement may be wholly tautological. But if it meant to imply that manufacturing industry is the sole source of national wealth, it is not correct.

[2] M. Moseley, *Irascible Genius* (Hutchinson, 1964), p. 145.

[3] M. Josephson, *Edison* (Eyre and Spottiswoode, 1961), p. 186.

be applied commercially. Prior to the invention of the transistor, it was being said by some scientists that there was nothing more to be discovered in the field. Conversely, there are even more cases where new scientific ideas were ridiculed by those who might have made use of them much earlier. In medicine, the theory of the circulation of the blood was at first laughed at; the stethoscope and the laryngoscope were regarded as physiological toys; the use of chloroform was considered impious; hypnosis, massage, physiotherapy, plastic surgery and orthopaedic surgery were looked upon with deep suspicion.[1]

There is, finally, a simple economic reason why science and technology do not necessarily march together. It is that they are often in competition for the same resources of men and materials and, to that degree, the expansion of the one may imply the denial of the claims of the other. This lesson is now being learnt, although sometimes only through hard experience. In 1956, the British Advisory Council on Scientific Policy had declared:

> The correlation between increases in industrial output and increases in the number of scientists in employment is more than fortuitous. Modern science, whether basic or engineering science, is the source of almost all the ideas on which the development of modern industry depends, and certainly the source of all the more important ideas.[2]

But, by 1963, the same Council was forced to admit that:

> We know of no way to determine precisely what proportion of any country's gross national product should be devoted to the advancement and exploitation of science.[3]

By 1967 the official British attitude, having now turned a full circle, was that technology was being starved by the growth of science. The University Grants Committee was complaining that too many science students were staying at the universities to obtain higher degrees and conduct research. The President of the Royal Society was appealing to existing scientists to 'make Britain wealthy by diverting more of their effort to utilitarian channels'.[4]

IV

WHERE DO INVENTIONS ARISE?

In the original edition of *The Sources of Invention* the conclusion to which we attached the greatest importance, but which seemed to cause surprise,

[1] Reginald Pound, *Harley Street* (Michael Joseph, 1967), pp. 52 *et seq.*

[2] Committee on Scientific Man Power, *Scientific and Engineering Man-Power in Great Britain* (1956).

[3] British Advisory Council on Scientific Policy, 17th Annual Report (1963).

[4] *Daily Telegraph*, Dec. 1st, 1967. See also P. M. S. Blackett, *Technology, Industry and Economic Growth*, Thirteenth Fawley Lecture (1966).

was that the sources of invention were numerous, scattered and varied. Studies on this subject which have been carried on since 1958 fall into four groups: case histories of specific inventions, of individual firms, and of whole industries; and statistical analyses covering the whole or large parts of the industrial system.

V

THE INDEPENDENT INVENTOR

In the first edition we drew up a long list of twentieth-century independent inventors and provided details of some of the more outstanding personalities.[1] The works of later writers make it possible to add to this list. Enos, in his penetrating study of innovations in petroleum processing, has concluded:[2]

> The most novel ideas – cracking by the application of heat and pressure, continuous processing, fractionation, catalysis, the generation of catalysts, moving and fluidised beds – occurred to independent inventors, men like Dewar and Redwood, Ellis, Adams, Houdry and Odell, who occasionally have contributed their talents, but never their employment or loyalties, to the oil companies. The most capable inventors, with the notable exception of Eugene Houdry, have stopped at invention.

Miller and Sawers, from their detailed study of the history of the evolution of modern aircraft,[3] have concluded:

> Of all the major inventions made in the past half century, only those of two types of flap can wholly be credited to the employees of aircraft manufacturers. The institutions that have been most productive of inventions are the universities and government-financed research institutes – especially in Germany – whilst about a quarter of the inventions have come from inventors with no institutional backing.

In his study of the textile industry, Schon found a striking number of innovations which involved the activities of independent inventors.[4]

The case histories added in this edition bring to light other independent workers: Cockerell with the hovercraft; Moulton with his bicycle; Wankel with the rotary engine; Sheppard and Clarke and other scientists in the study and treatment of rhesus-haemolytic disease; Zuse, Eckert, Mauchly and von Neumann with the perfecting of the computer; Purdy and Mackintosh, Higonnet and Moyroud with photo-typesetting.

Other names which can be added to the roll are Jacob Rabinow (the clock regulator); Richard Walton (the shrinking of knit goods); Samuel Ruben (the mercury dry cell);[5] Charles Learned (emergency signalling

[1] See pp. 84–8.
[2] Enos, p. 234.
[3] Miller and Sawers, p. 246.
[4] Donald A. Schon, *Technology and Change*, p. 147.
[5] D. V. de Simone, Hart Committee, Part 3, pp. 1093 *et seq.*

mirrors); Charles Hedden (land mine locators);[1] Dennis Gabor, Emmett Leith and Juris Upatnieks (holography); Leonard T. Skeggs (auto-analysis of body fluids); F. Pfleumer (magnetic tape); Francis Bacon (fuel cell); Lachmann (slotted wing); Hele-Shaw, Beacham and Turnbull (variable pitch propeller); Busemann and Betz (swept back wings).

More general studies continue to provide evidence of the important role of the independent inventor. In the United States, although the percentage of patents issued to individuals continues to fall, it still remains about one-quarter of the total and the absolute number of patents issued to individuals has increased over the past thirty years.[2] It is relevant that a high proportion of all inventions seem to be made by persons who are not full-time inventors and, further, that although many inventors have enjoyed a high-class training, many inventions are still made by individuals who have not had a higher education.[3] A particularly interesting study reveals that there are many different types of inventors and they are not by any means all likely to be found in large industrial laboratories.[4]

Nelson, we feel, has struck the right balance in his summing-up on this subject:[5]

> The private inventor will probably continue to play an important role in inventive efforts which do not involve heavy expense and in areas where ingenuity and practical experience are more important than formal scientific and engineering training and probably will continue to be responsible for a large share of major inventive efforts (major in that they take lots of time, but not money) in these fields.

VI

INVENTIONS IN FIRMS OF DIFFERENT SIZE

In this century the idea has died hard that the bigger firm possessed inherent advantages over the smaller and must, therefore, finally oust it. Up to 1950 it was widely assumed that these economies of size lay in the sphere of manufacturing and marketing. The doctrine fell into disrepute, partly because no direct evidence could be found that the biggest firms did in fact show the lowest costs and partly because more careful enquiries revealed that industrial concentration was not on the increase, that the big firms were not supplanting the smaller.[6] But no sooner was the one myth dispelled than another sprang up to take its place. This was that size was essential to efficiency because innovation was bound up with size. In the

[1] D. A. Schon, Hart Committee, Part 3, p. 1393.

[2] Schmookler, p. 26.

[3] Schmookler, p. 268.

[4] D. W. Mackinnon, *The Rate and Direction of Inventive Activity* (Princeton University Press, 1962), pp. 361 *et seq.*

[5] R. R. Nelson, Hart Committee, Part 3, p. 1147.

[6] J. Jewkes, 'Are the Economies of Scale Unlimited' in *Economic Consequences of the Size of Nations* (Macmillan, 1963), p. 102.

first edition of *The Sources of Invention* we hazarded the opinion that this later doctrine was as much an over-simplification as that which it had replaced.

Many statistical studies have since been made of the connection between the size of firms and their inventiveness and, almost without exception, they fail to find a significant correlation between the two. The conclusions of Scherer[1] are typical of many other workers: '... giant firm size is no pre-requisite for the most vigorous inventive and innovative activity'. Statistical correlations of this kind are limited in value because they measure what the firm is spending and not what it is getting for its money, and they bulk together spending on research and spending on development although it is the former with which we are immediately concerned. But the tentative statistical results are supported by qualitative evidence.

First, the histories of big firms: Mueller,[2] in his study of du Pont, showed that of eighteen new products put on the market by the firm, the firm itself discovered five and shared in the discovery of one other. Of the seven most important product and process improvements, du Pont was itself responsible for five. Unilevers were, until quite recently, not the technical leaders in their industry, as the history of the hardening of fats, of methods of quick freezing and quick drying of foods, and of the synthetic detergents clearly brings out.[3] The case of IBM is of especial interest not only because of the extraordinary commercial success of the Company in all parts of the world but also because of the frank confessions by the Company of the occasions on which they have lagged behind technically. IBM turned down Carlson when he came to them with his inventions in xerography, which in other hands were to become one of the outstanding successes of the century.[4] An official of IBM has explained how 'during the really earth-shaking developments in the accounting machine industry, IBM slept soundly' and of how, with large electronic computers, the Company 'had to come from behind'.[5] One important discovery made by the Company came about through chance:[6]

> The disk memory unit, the heart of today's random access computer, is not the logical outcome of a decision made by IBM management. It was developed in one of our laboratories as a bootleg project – over the stern warning from management that the project had to be dropped because of budget difficulties. A handful of men ignored the warning.

[1] F. M. Scherer, Hart Committee, Part 3, p. 1200.

[2] W. F. Mueller, *The Rate and Direction of Inventive Activity* (Princeton University Press, 1962), p. 323.

[3] C. Wilson, *Unilever 1945–1965* (Cassel, 1968), chap. 4.

[4] D. V. de Simone, Hart Committee, Part 3, pp. 1110–11.

[5] C. Freeman, 'Research and Development in Electronic Capital Goods', *National Institute Economic Review* (Nov. 1965), p. 60.

[6] D. A. Schon, Hart Committee, Part 3, p. 1217.

They broke the rules. They risked their jobs to work on a project they believed in.

Conversely, the histories of small firms or groups of firms have revealed that they have been responsible for a considerable flow of inventions. In the United States great importance is now attached to a number of small research firms, clustered around such centres as M.I.T., Princeton and Stanford, which have been launched by small groups of scientists and technologists, have been responsible for many inventions and, sometimes with financial support from specialised agencies, have placed many new products on the market.

Despite the commercial dominance of IBM in the computer industry, smaller firms have played a significant part in the technical advance. Small firms, especially the Eckert-Mauchly Computer Corporation, stimulated the creation of the computer industry. Another small firm, Control Data, was established in 1957, with Seymour Cray as its brilliant technical director. Despite its limited resources, this firm produced the first solid-state computer in 1960, became the leader in the industry for scientific computer systems and in 1967 ranked in terms of profits second only to IBM.[1]

Many recent innovations in the printing industry have been made by smaller companies, which have later been taken over by larger concerns.[2] K. S. Paul and Associates in Britain developed the Paul-PM electronic photosetting machine, with aid from N.R.D.C., and were later bought out by Mergenthaler-Linotype. Dr. Rudolph Hell of Kiel, in a small firm, developed a very large electronic photosetter, the Digiset, the American rights of which were sold to R.C.A.

Other organisations which were small when they first began the development of an important invention, although they subsequently grew because of their success, are the Technicon Instrument Corporation with the Auto-Analyser; Velsicol with Chlordane, Aldrin and Dieldrin; Texas Instruments with the Silicon Transistor; VOEST, Alpine and McLouth Steel with Oxygen Steel-making; Ampex with Video Tape Recording.

Turning now to the studies of whole industries, Peck has shown that between 1946 and 1957 in the aluminium industry a large proportion of the inventions came not from the primary aluminium producers, which were very large firms, but mainly from equipment manufacturers which, on the whole, were much smaller.[3]

Enos,[4] speaking of the petroleum industry, has said:

> If the most radical departures in thought have been made by inventors on the periphery of the oil industry, the least radical have been made by

[1] T. A. Wise, 'Control Data's Magnificent Fumble', *Fortune* (Apr. 1966). 'Control Data's Newest Cliffhanger', *Fortune* (Feb. 1968).

[2] W. P. Jaspert, 'Lines of Development for Printers and Manufacturers', *Print in Britain* (Dec. 1966).

[3] M. J. Peck, *Competition in the American Aluminum Industry 1945–1955* (Harvard University Press, 1961).

[4] Enos, p. 235.

inventors working for the large integrated oil companies. Of all the major inventions in petroleum cracking only one – Burton's choice of the cracking charged stock with a narrow boiling range – was discovered by such an employee.

In the American iron and steel industry the evidence, recently confirmed by the case history of the oxygen steel process, points to the comparative technical backwardness of the biggest firms.[1] In electronics Freeman[2] concludes: 'The history of this industry amply demonstrates the value of new ideas coming from small firms as well as large ones'. In his study of aero-engines, Schlaifer[3] claims that 'The original development of many important innovations has been due to very small companies, and particularly to companies formed for the sole purpose of developing a single innovation'. Nelson, Peck and Kalachek[4] have pointed out that 'Jet engines and rocket engines were pioneered by relative newcomers'.

In the past, therefore, no one type of institution has enjoyed a monopoly of invention. But are there any indications that this generalisation is becoming less true and that the balance has recently been shifting in favour of the large corporations? In the past decade there have been some remarkable successes involving costs and risks which could hardly have been entertained except by very large firms: Pilkingtons with Float Glass and Beechams with Semi-Synthetic Penicillins are instances of this.

Freeman[5] has suggested that if the list of sixty-one inventions studied in detail in *The Sources of Invention* are divided into two groups – those which can be dated before 1928 and those coming after that year – the very big organisations show a better record in the second period than in the first. Our sample of inventions was not in any sense a scientifically balanced one (it is difficult to see how any such thing could be constructed). It was an attempt to be representative and it may have embodied a bias which more extensive studies will correct. But the most recent evidence does not support Freeman's suggestions. Hamberg,[6] in his study of inventions emanating between 1946 and 1955, finds that of a total of twenty-seven, twelve originated in the work of independent inventors. Among the case histories which we have added to this volume, there are a number of the more important – Oxygen Steel-making, the Hovercraft, Computers, the Wankel engine, the prevention of Rhesus Haemolytic disease, Chlordane and associated chemicals, Photo-typesetting – which cannot be attributed mainly to large institutions.

[1] See pp. 338–41.

[2] C. Freeman, 'Research and Development in Electronic Capital Goods', *National Institute Economic Review* (Nov. 1965).

[3] R. Schlaifer, Hart Committee, Part 3, p. 1235.

[4] R. R. Nelson, M. J. Peck and E. D. Kalachek, *Technological Advance, Economic Growth and Public Policy* (Brookings Institution, 1967), p. 71.

[5] C. Freeman, 'Science and the Economy at the National Level'; O.E.C.D., *Problems of Science Policy*.

[6] D. Hamberg, *Research and Development* (Random House, 1966), p. 16.

We think that this is also the correct interpretation with other recent innovations: the Maser and Laser, the Tunnel Diode, the Fuel Cell, certain of the Tranquilliser Drugs and the Auto-Analyser.

It seems significant that only of one industry has it been claimed that the greater part of the inventions emanated from the largest companies. Freeman, in an important study,[1] has suggested that this is true of the plastic industry where he finds that, up to 1945, I. G. Farben was the inventive leader in the sense that it spent more on R and D, took out more patents and was the innovator of a larger number of the materials which now constitute the bulk of the output of plastics. After 1945 the lead passed to the giant American corporations, and especially du Pont. We wonder, however, whether Freeman has given sufficient credit in this field to Ziegler for his work on polyethylene, to Giulio Natta on polypropylene, to Whinfield on terylene, to Kipping on the silicones. And it is worth noting that, although over the whole period up to 1955 the largest five firms in the industry were responsible for about one-third of the patents, another one-third of the patents came from firms which were not found among the thirty largest.

VII

THE NATURE OF INVENTION

In the first edition we suggested that it should not occasion surprise that the sources of invention are numerous and diverse, because the invention can easily be discouraged by an unfavourable environment and may often be frustrated in a large research organisation.[2] Later writers have emphasised other reasons for reaching the same conclusion.

VIII

SMALL AND LARGE INVENTIONS

The greater part of inventive advance seems to be accounted for by the accumulation of comparatively small changes. Major inventions seem to be less important than was hitherto thought.[3] A considerable part of all patents are used commercially (this is one matter on which our own earlier more impressionistic views have proved incorrect). Many patents are still taken out by individuals having no highly technical training or by part-time workers. The cost of invention continues in many cases to be modest.[4] The

[1] C. Freeman, 'The Plastics Industry', *National Institute Economic Review* (Nov. 1963).

[2] See pp. 107–15. A recent interesting case is that of Seymour Cray, whose brilliant work was largely responsible for the success of the American company, Control Data, in the computer industry. Cray is reported as detesting large organisations and administrative chores. He insists upon working with a small staff in a secluded laboratory one hundred miles from the headquarters of the Company. Moulton is clearly an inventor with the same outlook.

[3] Schmookler, pp. 18 *et seq.*

[4] D. Novick, Hart Committee, Part 3, p. 1243.

'preliminary activities' in the developing of new cracking processes for petroleum involved relatively small sums.[1] The first critical steps in many important inventions in the aircraft industry did not cost a great deal of money: 'Knowledge is rarely as important as thought in inventing'.[2] Proving the principle of the Hovercraft cost no more than £120,000. Du Pont started their research on moisture-proof cellophane in 1924 with an initial appropriation of $5,000–10,000. By 1927 a satisfactory process had been found and in 1929 it received its basic patents; up to then du Pont had spent about $250,000.[3]

IX

THE MOTIVES OF INVENTORS

The motives of the inventor are not wholly economic. We think that Schmookler tends to over-emphasise the importance of economic demand: 'The supposition that the inventor is a man possessed by an idea and driven for months or years to develop it regardless of its market value, probably holds for some inventors. It is certainly the kind of inventor imagined by cartoonists but it hardly describes the typical inventor.'[4]

No one is likely to claim that a steady flow of inventions will continue in respect of some commodity or service, the demand for which has wholly or largely disappeared. But the non-economic factors should not be underestimated.

(i) Many inventions have been produced not in response to a specific need but because of the sense of craftsmanship or the pleasure in contrivance in the inventor: 'technology for technology's sake'. This was true of a number of aeronautical inventions,[5] especially wing flaps, knowledge of streamlining and stressed skin metal construction. Wankel became interested in the rotary engine because he considered 'the shaking and pounding of the reciprocating piston engine unaesthetic as compared with the running of a turbine or an electric motor'. Inventors often have no idea to what purpose their inventions can or might be put. Edison and Berliner, for instance, did not think of the gramophone as a method of reproducing music. It is not easy to interpret the work of Whittle, of Cockerell or of Fleming as a response purely to economic incentives.

(ii) Many inventions are 'produced before their time'. This does not necessarily mean that the inventors have been indifferent to private economic gain. It may be that they have been bad judges of what is likely to make money. But there are enough cases where an inventor persists in the perfection of his branch of a new technique, although

[1] Enos, p. 238. [2] Miller and Sawers, p. 253.
[3] W. F. Mueller, Hart Committee, Part 3, p. 1458.
[4] Schmookler, p. 112. [5] Miller and Sawers, p. 52.

he knows that it can never be fruitful unless simultaneous progress is made elsewhere, to suggest that economic motives are secondary. Babbage designed calculating machines which needed greater precision in the machining of parts than was possible at the time. Aeroplanes were designed before suitable engines were in existence. Machines have been conceived of in which the mechanical processes were new and sound but which called for stronger alloys than had yet been found. Perhaps it is true that the greatest inventors, for example Leonardo da Vinci, are those in whom economic motives are weakest or economic judgment most unreliable. Necessity is not the only mother of invention.[1]

We are now more than ever convinced[2] of the benefits of having a large number of different outlets from any one of which innovations might issue at any time.

X

THE ADVANTAGES OF ATTACK FROM MANY ANGLES

There will always be a wide range of unsatisfied needs more or less precisely defined (a cure for the common cold; a silent aero-engine; a typewriter which will operate from dictation; a car which does not need oil, water or air; an electric battery of much greater endurance, and so on through an almost endless list) and an equally wide range of hypotheses about how they might possibly be met.

It seems inconceivable that any one type of institution could properly explore every suggestion holding out some promise of success. There are not, and never will be, enough large firms to follow up every feasible idea. This is why the history of every large firm presents instances of its apparent failure to take up some idea which ultimately proved to be highly profitable. The big firm may well be wise not to spread its energies too widely. In the aero-engine industry the bigger firms, and the firms which were ultimately more successful in the later developments, set to one side many tempting possibilities for radically new types of engines.[3]

Thus the way may be left open for other, small organisations to try their luck. The differences in the sizes of firms may very well be associated with specialisation in research. Some projects will threaten to be costly but offer a reasonable chance of success (as, for example, where 'systems' are concerned as contrasted with the possibilities of inventing smaller pieces of

[1] The case of goldmining is most intriguing and is well worth further study. The world price of gold has for long been fixed; costs of mining have steadily risen so that profits have fallen. In no industry have the prizes for invention been so potentially attractive. In fact, with the exception of the employment of the magnetometer, the improvements in technique in goldmining have been relatively meagre.

[2] See pp. 184–6.

[3] R. Schlaifer, Hart Committee, Part 3, p. 1235.

equipment). These may well be the lines of research which commend themselves to the bigger firms. Other projects will be more risky but explorable with much smaller outlays. This may be the proper field for the smaller firms or even firms brought into existence for the purpose of following up a single inventive idea.

A free economic system tends to circumvent resistances to change by bringing into action new agencies and institutions. A large firm which has already committed much capital to one method of production will naturally hope that its equipment will not quickly be made obsolete by radical new methods. It may, therefore, be disinclined to conduct research which would help to produce such upsets and it may be tempted, perhaps unwisely, to delay the introduction of new methods even when they have been employed elsewhere. This probably explains the unimpressive performance of the United States Steel Corporation in the matter of inventions, especially with the new oxygen steel process. Or a large firm, where it has already been highly successful because of one invention, may tend to rest on its laurels. This happened with the Standard Oil Company of Indiana over the Burton Process.[1] Alfred P. Sloan has recounted how one of Kettering's most imaginative efforts – 'the copper-cooled engine' – was not adopted by General Motors, not so much because of technical doubts or lack of capital but because, in a corporation of that size, so many rival views were pressed with great skill and strong conviction that the proposal was finally dropped.[2] Such conditions explain why not infrequently one industry will break into the field of another[3] – innovation by 'invasion' – or one firm, not overburdened with older equipment, may leap-frog into newer ways of production over the head of longer-established companies.

In a recent important book, *Technology and Change*, Schon has analysed the reasons why, within the corporation, resistance to innovation is only to be expected because change upsets everything that is orderly, uniform and predictable. It seems clear from his analysis that the resistances are more likely to be met with and stronger in big firms than in smaller.[4]

XI

THE DEVELOPMENT OF INVENTIONS

One criticism of the first edition of *The Sources of Invention* was that, whatever may be true of inventions themselves, in their development the big organisation normally has a great advantage and that, in failing to emphasise this view, we had provided a distorted picture of the process of innovation as a whole. There is now more information on the subject which we try to summarise below.

[1] Enos, chap. I.
[2] Alfred P. Sloan, *My Years with General Motors* (Sidgwick and Jackson, 1965), chap. 5.
[3] D. A. Schon, *Technology and Change*, p. 161.
[4] D. A. Schon, *Technology and Change*, chaps. III and IV.

The cost of development is by far the larger, and tends to be an increasing, part of the cost of research and development. It seems true that a large – some students think the greater – part of the costs of development are incurred in bringing into existence new or improved products rather than improved processes.[1] In turn, product improvement consists not of a few big changes but of a very large number of smaller ones.

The costs of specific developments are not easy to define. As they are often quoted, in addition to what is spent on perfecting a product or process before it is marketed at all, they will normally contain the cost of pilot plants, they may include the costs of market research and sales programmes[2] and even the capital costs of the first commercial manufacturing plants. In some instances they may include the long period costs of improving a product or process which already has a substantial market. The figures, therefore, often include expenditure which would normally be regarded as outside that properly attributable to technical improvements.

In Table IX a list has been drawn up of development costs in a number of important cases where the information appears to be reasonably reliable. The astronomical expenditures on the atom bomb, nuclear energy, guided missiles and supersonic aircraft may be set aside as having little or no relevance to normal economic activity. Most of these developments have been paid for by governments under conditions which almost inevitably have led to violent inflations of cost: the paramount demand for speedy results; free access to the pocket of the taxpayer; the absence of any market test as to the economic value of the product and the possibility of justifying virtually any scale of expenditure by emotional appeals to the need for national defence or for the maintenance of national prestige. The heavier the hand of government, the more likely an escalation of costs.

There still, however, remain in Table IX sufficient normal cases to illustrate how large development costs can be and this remains true even though sometimes the costs of later stages of development may be met partly from profits gained in the earlier stages.

Nor can there be any doubt that, at least in some sectors of the economy, development costs have been rising. Some obvious reasons for this,[3] already discussed, seem to have become more powerful than ever. Competition in the placing of new products may well be becoming sharper. For example, it is said that du Pont had by 1964 invested over $25 million in the development and marketing of Corfam, a substitute for leather, doubtless hoping that they would be far ahead of competitors. But a rival product was quickly put onto the market by Goodrich. When du Pont introduced their plastic

[1] E. Gustafson, 'Research and Development, New Products and Productivity Changes', *American Economic Review* (May 1962).

[2] Thomas J. Watson Snr., the dominant personality in IBM 'seldom admitted any distinction between sales or market research and technical research' which perhaps goes a long way to explain the figure of $500 million sometimes quoted as the cost of development of IBM's System 360.

[3] See pp. 155–60.

TABLE IX

Estimated Costs of Specific Developments

	Period	Cost	Comments
Petroleum Cracking Processes[1]			
Burton	1909–13	$92,000	'Development of New Process' but not including 'Major Improvements in the New Process'
Dubbs	1917–22	$6,000,000	
Tube and Tank	1918–23	$600,000	
Houdry	1925–36	$11,000,000	
Fluid	1938–41	$15,000,000	
T.C.C. and Houdriflow	1935–43	$1,150,000	
Fibres			
Du Pont Nylon[2]	1928–38	$1,178,000	Including Pilot Plant
Du Pont Orlon[2]	1941–7	$5,000,000	Including Pilot Plant and Market Development
Du Pont Dacron[2]	1947–50	$6–7,000,000	Including Pilot Plant
I.C.I. Terylene[3]	1941–9	£4,000,000	Including Pilot Plants
Television			
R.C.A.[4]	Up to 1939	$2,700,000	
E.M.I.[5]	Up to 1939	£550,000	
Moisture-proof Cellophane			
Du Pont[2]	1924–9	$250,000	Up to receipt of Basic Patents
Float Glass			
Pilkingtons[6]	1952–7	£500,000	Including Pilot Plant
	1952–67	£7,000,000	Including first Float Line
Surface Modified Float Process	1963–7	Nearly £1,000,000	
Video-Tape Recording			
Ampex[7]	1951–5	$1,000,000	Development costs of components 1956–60: $15 million
Diesel-Electric Locomotives			
G.M.[8]	1930–4	$4,000,000	

	Period	Cost	Comments
Aircraft[9]			
D.C.4	1939–42	$3,300,000	
D.C.6	1945–6	$14,000,000	
D.C.8	1955–9	$112,000,000	
XB. 70	1960–5	$1,500,000,000	
Concorde	Up to 1973	£730,000,000	Minimum quoted figure
Hovercraft			
S.R.N.1		£120,000	For design, construction and early testing
S.R.N.1		£350,000	Whole programme including early tests on flexible skirts
Up to and including S.R.N.4	Up to 1968	£17,000,000	Total expenditure of B.H.C. and its constituent companies
Photo-typesetting Electronic Machines			
Paul P-M	Up to 1967	£150,000	Including development cost of some peripheral equipment
Alphanumeric APS-2	1963–7	$518,000	
Photon	Up to 1953	$720,000	
Semi-Synthetic Penicillins			
Beecham Group	1947–57	£2,500,000	Up to discovery of penicillin nucleus
	1957–66	£9,500,000	
Transistor			
Bell Labs.	1946–50	$140,000	Up to patenting of the process of producing transistors
	1950–61	$28,000,000	
	1961–5	$28,000,000	
Computers[4]			
A.S.C.C.	1937–44	$400,000	
ENIAC	1942–6	$500,000	
LEO	1947–53	£129,000	

TABLE IX [*continued*]

	Period	Cost	Comments
Missiles[10]			
Atlas		Over $3,000,000,000	
Atom Bomb[11]			
Electro-magnetic separation plant	Up to 1946	$304,000,000	
Gaseous diffusion plant	Up to 1946	$253,000,000	

Sources: 1. J. L. Enos, *Petroleum Progress and Profits* (M.I.T. Press, 1962), p. 238.
2. W. F. Mueller, *Rate and Direction of Inventive Activity* (National Bureau of Economic Research, 1962).
3. See pp. 310–12.
4. C. Freeman, 'Research and Development in Electronic Capital Goods', *National Institute Economic Review* (Nov. 1965).
5. Lord Brabazon, *The Brabazon Story* (Heinemann, 1956), pp. 151–2.
6. Private communication.
7. R. Houlton, 'The Process of Innovation', *Bulletin of Oxford University Institute of Economics and Statistics* (Feb. 1967).
8. Testimony of H. L. Hamilton. Hearings before the Sub-Committee on Anti-Trust and Monopoly of the Committee on the Judiciary, U.S. Senate, 84th Congress, 1st Session, Part 6.
9. R. Miller and D. Sawers, *The Technical Development of Modern Aviation* (Routledge and Kegan Paul, 1968), p. 267.
10. R. R. Nelson, M. J. Peck and E. D. Kalachek, *Technology, Economic Growth and Public Policy* (The Brookings Institution, 1967), p. 158.
11. L. R. Groves, *Now It Can Be Told* (Harper, New York, 1962), pp. 97 and 117.

Delrin, they must have believed that they had something new; but Celanese shortly afterwards introduced a competitive plastic. As C. B. McCoy, the President of du Pont, put it, 'the more discoveries we make, the more rapidly our competitors move in against us'. If competitive products continue to tread more closely on each others' heels, then companies may either give up the struggle as involving too great cost or intensify their efforts to reduce the period of development, thus increasing their development costs. The emphasis now being placed on the safety of products by public authorities, especially in the case of pharmaceuticals and motor vehicles, has created the need for more and more expensive tests before a product is put onto the market. Where numerous components are bound together in 'systems' then the testing of the inter-operation of the different pieces may become progressively prolonged and complicated. And it may be truer now than ten years ago that if a new product is to be launched successfully it must be put onto the market in large quantities, which renders the manufacturer more anxious that all possible defects will have been removed from it beforehand.

There must be many occasions when large corporations, having already committed themselves to heavy expenditure on some particular development, are thereafter drawn on to further spending because they are reluctant to confess to error or failure or because, even with an unpromising development, it is still possible that something can be salvaged from past investment.[1]

Is it then true that fruitful development has become so costly that only the very largest firms can engage in it? There seem to be two reasons for doubting this proposition: the first is that cases of successful development by moderate sized firms are still to be found; the second is that of the total funds devoted to development a significant part is dispersed in relatively small amounts and this suggests that they were bringing returns.

XII

SUCCESSFUL DEVELOPMENT BY SMALLER FIRMS

In the first edition we provided cases, including the jet engine and the early stages of television, where important progress in development had been carried on at a relatively modest cost. Other instances can now be cited. Many firms, while still small, have been responsible for successful developments which led to rapid growth. Texas Instruments with their development of the Silicon Transistor; the Minnesota Manufacturing and Mining Company with its Masking Tape and Scotch Tape; the Haloid (now Xerox) Company with Xerography; the Polaroid Corporation with its revolutionary types of cameras – these are all well-known cases. Oxygen Steel-making was developed by small Austrian organisations and, in the United States, the process was first adopted by McLouth, a small steelmaker. The first development of computers for business use came from Mauchly and Eckert in the United States and Leo Computers, a new firm in England. Although IBM now dominates the field in computers, firms while still small, such as Control Data and the Digital Equipment Corporation, made their mark. Velsicol, then a small chemical company in the United States, developed Chlordane, Aldrin and Dieldrin. The experimental station of the University of Minnesota developed Taconite processing.[2] In the development of Electronic Photosetting at least three small firms have made important contributions.

A particularly interesting case is that of magnetic recording. An account has been given in other pages[3] of the early discovery of this revolutionary device and its improvements by independent inventors; of the long period

[1] D. A. Schon, *Technology and Change*, pp. 35–6.
[2] E. W. Davis, *Pioneering with Taconite* (Minnesota Historical Society, 1964), p. 89.
[3] See pp. 269–72.

during which it remained unexploited commercially and of the part subsequently played by institutions both large and small, profit- and non-profit-making. The history of the development and employment of magnetic recording since World War II has now been recounted in full by R. Houlton.[1] It is a revealing story of chance, the persistence of independent inventors, the ultimately successful risk-taking by a small manufacturing firm and by a lesser broadcasting network, and the conservative attitude of larger organisations.

A major invention or development will often open up the way for progress on a smaller scale by other firms. In computer technology, many small firms have undertaken the more limited tasks such as the development of computers for laboratory work or of components for computers, storage units, switching mechanisms or information-feeding devices. A new plastic gives smaller firms an opportunity of developing new uses and forms for it.

Even when the largest firms are employing in total a great number of research workers, these workers may often be deployed in small teams organised in separate departments or even separate companies. The American Telegraph and Telephone Company has the Bell Telephone Laboratories to do its research and development and the Western Electric Laboratories to carry out development in manufacturing. When Bell Telephone undertook the major project of developing the TH system for microwave transmission of telephone and television messages, it employed a team of 90 qualified men.[2] It is true that the Bell Laboratories employed in total about 11,000 people (of whom one-third were professional scientists and engineers) but this number must be thought of in relation to the wide range of research activities carried on. L. A. B. Pilkington, the inventor of Float Glass, has said 'large numbers of people are not an alternative to people of really good quality'.[3]

Very large development groups are found in the aircraft and computer industries. But, certainly with aircraft, there are widely differing views in the industry about the numbers of men which it is profitable to employ on one project: Dassault, the most successful firm in Europe at developing military aircraft, normally employs a design team numbered in tens rather than in hundreds. In the United States the design team headed by Kelly Johnson at Lockheeds has had a long run of successful designs of advanced military aircraft, and this team numbers only 200–300, operates essentially as an independent unit and has taken care to insulate itself by forming a small research company within the main company. In Britain the design team headed by the late Sir Sydney Camm of Hawker-Siddeley had a long record

[1] R. Houlton, 'The Process of Innovation: Magnetic Recording and the Broadcasting Industry in the United States', *Bulletin of Oxford University Institute of Economics and Statistics* (Feb. 1967).

[2] T. Marschak, 'Strategy and Organisation in a System Development Project', *The Rate and Direction of Inventive Activity* (Princeton University Press, 1962).

[3] L. A. B. Pilkington, 'Float Glass', *Advance* (Nov. 1966).

of success, yet Camm disliked employing as many as 200 or 300 men on one project and developed the P-1127, a vertical take-off machine, with a team of ninety men. In the electronics industry, the computer is virtually alone in requiring design teams which run into hundreds or thousands, and even here one finds differences between companies and between products. Freeman, in his list of 'notional' threshold development costs for electronic capital goods, suggests that the minimum effective annual R and D expenditure will range all the way from £40,000–75,000 for radio communication receivers to £2,000,000–4,000,000 for a range of E.D.P. computers, with soft ware and peripherals.[1]

XIII

THE DISPERSION OF RESEARCH AND DEVELOPMENT EXPENDITURES

Perhaps nothing has done more to create the impression that only the giant firms can usefully engage in development than the form in which the statistics are usually presented.

In the first edition, in Chapters VI and VII, we discussed the limited statistics which were then available and suggested that the figures could carry a somewhat different interpretation from that sometimes put on them.

TABLE X

United States, 1964

Full-time Equivalent Number of R and D Scientists and Engineers by Size of Company Having an R and D Programme

	Full-time Equivalent Number of R and D Scientists and Engineers		
Size of Company (Workers)	All Industry	Aircraft and Missiles Electrical Equipment and Communication	All Other Industries
Up to 1,000	34,700	7,500	27,200
1,000–5,000	36,500	11,000	25,500
Above 5,000	276,400	166,400	110,000
Total	347,000	184,900	162,700

Source: National Science Foundation, *Basic Research, Applied Research and Development in Industry* (1964).

[1] C. Freeman, 'Research and Development in Electronic Capital Goods', *National Institute Economic Review* (Nov. 1965).

TABLE XI

United States

Median and Interquartile Range of Funds for R and D Performance as Per cent of Net Sales in R and D performing Manufacturing Companies, by Industry and Size of Company, 1964*

Industry	Companies with Total Employment of:					
	1000 to 4999			5000 or more		
	Median	Lower Quartile	Upper Quartile	Median	Lower Quartile	Upper Quartile
Total	·8	·2	2·1	1·1	·4	3·3
Food and Kindred Products	·3	·1	·6	·3	·1	·6
Textiles and Apparel	·2	·1	·5	·3	·2	·9
Lumber, Wood Products and Furniture	·1	·1	·8	†	†	†
Paper and Allide Products	·6	·4	1·7	·7	·5	1·0
Chemicals and Allied Products	2·4	1·2	5·8	2·7	1·6	5·9
Industrial Chemicals	2·9	2·4	5·6	2·9	2·2	5·3
Drugs and Medicines	5·8	5·2	8·2	6·7	2·2	9·4
Other Chemicals	1·7	·6	3·1	1·5	1·2	2·3
Petroleum Refining and Extraction	†	†	†	·9	·5	1·0

Rubber Products	·6	·3	1·5	†	†	†
Stone, Clay and Glass Products	·7	·4	1·0	2·0	·8	3·4
Primary Metals	·6	·2	1·4	·5	·3	·9
Primary Ferrous Products	·5	·1	1·2	·5	·3	·7
Nonferrous and Other Metal Products	·9	·5	3·3	·9	·5	1·3
Fabricated Metal Products	·8	·4	1·6	1·4	1·0	2·4
Machinery	1·5	·8	2·6	2·5	1·2	3·9
Electrical Equipment and Communication	2·1	1·2	6·1	4·6	2·0	11·3
Communication Equipment and Electronic Components	4·1	1·8	16·4	7·8	4·1	15·2
Other Electrical Equipment	2·0	1·0	2·9	2·9	1·6	8·9
Motor Vehicles and Other Transportation Equipment	·5	·3	1·5	1·2	·8	3·0
Aircraft and Missiles	5·8	2·2	11·7	21·8	10·6	30·7
Professional and Scientific Instruments	3·4	1·9	5·2	6·0	4·5	18·2
Scientific and Mechanical Measuring Instruments	3·2	1·5	4·7	†	†	†
Optical, Surgical, Photographic and Other Instruments	3·9	2·7	7·0	†	†	†
Other Manufacturing Industries	·5	·2	1·7	·3	·2	1·0

* Data are not available for companies with less than 1,000 employees. † Not separately available but included in total.

Source: National Science Foundation, *Basic Research, Applied Research and Development in Industry* (1964).

Since 1958, statistics, especially in the United States, have been compiled more widely and systematically. These reinforce what was indicated by the earlier information.[1]

It is true everywhere that formal research and development is carried on only by a small fraction of all industrial companies and that the overwhelming proportion of total expenditure is incurred by a very small number of those firms. In 1964, for example, in the United States, 13,400 companies had R and D programmes but of these only 1,500 showed expenditure of more than $100,000. Of the total expenditure on R and D, 300 companies accounted for 92 per cent. But this bald statement calls for qualification in several ways.

First, the total figures are greatly influenced by the aircraft, missile and the electrical equipment and communication industries. Table X shows this clearly. In these two industrial groups the larger companies (i.e. those employing more than 5,000 workers) accounted for 90 per cent of the R and D scientists and engineers. But if we take the whole of industry excluding aircraft, missile and the electrical equipment and communication industries, such larger companies employed only 68 per cent of the scientists and engineers.

Second, as Table XI shows, firms of the same size, even in the same industry, carry on research and development to very varying degrees. Smaller firms, with a good R and D record, often spend a larger proportion of their total receipts in this way than the bigger firms with a comparatively poor research record. In seventeen out of twenty-one major and minor industrial groups the 'best' smaller companies show higher relative expenditure than the 'average' bigger companies. This confirms, what indeed has been suggested by studies of more limited coverage, that the extent to which a firm will embark on development depends upon many factors other than its size.

Third, although the figures cover too short a period to be conclusive, it does not appear that the firms with the largest R and D programmes are gaining much on those with more modest programmes. Table XII shows that, between 1959 and 1965, if companies are ranked by the size of their research programmes (the first four, the first eight and so on), the groups at the top of the Table do not show any increase in their proportion of total

[1] We would wish, however, to reiterate that the use of any statistics of research and development involves conceptual and practical difficulties. These figures usually do not separate expenditure on development from that on research; but development accounts for some two-thirds of the total in both Britain and the United States, so that the common assumption that the distribution of R and D spending reflects that on development alone is perhaps not too misleading. Beyond that, however, 'development' as defined for the purpose of the statistics covers a much wider range of activities than the development of new products and processes, as discussed in the previous section. Much of the expenditure on 'development' as reported by firms is accounted for by the cost of minor improvements to existing products and an increase in total expenditure under this heading is as likely to reflect an increase in the number of development projects as an increase in the cost of each project.

net sales, or total employment or total expenditure on R and D. (Incidentally, the table shows that the firms with the largest R and D expenditures account for a much smaller proportion of total net sales and of employment than of R and D expenditures.)

TABLE XII

UNITED STATES

PERCENTAGE OF R AND D FUNDS: NET SALES AND EMPLOYMENT ACCOUNTED FOR BY COMPANIES WITH THE LARGEST R AND D PROGRAMMES, 1959–1965

Companies Ranked according to Size of R and D Programmes*	Percentage of All Manufacturing Companies with R and D Programmes								
	R and D Funds			Net Sales			Employment		
	1965	1964	1959	1965	1964	1959	1965	1964	1959
First 4 Cos.	21	22	22	8	8	7	9	9	8
First 8 Cos.	35	35	33	11	10	10	13	12	10
First 20 Cos.	57	57	54	19	18	18	21	21	20
First 40 Cos.	70	70	68	27	25	24	28	28	26
First 100 Cos.	82	82	81	43	41	41	42	41	40
First 200 Cos.	89	89	88	55	54	53	54	54	52
First 300 Cos.	92	92	91	65	62	61	63	62	60

* Companies were ranked individually for each year. Therefore particular companies comprising the size groups may have changed from year to year.

Source: National Science Foundation, *Basic Research, Applied Research and Development in Industry* (1967).

Fourth, there is no very clear evidence that those companies with an R and D programme have grown more rapidly than those which had no such programme. Here the statistical exercises become more hazardous. Table XIII, however, suggests that, in terms of employment and sales, the R and D companies as a whole, the 300 companies with the largest R and D programmes in each year and all manufacturing establishments grew at about the same rate between 1959 and 1965.

It is, of course, crude to compare the activities and achievements in Development of firms of varying size by examining only what they spend in these ways. 'Development' covers a multitude of varying tasks: technical services and 'trouble shooting'; the utilisation of formerly wasted by-products; the improvement of existing products; the discovery and launching of new products; changes, minor or major, in manufacturing processes for existing or new products. It may be that development costs incurred in small packets by smaller firms are directed towards less radical technical improvements and product innovations than the larger sums spent by bigger firms. But we cannot be certain. Again, it may be true that larger

TABLE XIII

United States

All R and D Performing Companies, the 300 Companies with the Largest R and D Programmes and All Manufacturing

(Employment and Net Sales, 1959–1965)

	Employment (March)						Sales					
	All R and D Performing Companies		300 Companies with Largest R and D Programmes		All Manufacturing		All R and D Performing Companies		300 Companies with Largest R and D Programmes		All Manufacturing	
Year	Actual (m.)	1959=100	Actual (m.)	1959=100	Actual (m.)	1959=100	Actual ($b.)	1959=100	Actual ($b.)	1959=100	Actual ($b.)	1959=100
1959	10·5	100	6·3	100	16·7	100	240·1	100	146·5	100	356·4	100
1960	10·7	102	6·2	99	16·8	101	243·6	101	143·7	98	364·8	102
1961	10·4	99	5·9	94	16·3	98	249·1	104	149·5	102	369·6	104
1962	10·5	100	6·2	99	16·9	101	263·7	110	158·2	108	399·6	112
1963	10·5	100	6·3	100	17·0	102	273·3	114	169·4	116	417·6	117
1964	10·7	102	6·6	105	17·3	104	287·2	120	180·9	123	445·6	125
1965	11·2	107	7·1	108	18·0	108	320·2	133	208·1	142	483·3	136

Sources: National Science Foundation, *Basic Research, Applied Research and Development in Industry* (1967).
Annual Statistical Abstract for the United States.
U.S. Dept. of Commerce, *Survey of Current Business.*

firms derive a larger return for a given expenditure on development than do smaller. So far, however, the evidence does not point clearly in that direction.[1]

To sum up on the subject of Development. Total expenditure has certainly been increasing rapidly in most industrial countries. There are undoubtedly many larger schemes which could not be handled except by spending on a scale formerly unthought of. In some industries, for various reasons, the general trend may well be for development to cost more. But this does not rule out the possibility, which we consider a probability, that a large number of ultimately valuable projects continue to be launched and brought to fruition at a much more modest cost. A reasonable surmise would be that the *range* of costs of development projects has been widening.

XIV

CONCLUSIONS

In the face of the facts uncovered in the past decade, ideas about innovation have been changing and much thinking, based upon an oversimplified conception of the working of the economic system, is being replaced by doctrines which, although less sweeping and spectacular than the old, are nearer to the truth.[2] It can no longer be claimed that supremacy in science

[1] J. Schmookler (Hart Committee, Part 3, p. 1257) and D. S. Watson and M. Holman ('Concentration of Patents from Government Financed Research in Industry', *Review of Economics and Statistics*, Aug. 1967) have provided evidence that the cost for each patent is higher for larger than for smaller firms. A. C. Cooper (Hart Committee, Part 3, p. 1293) has suggested that because they are more cost conscious, enjoy better communications between their production and research workers and have a shorter chain of command, smaller firms may be able to make correct decisions in development more quickly than larger. E. Mansfield ('Industrial Research and Development Expenditures', *Journal of Political Economy*, Aug. 1964) has suggested that, in the chemical, petroleum and steel industries, 'the inventive output per dollar of R and D expenditure in most of these cases seems to be lower in the largest than in the large or medium-sized firms'.

[2] To quote briefly from among the writings:

'There is in fact no positive linear correlation between size of firm and research intensity. ... In technically advanced industries ... there seems to be strong inducements to engage in research irrespective of size. ... Historically, credit for most significant inventions is not attributable to the very large companies. ... The large firms ... may, in fact, as a consequence of their size resist change. ... Small firms might be able to play a much bigger part in research than believed heretofore and may even sometimes provide the major stimulus.' C. Freeman, Poignant and I. Svennelson, O.E.C.D., *Ministers Talk about Science* (1965).

'If we were to devise a recipe for innovation, we would find diversity to be a necessary ingredient.' Enos, p. 262.

'If the trend were towards industrial sectors totally dominated by a couple of corporate giants with large research and development facilities – with formidable barriers to entry – I suspect that this is not an optimal industrial structure for facilitating technical advance.' R. R. Nelson, Hart Committee, Part 3, p. 1152.

'Both big business and small business play an absolutely essential role, and ... it would be a shame to interfere with the natural world of either one and try to get it

guarantees maximum economic growth when, in fact, the leading countries in science often show lower rates of economic growth than other countries. Or that expenditure on industrial research and development will bring corresponding increases in industrial production when it is undeniable that such expenditure has been increasing much more rapidly than production. Or that industrial research is a pre-requisite of growth in an industrial company when it is indisputable that very many firms succeed and grow by improving their methods of manufacturing and by selling products already long established. Or that success in invention and development will inevitably go to the larger firms with the larger research organisations when it is becoming increasingly apparent in the United States that small research firms, such as those found on 'Route 128', backed by appropriate financial agencies are a most fertile source of technical innovation. Or that government-operated research will always produce 'miracles', such as the atom bomb and space travel, when it is now clear that government activities provide some of the most outstanding cases of failure and waste. It is, therefore, not surprising that what were formerly considered as almost self-evident truths are now being challenged.

The interactions between science, technology and economic growth are much more complicated than was ever imagined by those who have dominated opinion and influenced public policy upon these matters in recent years. Each and every route to innovation has its impediments. The independent inventor is at a disadvantage because he works in comparative isolation, conscious that society is always ready to regard him as a charlatan. The small firm may have the *will* to innovate; its birth, survival and growth may often depend upon the exploitation of a new idea, but it may lack the power to innovate because it cannot find sufficient capital. The big firm has large capital resources and thus possesses the *power* to innovate, but it may lack the corporate *will* to do so because, in a situation inevitably of great uncertainty, the conflicting opinions within the corporation may lead to stalemate. Governments which themselves conduct research and development directed to economic ends tend, for political reasons, to recognise error and failure only reluctantly and, because of their great power

performed by the other who is less suited.' R. Schlaifer, Hart Committee, Part 3, p. 1238.

'It may be that the environment which typically evolves in a small company is remarkably suited to encourage efficient pursuit of product development projects, while other activities of the small company may or may not enjoy certain efficiency advantages.' A. C. Cooper, Hart Committee, Part 3, p. 1307.

'The case for bigness and fewness as a stimulus to industry R and D appears... quite weak.... Perhaps the sensible conclusion is that each industry should be treated as an individual entity.' D. Hamberg, *R and D*, p. 68.

'The qualities I am concerned about in corporate life are not related to bigness or smallness as such. There are small companies that are not orientated towards thoughtfulness and profundity, and there are a few large corporations in which they are encouraged.' Edwin H. Land, *U.S. Patent System 1790–1965*. Proceedings, vol. II, The Patent Office Society (1966).

and resources, to exaggerate the virtues of large-scale efforts in research. Governments which devote public funds to the encouragement of innovation by private institutions have not yet formulated the principles by which the scale and the distribution of this aid can rationally be determined.

Our main conclusion is, therefore, a simple one: that the path of innovation is always thorny, that there are no short cuts to success, no infallible formulae. This view, with all that it implies for public policy, seems now to be gaining ground. It is true that some economists seem still to cling to conventional error.[1] But not a few of the scientists who have been concerned with policy-making seem now to have recognised that there is no direct connection between science and economic growth, or even between research expenditure and economic growth.[2] It is, however, much to be hoped that they will not now switch their efforts towards encouraging scientists to devote themselves increasingly to 'utilitarian tasks'. For to invite scientists, who because of their special abilities and their sense of vocation have been entrusted by society with the task of discovering the ultimate meaning of things, to turn their minds to increasing the national export trade or to some such immediate purpose, would be to undermine the foundations of science itself.

In many instances, large companies are adjusting their earlier ideas about research and development: by stressing the benefits of sometimes being second in the field; by recognising the advantage of buying new ideas from

[1] Among economists, perhaps no one has exercised more influence than Professor Galbraith with his statement that:

> 'A benign Providence . . . has made the modern industry of a few large firms an almost perfect instrument for inducing technical change. . . .'

Since then nearly all the systematic evidence has run counter to any such doctrine. Yet, so far as we are aware, Professor Galbraith has said nothing in defence, or in modification, of his views. In his latest book, *The New Industrial State*, he merely repeats his unfounded assertions and dogmatically dismisses anyone who presumes to differ from him.

> 'It is a commonplace of modern technology that there is a high measure of certainty that problems have solutions before there is knowledge of how they are to be solved.' (p. 19.)
>
> 'Technology, under all circumstances, leads to planning; in its higher manifestations it may put the problems of planning beyond the reach of the industrial firm. Technological compulsions . . . will require the firm to seek the help and protection of the state.' (p. 20.)
>
> 'By all but the pathologically romantic, it is now recognised that this is not the age of the small man.' (p. 32.)

[2] A recent noteworthy statement to this effect by a scientist who has not always held such views is that of Sir Solly Zuckerman (*Scientists in War*, p. 105).

> 'I myself am somewhat sceptical about our chances of ever finding a set of universal principles which will tell us how much support pure research should receive, whether we accept as a working assumption that the cultivation of basic science should be regarded as an overhead cost to the economic exploitation of scientific knowledge in general, whether it is something which should be supported as one of man's cultural activities or whether the justification is a mixture of both these propositions.'

outside; by buying up small companies which have produced promising inventions and developments; by setting up, within their own organisation, semi-autonomous 'venture corporations' to handle new ideas which otherwise might be stifled at birth; by appointing special 'product champions' to combat resistance to change within the corporation.

Governments may not find it so easy to retrace their steps. In some instances their financial commitments are enormous. Catch phrases such as 'a national science policy', 'the technology gap' or the value of 'spin-off' may be politically embarrassing to discard. But even with governments it will not be possible for ever to ignore the relatively meagre economic benefits which have flowed from their massive expenditures on research and development.

The task which we set ourselves in this volume was a limited one of fact-finding and it is certainly no part of our purpose to draw up detailed blueprints of how western countries can best encourage innovation. Certain conclusions from our own work and that of other scholars in this field can, however, be put forward with some confidence:

1. The forces which make for innovation are so numerous and intricate that they are not fully understood. They are perhaps still as dimly comprehended as was the working of the human body five hundred years ago.
2. Governments, therefore, in seeking to encourage innovation should set down as their first aim the avoidance of harm, of inadverently checking what they are seeking to stimulate.
3. There can be no doubt that some of the ideas which have been highly influential in the last two decades have been unsound. They must have done damage either by obstructing innovation altogether or by encouraging it in one form but only at the expense of doing harm to it elsewhere.
4. It cannot be disputed that inventions and discoveries have had, and continue to have, many sources. It may be tempting to argue that one or other of these sources is more fruitful than others and should be stimulated even at the expense of the rest. Our impressions are that, given the present state of knowledge, it is safer to strive to keep all the sources open since competition strengthens the total flow of new ideas.

PART II

SUMMARIES OF ORIGINAL CASE HISTORIES

AUTOMATIC TRANSMISSIONS

THE story of the attempts to replace the manually operated gear-box and clutch of the motor-car by automatic transmissions giving smoother and simpler driving without an unacceptable loss of efficiency is a long and intricate one and only the bare outlines can be given here. The two main types of automatic transmission now used involve:

(*a*) the combination of a hydraulic coupling with an automatically controlled epicyclic gear-box. This was the first fully automatic transmission to be introduced commercially, as the Hydra-Matic gear, by the General Motors Corporation of the United States;

(*b*) the combination of a hydraulic converter-coupling (i.e. a torque converter in which the reactor is mounted on a free wheel, so that the features of a torque converter and a hydraulic coupling are combined) also with an automatically controlled epicyclic gear-box.

H. Föttinger in 1904 invented the separate torque converter and hydraulic coupling. He was a trained electrical engineer and chief constructor to the Vulkan Shipyard, Hamburg, and intended the torque converter to replace electric transmissions in ships. Numbers were built for this purpose; the hydraulic coupling also found some marine uses later.

The hydraulic coupling was first applied to vehicles by Harold Sinclair, an American engineer who interested the London General Omnibus Company in his idea. The Company first fitted the coupling to buses in 1926. As the 'Fluid Flywheel' it was used in 1930 in Daimler cars, combined with the Wilson epicyclic gear-box.[1] The Hydra-Matic, invented by Earl A. Thompson, is basically similar to this Daimler lay-out with one vital innovation, that of effective automatic control. Thompson was a brilliant engineer who had joined the Cadillac Company as a consultant in 1926 to develop his earlier invention of the synchro-mesh gear. As the director of a group of Cadillac engineers he later built a two-speed automatically controlled epicyclic transmission in the early 1930's, and remained in charge of its development into the Oldsmobile semi-automatic transmission, introduced in 1937, and the Hydra-Matic itself, introduced in 1939. The later work was done by the General Motors central research organisation, Thompson remaining in charge of Hydra-Matic development until he left the Company in 1941.

Although in the 1930's British individual inventors, notably A. W. Hallpike and A. A. Miller, produced automatic control systems for epicyclic gear-boxes, the British motor industry showed no interest and these transmissions were not manufactured.

Turning now to the torque converter: the original Föttinger torque converter was efficient only at one speed ratio and its development for vehicles consisted of widening the range of efficient speed ratios. Föttinger himself, on behalf of Belgian and German companies, studied its application to motor-cars, although the first

[1] The epicyclic gear-box had been first used on cars by F. W. Lanchester in the 1890's but it had been much improved by W. G. Wilson, a British consulting engineer, in the early 1920's.

person in Germany to apply a torque converter to a car, in 1926, was the individual inventor, Rieseler. In 1928 in Sweden, Alf Lysholm, chief engineer of the Ljungstrom Steam Turbine Company, produced the Lysholm-Smith torque converter; this was adapted by Leylands, the British firm, for buses and railcars in 1933, the first commercial application of the converter.

The converter-coupling was invented by Allan Coats, a Scottish individual inventor, in 1924. He appears to have made only one converter-coupling, although they were also built under licence in 1933 in America and France. He later turned to work on a normal converter, in which the reactor blades were pivoted; this he developed for some years after 1928 in association with Vickers-Armstrongs and the English Steel Company.

P. M. Salerni, an Italian living in England, has also worked on applying the converter-coupling to cars. After twenty-five years' work, he has developed the present Ferguson transmission, in which epicyclic gearing is placed before the converter-coupling. But the first converter-coupling to be produced commercially was the German Trilok, announced in 1934. It was the work of the Trilok research society, consisting at first of Professors H. Kluge, W. Spannhake and von Sanden; it had been formed and financed by a semi-governmental agency for the furtherance of scientific development, and given a contract to investigate the application of the diesel engine to locomotive drives. A converter-coupling was designed for this application; it was developed and manufactured by Kleine, Schanzlin and Becker of Frankenthal, the members of Trilok acting as consultants; its chief application was in heavy military vehicles.

The chief contribution to the development of the torque converter and converter-coupling in the United States was made by A. and H. Schneider and E. W. Spannhake[1] who began work in 1935. Though holding engineering positions in various companies, including the Warner Gear Division of Borg-Warner, the American Locomotive Company and the General Machinery Corporation, which financed them, their work on the torque converter was done in complete independence. Their contribution was to raise the efficiency of the converter and converter-coupling. Some of their torque converters and converter-couplings were used on military vehicles during the war; the first commercially successful civilian application of the converter-couplings in the U.S.A. was on White buses in 1946, under licence from the Schneiders. Some motor manufacturers had meanwhile begun to develop their own converter-couplings; the first to be introduced, in 1948, was the Buick Dynaflow, an ingenious combination of the converter-coupling with epicyclic gearing. Others have since followed until most automatic transmissions now embody a converter-coupling.

Some engineers believe that the converter-coupling is too inefficient for small cars; alternative types of automatic transmission have therefore been sought. Two recent British examples are the Hobbs, a purely mechanical design invented and developed as a solo effort by H. F. Hobbs; and the Smiths, which combines a gearbox with the electro-magnetic clutch invented by Jacob Rabinow.

The inventions here have, therefore, come from very different sources. The outstanding original inventions of the torque converter and the hydraulic coupling were made, with an eye on marine uses, by an electrical engineer employed by a

[1] E. W. Spannhake was the son of W. Spannhake, of the Trilok Research Society. H. Schneider had worked with W. Spannhake when the latter had been at the Vulkan Works at the time of the early development of the torque converter by Föttinger.

shipbuilding company. The improvement of the epicyclic gear-box was largely the work of a consulting engineer. Success in applying torque converters to vehicles was largely the work of individuals, although the Lysholm-Smith converter arose in a company. Individuals had much to do with the invention and development of a satisfactory converter-coupling which has proved of the greatest importance. On the other hand, the combination of the hydraulic coupling and automatic epicyclic gear-box, in the form of the Hydra-Matic, arose through the co-operation of a gifted individual and a large corporation.

REFERENCES

1. Föttinger, H., contribution to discussion on paper, 'Recent Developments in Hydraulic Couplings', *Inst. Mech. Eng. Proc.*, 1935, pp. 158–61.
2. Heldt, P. M., *Torque Converters*, 1942.
3. Shorter, L. J., 'Transmission Gear Developments', *Inst. Automobile Eng. Proc.*, 1937–1938, p. 337.
4. M'Ewen, E., 'Recent Developments in Automobile Transmissions', *Inst. Mech. Eng. (Automobile Division) Proc.*, 1947–48, p. 97.
5. Letter from C. A. Chayne, Vice-President, General Motors, Jan. 25, 1956.
6. Letter from A. W. Hallpike, May 30, 1956.
7. Letters from J. N. Fieldhouse, Vickers-Armstrongs Ltd., June 13 and 14, 1956.
8. Letter from E. W. Spannhake, June 26, 1956.
9. Chayne, C. A., 'Automatic Transmissions in America', *Inst. Mech. Eng. (Automobile Division) Proceedings*, 1952–53, Pt. 1.

BAKELITE

THE foundations of the modern plastics industry were laid by Leo Hendrik Baekeland when he invented bakelite, the first thermosetting plastic. Earlier known plastics hardened only upon cooling, softened when heated and were too soluble. Bakelite does not suffer from these defects and now has an extremely wide range of uses.

Although Baekeland benefited from the successes and failures of his predecessors, it was not until he had combined much scattered knowledge and experimented along unconventional lines that he perfected his product. He was born in Belgium and became a professor of chemistry and physics in Bruges. While visiting the United States he was persuaded by a photographic film and paper manufacturer to work in their laboratory, where he remained for two years. After leaving this company he invented 'Velox', a new type of photographic paper which made a print instantaneously, and he formed a partnership with a financial backer to manufacture the paper. He kept the process secret and refused to patent the idea. The Eastman Kodak Company made repeated offers to buy the firm and finally Baekeland agreed to sell.

In a converted barn at his home in Yonkers, New York, he then began to experiment widely. In 1904, when there was a sudden rise in the price of camphor, he sought unsuccessfully for a substitute. He next turned to the possibility of producing a synthetic shellac through the reactions of formaldehyde and phenolic bodies. It was common knowledge that these two materials react, but with varying results depending upon the conditions of the experiment and the proportions in which the materials were used. Other workers, notably Kleeberg, A. Smith, A. Luft and H. Story, had produced plastics which, for one reason or another, were not practical.

Baekeland's invention lay in the discovery of a hard, infusible, chemically resistant plastic, and of the process by which this product is obtained. All earlier patents indicated that such a plastic could be obtained only if the temperatures were kept below 100° C. Baekeland ignored the traditional method and proved that much higher temperatures are needed to create a plastic of exceptional quality. He found, too, that the quality and quantity of the condensing agent have a significant effect upon the nature of the product, and that bases are more suitable than acids as condensing agents. After developing his process into a three stage reaction,[1] he received patents in 1909 and, with his own money and financial help from his friends, he organised the Bakelite Corporation in 1910. He made many subsequent improvements both in the plastic and the methods of manufacture.

In this case an individual inventor, undeterred by traditional ideas, conceived and developed an important new product which also opened up the way for many later developments.

REFERENCES

1. T. N. E. C. Hearings, Part 3, Patents, pp. 1077–1104, 1939.
2. Baekeland, L. H., 'Bakelite', *Scientific American* Supplement, Nov. 20, 1909; Nov. 27, 1909.
3. Craig, L. H., *World's Work*, Apr. 1916.
4. 'Plastics', *Fortune*, Mar. 1936.

BALL-POINT PEN

THIS type of pen has a ball-bearing for a point which rolls the ink on to the paper and, with its special ink, is so constructed that it needs refilling only at long intervals. Although the idea that the writing point of a pen should consist of a revolving ball goes back much earlier, the modern form of this pen was the invention of two Hungarians, Ladislao J. Biro, who at various times had been a sculptor, painter and journalist, and his brother Georg, a chemist. The brothers conducted their original experiments in Hungary and patents were applied for in 1938. At the outbreak of war they moved to the Argentine and there, with the help of financial backers, especially H. G. Martin, a company was formed to perfect and produce the pen. In 1943 a defect in connection with the piston reservoir became apparent: ink was forced out by the piston whether the pen was being used or not. Biro and his colleagues found the solution in the provision of a ball-point in which the ball rested on a base seat intersected by feed channels and the employment of a reservoir

[1] George Baekeland described the operation of this process: '[The material] is placed in a hot mold; the heat of that mold begins to fuse or soften this plastic so that when pressure is applied . . . to the mold, the plastic flows through the mold and takes the form and shape of a mold, but continued heating in that mold does something that hadn't happened before. Continued heating brought on a chemical reaction within the material itself in the mold and it set up hard, and then having reached that point the mold could be opened, the piece taken out at a temperature so hot that it isn't convenient to handle, and there was no deformation and no more change, and any further heating would never soften that material again.' (T. N. E. C. Hearings, Part 3, Patents, p. 1079.)

consisting of a tube in which the ink was maintained as an uninterrupted column by capillary forces.

The patents on the Biro inventions were licensed to companies in a number of countries. In the United States the patent rights were acquired by the Eversharp Company and the Eberhard Faber Company. But, in fact, the first ball-point pens were put on to the American market by a shrewd businessman, Milton Reynolds. He had seen the Biro pens on sale in Buenos Aires in 1945 and had brought a number of them back to the United States. On the advice of an engineer and a patent attorney he concluded that he could develop a pen which would not infringe the Biro patents. Reynolds discovered that the idea of the ball-point was not new; an American, John J. Loud, had invented and patented a ball-point pen in 1888 for marking rough surfaces, but the patent had expired without its having been used for writing (although Loud's pen was later used for marking leather and fabrics). Although it had not been a commercial success, Loud's prior patent effectively limited the value of the Biro patents; Reynolds had only to re-design the feeding system of the pen in order to evade the Biro patents. Reynolds and an engineer quickly developed a successful pen with a gravity feed which, introduced in 1945, immediately became popular. The Eversharp pen came on to the market in the following year.

A further advance in ball-point pen design was made possible by the invention by an Austrian chemist, Fran Seech, of a new type of ink based on glycol as a solvent. He lived in California and produced this ink for some time in his kitchen. This ink tends to form a skin at its surface when exposed to the atmosphere, and so dries up quickly on paper. This self-sealing property greatly assisted in producing a nib which remains clean in use and is therefore suitable for retractable ball-point pens. The Papermate ball-point pen of this type, introduced by the Frawley Manufacturing Corporation, was the first to make use of this ink and proved a great success. Other companies have since introduced similar types of ink.

The successful ball-point pen is thus entirely the work of individual inventors; L. J. Biro made the major contribution to its invention, while the greatest subsequent improvement is attributable to another individual inventor, Seech.

CATALYTIC CRACKING OF PETROLEUM

METHODS of cracking petroleum by the use of heat had been known and employed in the nineteenth century, but it was not until large quantities of gasoline were needed that serious commercial interest was taken in them. Dr. Burton of the Standard Oil Company pioneered the first practical thermal cracking process in the early years of the present century and his and other methods were widely adopted by the oil companies.

An earlier catalytic cracking process, using anhydrous-aluminium chloride as catalyst, had been invented in 1915 by A. M. McAfee. A commercial unit was built in Texas by the Gulf Refining Company in 1916 but it proved unsuccessful; it was expensive and the catalyst remained effective only for a few hours and no way of regenerating it could be found.

New and improved catalytic techniques appeared in the 1930's which had important advantages over the thermal method: they yielded higher percentages of improved products and less of the residual oils, and they helped to meet the increased demands for higher octane petrol. After the failure of the McAfee process, a considerable amount of work was carried out between 1920 and 1930 in Germany on the use of metal oxides as catalysts. But the first commercially practicable system emerged from the work of Eugène Houdry. The son of a wealthy Parisian steel manufacturer, Houdry graduated from an engineering school. After the First World War he took up motor-racing as a hobby and was shown gasoline made from lignite by a process invented by E. A. Prudhomme, a pharmacist in Nice. Houdry provided Prudhomme with a laboratory and the services of three chemistry professors to try to perfect the process. The Société Anonyme Française pour la Fabrication d'Essences et Pétroles was subsequently organised, but it appeared that the Prudhomme process could not be commercialised because there was no way of preventing the poisoning of the catalyst: 'catalytic research men were absolutely certain that catalysts could not be regenerated'.[1]

At that point Houdry began his own researches and, after two years, he was able to show that with proper temperature control and the use of air and hydrogen, the nickel catalyst could be regenerated.

> 'I began to study catalytic cracking, knowing three important factors: first, the anti-knock quality of gasoline was most important; secondly, catalysts could do unbelievable things; and thirdly, catalysts could be regenerated and kept sharp for a long period of time.'

His search for a satisfactory catalyst led him to examine over one thousand substances and he finally fell upon activated clay – a material which chemists used as a bleaching medium for lubricating oil, and one which the German producers had warned their customers could not be reactivated.

Houdry offered to sell his process to the Anglo-Iranian Oil Company but they were not interested. In 1930 H. F. Sheets of the Vacuum Oil Company asked him to demonstrate it at their refinery. The Houdry Process Corporation was organised, Vacuum receiving one-third of the shares and Houdry and his associates the rest. For two years Houdry demonstrated his invention to Vacuum and Socony-Vacuum,[2] but they concluded that the process had no commercial future. He then obtained permission to negotiate with the Sun Oil Company, which took out a licence under the process patents and agreed to finance the subsequent development. For two years intensive development work proceeded; the technical staff at the Sun Oil Company established the basic engineering principles for the design and construction of large-scale equipment and perfected Houdry's method of regenerating the catalyst, at an estimated cost of over $2 million. In 1935 the Sun Oil Company offered to exchange its patents for the right to purchase a one-third interest in the process and Socony's promise to pool later research results. Socony acquired a world licence to manufacture under the Houdry process, the value of which was much enhanced by the unexpected discovery that it could produce an aviation fuel of high quality.

Houdry's discoveries spurred on other oil companies to produce better catalytic cracking methods. The Standard Oil Company of New Jersey had begun experiments in 1930, at first with the 'fixed bed' method (that used by Houdry) but later

[1] Eugène Houdry, 'Discovery of the Catalytic Cracking Reaction', *Journal of the Patent Office Society*, May 1954, pp. 375–6.

[2] Vacuum and Socony-Vacuum merged in 1931.

E. J. Houdry: catalytic cracking of petroleum

S. Junghans: continuous casting of metals

J. B. Tytus: continuous rolling of steel

with the 'fluid' method which was simpler mechanically and was continuous.[1] Standard developed this process, took out patents on it and introduced it in 1941. Since then important improvements of many kinds have been introduced by a number of firms.[2]

In this case an individual who had no immediate connection with the oil industry made feasible the first commercial catalytic cracking process by solving the critical problem of regenerating the catalyst. One large oil firm developed his ideas after another had decided the process would never be practical. Other large oil firms, which had simultaneously been studying catalytic techniques, later introduced much improved processes.*

REFERENCES

1. Murphree, E. V., Brown, C. L., Fischer, H. G. M., Gohr, E. J., and Sweeney, W. J., 'Fluid Catalyst Process', *Industrial and Engineering Chemistry*, July 1943.

2. Sittig, Marshall, 'Catalytic Cracking', *Petroleum Refiner*, June, Aug., Oct. 1947.

3. Houdry, Eugène, 'Discovery of the Catalytic Cracking Reaction', *Journal Patent Office Society*, May 1954.

4. McKnight, David, Jr., *A Study of Patents on Petroleum Cracking*, 1938.

5. 'Monsieur Houdry's Invention', *Fortune*, Feb. 1939.

6. Houdry, E., Burt, W. F., Pew, A. E., Jr., Peters, W. A., Jr., 'Catalytic Processing of Petroleum Hydrocarbons', *American Petroleum Institute Proceedings*, 1938.

7. Brooks, Benjamin T., 'The Petroleum Industry in America', *Journal of the Society of Chemical Industry, Transactions*, Aug. 24, 1928.

8. Brooks, Benjamin T., 'Development of Petroleum and Petrochemical Processing', *Petroleum Engineer*, Jan. 25, Feb. 5, 1954.

9. 'Catalytic Cracking and Solvent Refining', *Oil and Gas Journal*, May 31, 1951.

10. Wilson, Robert E., 'Petroleum Industry', *Industrial Science – Present and Future*, 1952. Collection of papers presented in Philadelphia at section on Industrial Science of the American Association for Advancement of Science, Dec. 28–30, 1951.

11. Correspondence with Mr. Eugène Houdry.

CELLOPHANE

PLAIN 'Cellophane' is a thin, flexible, non-fibrous film of regenerated cellulose containing glycerol as a softener. Its transparency, strength and flexibility make it a useful and attractive material for wrapping and for other purposes. Jacques Edwin Brandenberger, a Swiss-born French chemist, usually receives credit for its invention but the courts have disagreed much about the case. The weight of the evidence

[1] Several American and European companies were engaged in a co-operative research effort to perfect catalytic refining techniques. In this group were the Anglo-Iranian Oil Co.; the Standard Oil Co. of New Jersey; M. W. Kellog Co.; the Shell Oil Co.; the Standard Oil Co. of Indiana; the Texas Co.; the Universal Oil Products Co. and I. G. Farbenindustrie.

[2] Other oil companies perfected their own catalytic cracking units, although the fixed-bed, moving-bed and fluid units remain the three basic types. Phillips Petroleum developed their Cycloversion Process and the Houdry Process Corporation introduced a better bead catalyst moving-bed type called the Houdriflow, which was developed by Socony. These are only a few examples of the improved processes.

* [For a later definitive study of the history of petroleum refining see John L. Enos, *Petroleum Progress and Profits*.]

suggests that Brandenberger, by transforming an unmanageable film into a practical product, was responsible for real pioneering work.

Brandenberger was a dye chemist who had been trying, from 1900 onwards, to produce a protective covering which would remain clean or could be cleaned easily. His first cloth coated with liquid viscose was too stiff; he next prepared a thin sheet of viscose joined to the cotton fabric which also proved unsatisfactory. He came to the conclusion that cellulose film would have value if manufactured separately and to this task he applied the experience he had gained in the textile industry about continuous cellulose tissue processes.[1] The first material he made was thick and brittle but by 1912 he had produced a thin, supple film and he obtained European and United States patents on the process and on the manufacturing machinery. The Comptoir de Textiles Artificiels, the largest French rayon producer, became interested in his work and agreed to finance him. This French firm formed La Cellophane, transferred the Brandenberger patents to it, and employed Brandenberger to direct the development. La Cellophane was the first company to manufacture plain 'Cellophane' on a commercial basis.

Du Pont, the American chemical company, had been associated with Comptoir in the manufacture of rayon through their jointly owned Fibersilk Company and, after the First World War, du Pont, anxious to diversify their products, became interested in 'Cellophane'. Brandenberger and two officials of Comptoir visited the United States in 1923 and an agreement was reached to form a new company, the du Pont Cellophane Company, to which was assigned the United States rights to the patents, processes and 'know-how' of La Cellophane. In 1924 du Pont produced the first 'Cellophane' made in the United States.

In January 1925 du Pont commenced the search for a moisture-proof 'Cellophane' and in 1926 two of their employees, William Hale Charch and Karl Edwin Prindle, found ways of applying a suitable thin waterproof coating to both sides of the 'Cellophane' film. Du Pont received patents both on the process and on methods of manufacture; a licence to manufacture was granted to the Sylvania Industrial Corporation, which commenced production in 1933.

The crucial 'Cellophane' inventions were those of an individual experimenter, Brandenberger. A large French textile firm backed him and, with its help, 'Cellophane' was further developed. One of the largest American firms then took up the basic idea, carried on the development and discovered in its own laboratories a new and valuable type of 'Cellophane'.

REFERENCES

1. U.S. *v.* E. I. du Pont de Nemours & Company, 118 F. Supp. 41 (D.C. Del., 1953).
2. Du Pont Cellophane Company *v.* Waxed Products Company, 6 F. Supp. 859 (D.C. N.Y., 1934).
3. Du Pont Cellophane Inc. *v.* Waxed Products Company, 85 F. 2d 75 (C.A. 2d, 1936), cert. denied 57 S. Ct. 194.
4. 'Just About All About Cellophane', *Fortune*, Feb. 1932.
5. Hyden, William L., 'Manufacture and Properties of Regenerated Cellulose Films', *Industrial and Engineering Chemistry*, May 1929.
6. Brandenberger, Jacques E., 'Notes on Cellophane', *Journal of Franklin Institute*, Dec. 1938.

[1] J. E. Brandenberger, 'Notes on Cellophane', *Journal of Franklin Institute*, Dec. 1938, pp. 797–8.

CONTINUOUS CASTING OF STEEL

ATTEMPTS to cast steel continuously have been made for the past century, yet it is only in the last few years that the first commercial plants have been installed.[1] The economic potentialities of this process encouraged inventors to persist in the face of failures which made the steel industry in general sceptical of its practicability. By cutting out the soaking pit and blooming mill of the normal steelworks, it can save both capital and labour costs. It may also reduce the minimum economic size of a steel-works. A greater yield of high-quality metal is obtained, which makes the process especially attractive for the more expensive steels; but it is now beginning to be used to cast lower-grade steels.

The basic idea for the processes used, in which molten steel is poured through a brass or copper water-cooled mould and emerges with a solid outer skin formed on the billet, dates back to the late nineteenth century.[2] Many attempts to solve the problems involved in operating such a process have since been made; success seems to have been achieved by painstaking study and attention to detail, and commercial processes only differ from the original suggestions in employing a shorter mould which moves up and down.

The continuous casting of non-ferrous metals became a commercial success before that of steel. Their lower melting-point makes them easier to cast, and it has been from success there that most processes for casting steel continuously have developed. The most widely used process for non-ferrous metals is the Rossi-Junghans, invented by Dr. Siegfried Junghans in 1927. Dr. Junghans was a member of the family which founded Gebrüder Junghans, one of the largest firms in the German watch and clock industry, and his process was the outcome of a thorough study he and his assistants had made of the problems involved in casting metals continuously. Junghans did his first experiments at the brassworks he ran for his family firm. They were taken over by Wieland-Werke A.G. in 1931, and he worked with this firm until 1935, when he left to work on his own to apply his process to all metals. Its unique feature is the reciprocating mould, intended to reduce the risk of the molten metal adhering to the mould as it solidifies; this has been the chief problem to overcome in developing a successful process. Irving Rossi of New York was responsible for the commercial exploitation of the process, which soon came into wide use both in Europe and America.

Virtually all aluminium is now continuously cast, the largest proportion of it by the Aluminium Company of America's 'Direct Casting' process. This process employs aluminium moulds, and molten aluminium is poured through them in intermittent rather than continuous casting, to give ingots some ten feet long.

[1] From about 1890 to 1910 B. Atha in the United States produced file steel in a continuous casting machine. A single crucible of molten steel was poured through the mould at a time; the mould was vertical and did not incorporate any elaborate cooling arrangements. This work proved abortive and no such good results seem to have been obtained again until the 1940's.

[2] Another process, originated by Sir Henry Bessemer in 1858, consists in pouring steel on to two rollers, which is cast as a sheet when it passes between them. Much effort has been devoted since to making the process practicable; C. W. Hazelett, an individual inventor, obtained some promising results in America during the 1930's with such a process. More recently he has been working on a modification of the process, but commercial production has not yet been achieved.

This process was invented and developed during the early 1930's and put into general use by 1939. Another successful American process for the continuous casting of non-ferrous metals is that of Byron E. Eldred, a professional individual inventor, invented around 1925. This employs a graphite mould; the metal is cooled slowly, and the product is of high quality. The Aluminium Company took a licence under Eldred's patents, which were later acquired by the American Smelting and Refining Company and the Chase Brass and Copper Company. These companies were using a process almost identical to Eldred's patented in 1935 by F. F. Poland and F. A. Lindner; and they have since built further plants using the Poland-Eldred process.

The most commonly used process for the production of steel is the Rossi-Junghans, which has been adapted to cast steel since the end of the war. Dr. Junghans had long been considering casting steel in his machine but he was not able to restart work after the war until 1948; he produced his first cast of steel in March 1949 at Schorndorf, where he had established his works during the war. After carrying on his experiments privately, despite the heavy cost, he made an agreement in September 1950 with Mannesmann A.G. An experimental plant was set up at their Huckingen works, where the process was developed in association with Dr. Junghans. Other steel firms soon made agreements with Dr. Junghans; a consortium has been formed in Germany to exploit the process and a number of commercial plants installed.

Similar work on the casting of steel in the Junghans machine has been done in America by Irving Rossi, apparently working independently of Dr. Junghans. He has formed the Continuous Metalcast Corporation to exploit the process. The first experimental plant to his design was installed at the Allegheny Ludlum Steel Company's Watervliet plant in 1950. The first commercial plant was installed at the Atlas Steel Company's plant at Welland, Ontario, and started production in 1954.

Various other processes are under development in Europe and America. The Austrian firms of Gebrüder Böhler started experimental operation of a plant using a stationary mould in 1947, and commercial operation began in 1952. Since then, however, Böhler appear to have adopted the Junghans reciprocating mould. In France the firm of Jacob Holtzer began work on a process using a rapidly oscillating mould in 1952 with which they are producing limited quantities of steel.

In Great Britain numbers of individuals and companies have tackled the problem, especially the United Steel Corporation and the British Iron and Steel Research Association. The latter began to study continuous casting in 1947 and set up an experimental plant at William Jessop's steelworks in Sheffield in 1954. This machine had a mould that was free to move down against the action of a spring, so that if the steel stuck, the mould could move with the billet and the solid metal skin on it would not be ruptured. This feature has now been abandoned for a variant of the reciprocating mould introduced by Dr. Junghans. A number of different types of machines incorporating the reciprocating mould have been installed in Britain.

Much work has been done on the continuous casting of steel in Russia, though few original contributions have been made there; a recent Russian study of the subject acknowledges that it is the Junghans reciprocating mould that has made it possible to cast steel continuously under production conditions.

In America Edward R. Williams has made a major contribution; starting work in the later 1920's, he developed a process with a thin-walled stationary mould to obtain rapid cooling of the steel. After prolonged experimentation, in 1942 the Republic Steel Corporation began a development programme under his patents. The Babcock and Wilcox Tube Company joined in this work in 1944, and in 1946 the two companies acquired exclusive rights under the Williams patents. A pilot plant began operating in 1948.

Credit for the successful introduction of the continuous casting of steel thus belongs chiefly to men working outside the steel companies. The expense of experimenting with these processes, and the fact that much of the work resembled development more than invention, make it surprising that this should have been so. The steel companies only seem to have shown a serious interest in the processes after the continuous casting of non-ferrous alloys had become established; though they have contributed to recent development work, it is the persistence and ingenuity of a comparatively small number of individuals, notably Dr. Junghans, which have made the continuous casting of steel a reality.

REFERENCES

1. Morton, J. S., 'Continuous Casting of Steel', *Iron and Steel*, May 1955.
2. Lippert, T. W., 'Continuous Casting', *Iron Age*, Apr. 4, 1940.
3. Lippert, T. W., 'Continuous Casting of Semi-Finished Steel', *Iron Age*, Aug. 19, 1948.
4. Williams, E. R., 'Continuous Casting of Metals: the Williams Process', *Steel*, Mar. 6, 1944.
5. Goss, N. P., 'Continuous Casting of Metals: the Goss Process', *Steel*, Mar. 6, 1944.
6. Rossi, I., 'Continuous Casting of Steel', *Journal of Metals*, Mar. 1951.
7. Speith, K. G., and Bungeroth, A., 'The Junghans Method of Continuous Casting of Steel', *Metal Treatment and Drop Forging*, May 1953.
8. Hruby, T. F., 'Fewer Steps to Finished Steel', *Steel*, Nov. 8, 1954.
9. Lippert, T. W., 'Continuous Casting', *Iron Age*, Feb. 24, 1944.
10. Communication from Frau Junghans, Nov. 1956.
11. Boichenko, M. C., *Continuous Casting of Steel*, Butterworth, 1961.

CONTINUOUS HOT STRIP ROLLING

HOT continuous wide strip[1] rolling is recognised as one of the most important twentieth-century innovations in the steel industry. Its introduction greatly improved the quality of steel sheets and substantially reduced costs.

The idea of continuous rolling is an old one: but until the early part of the twentieth century efforts to render it practicable for sheets over twenty inches wide met with failure. In 1892 a mill in Teplitz, Bohemia, began to roll sheets up to

[1] Since the introduction of the wide strip mill, the terminology has become somewhat mixed, the terms sheets and strip being frequently used interchangeably. In earlier days strip ordinarily referred to sections narrower than about 12 inches and sheets to those wider than this. Today both are rolled on the hot continuous wide strip mill. In this account the wide material is referred to as sheets.

50 inches in width and lengths up to 60 feet by a continuous process, but difficulties with the maintenance of gauge uniformity led to its abandonment in 1907. Charles W. Bray designed a continuous sheet mill in 1902 for the American Sheet and Tin Plate Company which, after operating experimentally for a few years, shut down.[1]

John B. Tytus was the driving force behind the invention of a practical hot wide strip continuous rolling process. The son of a prominent paper manufacturer of Middletown, Ohio, he graduated in arts at Yale and, finding the paper business unattractive, in 1904 he took a job in the American Rolling Mill Company. He was immediately struck by the apparent inefficiency of the methods of rolling steel sheets.[2] In eighteen months he acquired a working knowledge of the conventional rolling methods and was promoted to assistant plant superintendent. He slowly formulated his ideas for a novel wide strip continuous rolling process and, by 1920, had embodied them into sketches and charts. Charles Hook, now chairman of the Board of Armco, worked with Tytus and gave him the benefit of his long practical experience. Hook was convinced that Tytus's ideas were sound and that the time was ripe to introduce them in view of the ever-increasing demands from the automobile industry for steel sheets. At the time Armco was flooded with new orders for sheets and, instead of gambling on a radically new process, chose to install four new standard mills to meet the demand; but in 1921 Armco purchased the Ashland Iron and Mining Company where it was decided to produce steel sheets and, instead of using the conventional methods, to erect a mill embodying the ideas of Tytus. At the end of 1923 the first continuous mill was completed. Within six months the Armco management expressed their satisfaction with the mechanical soundness of the mill, patented the process and licensed the patents to other steelmakers.

Armco was not alone in its research on continuous wide strip rolling. At about the same time two engineers, H. M. Naugle and A. J. Townsend, also made substantial contributions to the continuous rolling process, working at the Columbia Steel Company in Butler, Pennsylvania. Unlike Tytus, who had constructed the Ashland mills to roll short wide sheets which followed one another through the succession of stands, Naugle and Townsend attempted to roll wide strip several hundred feet in length. Columbia completed plans for production in 1925 and their plant commenced work in 1926. Since Armco patents covered continuous processes for rolling sheets twenty or more inches wide, Columbia was notified in 1927 that they were in all probability infringing Armco patents. Columbia finally agreed to sell all their plants and patents to Armco, which reorganised the Butler plant and incorporated in it the best features of both processes.

Tytus's success arose out of the discovery that, contrary to traditional ideas, a true cylinder could not possibly roll wide, thin sheets; that the steel must have a slightly convex cross-section, and that it must have progressively less convexity with each successive pass towards the end of the mill. Under proper control this convexity forms a guide for the sheet by surface contact, preventing any lateral movement and insuring its regularity of shape. Tytus also isolated the five most important variables that required precise control for successful operation: the prepared contour of the rolls; the temperature of the rolls; the composition and springiness of the rolls; the spacing of the rolls; and the shape, composition and

[1] Frank Fanning, 'Wide Strip Mills – Evolution or Revolution?' Paper read before General Meeting of the American Iron and Steel Institute at New York, May 21–22, 1952.

[2] Christy Borth, *True Steel*, 1941, pp. 146–7.

temperature of the sheet. The development of the four-high mill at about the same time enabled these variables to be more easily controlled.[1]

Thus the credit for this important new technique would seem primarily to belong to Tytus while working in what was then one of the smaller American steel companies, and secondarily to Naugle and Townsend of the Columbia Steel Company.

REFERENCES

1. Knox, J. D., 'Rolls Sheet Steel Direct from Ingot Without Reheating', *Iron Trade Review*, June 16, 23, 30, 1927.

2. Malborn, J., 'The Development of the Continuous Rolling of Sheet and Tinplate', *Blast Furnace and Steel Plant*, Oct. 1938.

3. Eppelsheimer, D., 'The Development of Continuous Strip Mills', *Journal of Iron and Steel Institute*, vol. 138, 1938.

4. Badlam, S., 'The Evolution of the Wide Strip Mill', *Yearbook of American Iron and Steel Institute*, 1927.

5. Borth, Christy, *True Steel*, 1941.

COTTON PICKER

THE removal of cotton from the plant by mechanical methods is by now well established; a quarter of the total American crop is harvested in this way. Satisfactory machines for this purpose, however, have only recently made their appearance and there is a long history of inventive trial and development behind them. Many types of machines were experimented with – in some the cotton was plucked by revolving spindles on prongs, by pneumatic apparatus or electrically charged belts; in others it was stripped from the plant by combs or rollers; in others again the whole plant was cut off and the cotton threshed from the vegetable matter. The picker type, employing revolving spindles, and the stripper have proved the most valuable. This note is concerned with the former.

Patents carrying the names of many individual inventors were being taken out on cotton pickers from 1850 onwards, but the machines in general employment today can be traced back to the work of a few individual inventors and certain companies which undertook the prolonged and intricate tasks of development. Angus Campbell was the pioneer of the Price-Campbell pickers, finally perfected and marketed by the International Harvester Company. John and Mack Rust first invented the type of machine still known under their name, which was partly developed by the Rust brothers themselves and was later taken up for subsequent development by the Allis-Chalmers Manufacturing Company and other companies. The patents of Hiram M. Berry, another individual inventor, were taken up by Deere and Company, but here the development proved unsuccessful.

Angus Campbell, a pattern-maker from Chicago, first became aware of a need for such a machine in the 1880's when he observed the slow and tedious process

[1] The history of the four-high mill goes back to the nineteenth century, but the four-high mills in use then had no roller bearings like their modern counterparts. The Rome Brass and Copper Company installed the first modern four-high mill with roller bearings in 1925, and it has served as a model for the later mills of this type.

of picking cotton by hand. In 1889 he placed his first spindle-type picker in the field, but the machine left cotton on the ground and caused damage to the bolls and blooms. In each of the next twenty years Campbell transported models for testing from Chicago to the southern cotton fields. The lack of adequate machine-shop facilities near the cotton fields hampered development. Campbell received little encouragement from others; his ideas were thought foolish and he concealed his work to avoid ridicule. He tried various types of spindles, cylinders and drives in an effort to find a satisfactory combination. When the gasoline-engine made its appearance, Campbell adapted it for use with his cotton picker.

Financial help was hard to come by, but in 1908 Campbell got the support of Theodore H. Price, a wealthy cotton dealer. The results of tests made with Campbell's spindle-picker in 1910 were encouraging, yet the machine was still not practical. From 1910 until his death in 1922, Campbell, with others at the Price-Campbell Cotton Picker Corporation, strove to perfect the picker.

The International Harvester Company had taken a keen interest in cotton-picker machines and they had tested many types. When their experiments with a pneumatic, and air-suction picker proved unsuccessful in the early 1920's, they turned to an investigation of the spindle-type pickers. They purchased outright the Price-Campbell patents in 1924 and built their first spindle machine, a single-row self-propelled type which performed poorly. The depression caused the Harvester Company to postpone the introduction of a simplified spindle-picker; labour-saving devices were by no means in demand at that time. Further improvements were necessary and it was not until 1942 that the Company began to produce commercial machines on a small scale; by 1948 the scale of production had increased substantially.

John and Mack Rust began their work long after Campbell had started. They grew up on a farm in Texas and earned their first wages in cotton picking. John displayed inventive talent at an early age. In 1922 he joined a firm organised to develop a novel design for a wheat combine. His ideas about a cotton picker, which, like Campbell's, was of the spindle type, crystallised in 1924, when he completed the first sketch of his machine. In 1927, when working as a maintenance man for the Gleaner Combine Company in Missouri, he devised a solution for the problem of stripping the cotton off the spindles; he substituted a smooth for a barbed spindle and moistened the spindle before bringing it into contact with the cotton. Previous inventors had explored the possibilities of smooth spindles, but had found them impractical.

John Rust returned to his sister's farm to develop his ideas, setting up a shop in the garage, where his brother Mack, a graduate engineer, joined him in 1928. They were short of money; John borrowed $4000 from his friends and relatives. He refused to take his patent to a corporation because he feared that he would lose direct control of the development. The brothers constructed their first complete model in 1930. In 1933 they tested their machine at the Experimental Station of the University of Mississippi, where it performed very well.

> 'Now for the first time the Rust brothers had the opportunity to see the machines developed by competing inventors. Although they were aware that since the Civil War 750 patents had been issued by the Patent Office in Washington to inventors of cotton-picking machines, neither John nor Mack Rust had felt, until they had completed their first model, it was of much use to study the failures of other inventors. They say now that if they had studied this long

history of disappointments they would have been so discouraged, that they would never have got as far as they have.'[1]

After moving to Memphis in 1934, the Rusts conducted successful runs in the Arkansas fields. They made ten machines for use during the 1936 picking season, sold two machines to the Russian Government and enlisted some support from New York financiers. But their troubles were by no means at an end. The smooth working of their machines depended upon favourable conditions, and their financial backers were inconstant. In 1940 they were compelled to sell their equipment to meet their obligations, and the partnership between the two brothers was dissolved. In 1941 John Rust re-designed several parts of the machine and filed new patent applications. Allis-Chalmers took licences under these patents, employing Rust as a consultant, and developed a machine which went into commercial production. Ben Pearson, Inc., a smaller firm in Arkansas, also secured a licence from Rust.

Hiram M. Berry is a third individual inventor who perhaps merits mention. As early as 1925, in Greenville, Mississippi, he was attempting, with very limited resources, to design a picker with barbed spindles. The Berry patents were in 1945 bought by Deere & Company, who themselves had for some years been trying unsuccessfully to develop a spindle picker. Deere & Company, however, found it impossible to perfect the Berry picker and they later placed on the market a machine based on the Price-Campbell principles subsequently developed by the International Harvester Company.

The basic ideas of the cotton picker may, therefore, be said to have arisen with individual inventions. The Campbell picker led to one main stream of development, carried through to a successful conclusion by the International Harvester Company. The work of the brothers Rust led to a second successful type of machine; in this case development, at the early stage, was carried through by the Rusts themselves but at a later stage by the Allis-Chalmers Company.

REFERENCES

1. 'Mechanical Cotton Pickers', *Manufacturer's Record*, Mar. 1945.
2. Strauss, R. K., 'Enter the Cotton Picker', *Harper's Magazine*, Sept. 1936.
3. Hagen, C. R., 'Twenty-Five Years of Cotton Picker Development', *Agricultural Engineering*, Nov. 1951.
4. 'The Revolution in Cotton', *American Mercury*, Feb. 1935.
5. 'Mr. Little ol' Rust', *Fortune*, Dec. 1952.
6. Day, W., 'Picking Cotton by Machine', *Scientific American*, Mar. 4, 1911.
7. 'A Mechanical Cotton Picker', *Scientific American*, Jan. 17, 1891.
8. Smith, H. P., Killough, D. T., Byram, M. H., Scoutes, D., Jones, D. L., 'The Mechanical Harvesting of Cotton', *Texas Agricultural Experiment Station Bulletin No. 452*, Aug. 1932.

CREASE-RESISTING FABRICS

A SERIOUS defect of cotton, linen and artificial silk fabrics is their tendency to crease. It was not until the late twenties that an effective remedy was invented in the laboratories of the Tootal Broadhurst Lee Company Ltd., a medium-sized

[1] R. K. Strauss, 'Enter the Cotton Picker', *Harper's Magazine*, Sept. 1936.

Lancashire textile firm.[1] Their crease-resisting process constitutes the outstanding, perhaps the only, major non-mechanical advance conceived of, and fully exploited within, the textile industry proper in this century.[2] The process is remarkable in other ways. While not wholly split off from earlier knowledge, it had no extensive scientific background. It has continued to hold the field since its discovery. The inventing company was the first to seize upon the problem and then to pursue its researches stubbornly to a successful conclusion.

Immediately following the First World War, Tootal Broadhurst Lee, largely through the initiative and imagination of Mr. (later Sir) Kenneth Lee, one of the directors and soon afterwards Chairman, set up a small research group of physicists and chemists under the direction of Dr. R. S. Willows. The step was taken in the belief that the cotton industry had sadly neglected scientific research in the preceding half-century, and that this in itself constituted grounds for presuming that valuable results were to be obtained for the seeking. Sir Kenneth has himself described the principles which were to govern the operation of the research unit.[3] No attempt was to be made to gather together scientists with a special knowledge of the cotton industry; indeed fresh minds were sought for 'since such workers would not have got into ruts, and would be more likely to bring a new outlook to the problems proposed'. The scientists were to work in close contact with the normal routine testing carried on by the firm, for in this way it was supposed that they 'would become acquainted with the commoner defects in fabrics, which might form a starting point for further research'. Even more interesting, the work of the scientists was to be 'directed' in that a specific problem was set before them: they were to seek for ways of reducing the propensity of cotton fabrics to crease, so that cotton would behave much like wool in that respect, without otherwise altering its character as a textile material. The set problem was successfully resolved but, as will be explained below, the methods finally devised proved to have other extremely valuable applications.

The research work at Tootal's began in 1918. Patents applied for in 1926 and 1927 under the names of Foulds, Marsh, Wood, Boffey and Tankard were granted in 1929. In 1932 the firm was able to announce that their invention was ready for commercial application. In the fourteen years covered by the research and development the cotton industry passed through a long period of depression. It would not perhaps have been surprising if, in view of the great difficulties and repeated disappointments of the scientists, the task had been abandoned. The final outcome represented, therefore, as much an act of courage on the part of those responsible for the solvency of the business as a work of great pertinacity and ingenuity on the part of the inventors themselves.

The essential novelty of the invention lay in forming a synthetic resin *inside* the cotton, linen or artificial silk fibres so that the elastic property of the resin would be imparted to the material. In 1906 Eschalier had made some empirical observations about the behaviour of artificial silk towards formaldehyde. But before the Tootal Broadhurst Lee invention there had been very few proposals to treat

[1] Tootal Broadhurst Lee does not rank among the one hundred largest companies in British industry. In 1954–55, with assets of £9 million, it ranked as the sixth largest company in the cotton industry.

[2] J. R. Whinfield, 'Textiles and the Inventive Spirit', Emsley Lecture, Textile Institute, Oct. 1955.

[3] Kenneth Lee, 'Industrial Research: A Business Man's View', Royal Institution, Dec. 15, 1933.

textile materials with resins, and then only so as to coat or embed the textile material and thus totally alter its character and appearance. Thus the external application of synthetic resins of one kind or another to fibres and fabrics had been proposed previously, but this, while creating stiffness in the materials, gave them no powers of crease resistance. In the Tootal Broadhurst Lee process, the internal impregnation of the fibres was brought about by first placing an aqueous solution of the chemicals, which will form a resin, within the fibres; next drying the fabric and then subjecting it to high temperatures which, with the aid of a catalyst, transform the chemicals, where they lie within the fibres, to resin. The fabric is then washed to remove any uncombined chemicals or resin formed on the outside of the fibres. That is to say, the resin is manufactured, from its chemical constituents, inside the fibres.

This method would perhaps well have justified itself by performing only the task originally set out – i.e. the imparting of crease-resisting properties to cotton cloth. In fact, it had another, unexpected property – its power to increase the strength of viscose rayons, and particularly spun rayons when wet. Rayon is naturally a weaker fibre than cotton or linen and, in particular, rayon when wet is extremely fragile and subject to damage during laundering. The crease-resisting process increased substantially, both in the wet and the dry state, the tensile strength of rayon fabrics, in addition to its effects in rendering them more resistant to creasing. In this way a new discovery, that of crease-resisting, greatly enhanced the value of an old one, that of viscose itself, and opened up much wider markets for rayon as a dress fabric. Crease-resisting is a case of success in 'directed' research, but it also provides a very good example of the unexpected and unpredictable windfalls that arise in the course of experimentation.

Although the period of research, before success was achieved, was prolonged, and its costs, having regard to the moderate size of the firm, were far from insignificant, the crease-resisting process was produced with relatively modest expenditure. In the first stage, that of invention, the research staff consisted of about half a dozen scientists with assistants. In addition, however, Tootal Broadhurst Lee carried out the development of their invention. The designing of full-scale plant for the preparation of the resin-forming mixture and for impregnating the cloth with this mixture presented no great difficulties. A more formidable task was that of designing equipment which would transform the chemicals into resin within the cotton fibres by heating the treated fabric at a higher temperature and for a longer time than was customary at that date. Moreover the equipment had to give the fabric an exact amount of heating which could be exactly repeated so as to give a constant result in all batches of material treated. The development stage, while of course considerably more costly than the stage of invention (Sir Kenneth Lee spoke of the development cost as 'scores of thousands of pounds') involved sums which were only a fraction of those which seem to have been spent on some other inventions, especially chemical inventions, of the twentieth century.

A medium-sized firm, therefore, took the lead both in appreciating the problem to be solved, in possessing sufficient confidence that it could be solved and of finding the proper lines for its solution. No one of the large chemical firms seems to have been pressing towards the same answers. I.G. Farbenindustrie was apparently showing some interest in the field; in 1929 they held a patent connected with the use of urea as an impregnating agent to strengthen rayon in the wet state. But they seemed to be unaware of the crease-resisting potentialities of this approach; they were concerned with the treatment of fibres rather than fabrics and they treated fibres with small quantities of resin which induced no crease-resisting characteristics.

To sum up. The discovery of the crease-resisting process occupies an important place in the history of invention in the twentieth century, partly because it is a rare case of success being achieved in a 'set problem' (although the solution of that problem yielded other unpredictable gains) and partly because the discovery was made, and the development carried through, without the enormous expenditures sometimes assumed to be inseparable from important progress in the chemical field.

REFERENCES

1. Lee, Kenneth, 'Industrial Research: A Business Man's View', Royal Institution, Dec. 15, 1933.
2. Wood, F. C., 'Research, Textiles and the Future', *Textile Institute Journal*, June 1939.
3. 'Triumph After Years of Research', *Chemical Age*, Aug. 13, 1932.
4. Amick, C., 'Crease Resisting Fabrics', *American Dyestuff Reporter*, Oct. 7, 1935.

CYCLOTRON

In recent years very large, intricate and costly cyclotrons have been constructed to aid scientists in their study of matter, but they had humble beginnings. The creator of the cyclotron, Ernest O. Lawrence, was an academic scientist who combined a natural scientific curiosity with a flair for practical application. He was born in South Dakota in 1901, received his Ph.D. from Yale in 1925 and remained there until he accepted a position at the University of California.

In 1929 Lawrence found the clue which led to the invention of the cyclotron. He had, in the spring of that year, read a paper written by R. Wideroe, a German physicist, and his attention was drawn particularly to the apparatus employed. Wideroe had put together two long vacuum tubes in such a way that he could increase the speed of electrified particles even though he employed a relatively small voltage. Lawrence thought it might be possible to put together a great number of these tubes; the particle acceleration achieved might be sufficient to break up atoms. He thus had formulated the basic notion of a cyclotron; he was confident that he could generate a tremendous acceleration by giving the particles a series of perfectly timed electrical pushes, each of which would be relatively small. He discarded the idea of hooking up tubes in a long line, as Wideroe had done, because this type of apparatus could not at that time produce a very great acceleration in a small space, and seized upon that of using a curved path for the particles, so that the particles could circulate continuously, travelling long distances in a relatively small volume and using the same accelerating system over and over again. An electrically charged particle entering a magnetic field directed at right angles to the motion of the particle, proceeds to move in a circle with constant speed; as the particle speed is increased, the radius of the circle in which the particle moves also increases. Further acceleration occurs at each revolution. Lawrence wrote out a simple mathematical relation to describe these facts and his apparatus followed these lines.

Although Lawrence's associates agreed that he was right, they were pessimistic about the possibility of constructing a machine that would operate successfully. When Dr. W. D. Coolidge presented the Comstock Prize of the National Academy of Sciences to Lawrence in 1937 he declared:

> 'Dr. Lawrence envisioned a radically different course – one which did not have those difficulties attendant upon the use of potential differences of millions of volts. At the start, however, it presented other difficulties and many uncertainties, and it is interesting to speculate on whether an older man, having had the same vision, would have ever attained its actual embodiment and successful conclusion. It called for boldness and faith and persistence to a degree rarely matched.'[1]

Lawrence and a graduate student at California, N. E. Edlefsen, built the first cyclotron with improvised means:

> 'There was an ordinary wooden kitchen chair on top of a physics laboratory table at the University of California in Berkeley. On either side of this chair stood a clothes tree, with wire hanging on the hooks which normally would hold hats and coats. Between the two poles the wire was suspended in loose hoops. The loops went all around the chair, on the seat of which was an object about the size and shape of a freshly baked pie. It was made of window glass, sealing wax and brass.'[2]

After constructing a metal cyclotron of the same size, the next step was to build a larger one. A sum of $10,000 was collected for the purpose and Lawrence was fortunate enough to obtain the use of a very large magnet, originally built for the Chinese Government but never delivered.

The building of this larger cyclotron presented formidable engineering problems, but these were overcome largely by the drive and energy of Lawrence himself. Although he explored the possibilities of other atom-smashing methods, he found none as effective as his cyclotron accelerator.

An individual conceived this highly important device and, with limited resources, transformed his ideas into a practical instrument.

REFERENCES

1. Blakeslee, Howard, 'Atomic Slingshot', *Science Digest*, Apr. 1949.
2. Schuler, L. A., 'Maestro of the Atom', *Scientific American*, Aug. 1940.
3. Presentation of the Nobel Prize to Professor Ernest O. Lawrence by Professor R. T. Birge, *Science*, Apr. 5, 1940.

DDT

DDT is a white crystalline substance made by reacting monochlorobenzene and chloral in the presence of sulphuric acid. Its amazing insect-killing power was discovered by chemists in the Swiss firm of J. R. Geigy in 1939. This firm had an

[1] Presentation of the Nobel Prize to Professor Ernest O. Lawrence by Professor Raymond T. Birge, *Science*, Apr. 5, 1940.

[2] Howard Blakeslee, 'Atomic Slingshot', *Science Digest*, Apr. 1949.

established reputation for synthetic dyestuffs, but before the discovery of DDT was relatively unknown for insecticides. For about twenty years Geigy chemists had searched for a moth-proofing agent, which would be odourless, colourless, non-toxic to humans and resistant to deterioration by light. They finally marketed such a material under the name 'Mitin FF'.

It was next decided to broaden the scope of the research and to look for an insecticide that acted against a broader group of insects. The chemists first examined a group of natural insecticides, such as pyrethrum and rotenone, but found that these lost their power in the presence of light. Paul Müller, one of the Geigy chemists, synthesised DDT which turned out to be the crucial compound. In the course of his work he made diphenyltrichloroethane, an organic compound that showed promise as an insecticide. He next prepared a number of compounds, using the phenyl group, and one of these was DDT. Its chemical structure gave no hint of its extraordinary insect-killing properties. In fact, the same compound had been prepared by a student, Othmar Zeidler in 1874, but he had not suspected its insecticidal properties. Most of the DDT produced commercially, however, is made by Zeidler's method.

Müller tested his discovery on flies and was amazed by its effectiveness; he found that DDT was particularly deadly when used in a water emulsion; he was surprised to find that, after spraying DDT on a window and allowing it to dry, it retained its insect-killing powers for days. DDT was later found to possess the power of killing by contact the larvae of the Colorado beetle. Geigy informed the British Legation of the discovery in 1942, and the Allies used DDT extensively during the war, when it was particularly useful against lice. The Geigy patents were widely licensed.

This is an instance of a discovery by a chemical firm in a field which was relatively new to them.

REFERENCES

1. West, T. F., Hardy, J. Eliot, and Ford, J. H., *Chemical Control of Insects*, 1951.
2. 'Patent Status of DDT', *Chemical and Engineering News*, Sept. 10, 1945.
3. Froehlicher, Victor, 'The Story of DDT', *Soap and Sanitary Chemicals*, July 1944.
4. Frear, Donald E. H., *Chemistry of Insecticides, Fungicides and Herbicides*, 2nd ed., 1948.
5. West, T. F., and Campbell, G. A., 'The Story of DDT and its Role in Anti-Pest Measures', *Chemistry and Industry*, May 19, 1945.
6. 'DDT', *Society of Dyers and Colourists Journal*, Dec. 1945.

DIESEL-ELECTRIC RAILWAY TRACTION

In the past twenty-five years the diesel-electric locomotive has largely replaced the steam-locomotive in the United States and has gained much ground elsewhere. Diesel-electric traction represents the use of the diesel engine with the petrol-electric system of traction, both of which were invented in the 1890's. The first experiments with petrol-electric traction were unsuccessful owing to the crudity of the available engines, but in 1903 a system was successfully applied to some railcars of the British North-Eastern Railway. Railcars of this type enjoyed considerable popularity on the continent of Europe and in America before 1914.

The first use of the diesel-electric system was in 1913 on a railcar built in Sweden by the Swedish General Electric Company and the A.B. Atlas Diesel. A number of railcars and locomotives were built which proved to be a financial success. Sulzer Bros. in Switzerland also pioneered with diesel-electric railcars. Five were built for the Prussian and Saxon State Railways, and put into service in 1915. By 1925 the diesel-electric system was being developed in many parts of the world, the chief object being to reduce the weight of the units employed. Diesel railcars entered into service on private railways in Denmark in 1926; they were used in Germany in the early 1930's. But despite this progress, diesel-electric railcars and locomotives did not come into general use. The countries which used them most extensively were those lacking their own supplies of coal.

The use of diesel-electric traction on main lines began in the United States during the later 1930's. In 1923, indeed, a diesel-electric switching locomotive had been built by General Electric, Ingersoll-Rand and the American Locomotive Company. In 1925 the Canadian National Railways put diesel-electric railcars into service. But it was the General Motors Corporation which finally placed on the market in 1934 a light powerful diesel engine which established the success of this system of traction. Charles F. Kettering, the Research Director of General Motors, had been interested in the possible uses of the diesel engine for some years, and in 1930 General Motors purchased two companies, the Winton Gas Engine and Manufacturing Company and the Electromotive Company. The Winton Company had been established by Alexander Winton and had constructed the first diesel to be built completely within the United States. Winton became associated with a highly talented engineer, George Codrington, and the company was producing diesels for ships during the First World War. The Electromotive Company had between 1922 and 1929 constructed a large number of railcars with a petrol-engine and electric transmission. The moving spirit in the Electromotive Company was Harold L. Hamilton. General Motors thereby brought under their control some of the most experienced persons in this branch of engineering.

The development of the diesel engine after 1930 moved swiftly. In 1934 the Burlington and Union Pacific bought electro-motive models to power their trains to the west coast of America, and on these long runs across the Rocky Mountains the diesel-electric soon acquired a reputation for performance and economy. After the war the American railways found themselves with locomotives that were for the most part old and worn-out after the war and the slump. Most railways decided to take the opportunity to change their motive power to diesel-electric, and General Motors, as the largest company and the pioneer in this field, was able to take a substantial part of the locomotive market from the established companies. Recently, however, doubts have been expressed about the economic gains provided by the diesel locomotive; it has even been suggested that modern steam locomotives would be cheaper to run in America.

The final commercial success of diesel-electric traction was much more a matter of development than of invention. The combination of the diesel-engine and electric-traction equipment was not new, but General Motors, by acquiring two small pioneering firms, by placing the large resources of their research organisation to the task and by taking advantage of a large potential domestic market, finally established the great commercial advantages of the system.

REFERENCES

1. Hamilton, H. L., *Diesel Engine Development*, 1944.
2. Foell, Charles F., and Thompson, M. E., *Diesel-Electric Locomotive*, 1946.
3. Munger, W. P., Jr., 'The Development of the Diesel Locomotive in America', *Diesel Power and Diesel Transportation*, Nov. 1942.
4. Berge, Stanley, *Self-Propelled Diesel Cars and Multiple-Unit Trains*, 1952.
5. Codrington, George W., *Shadows of Two Great Leaders – Rudolph Diesel and Alexander Winton*, 1945.
6. Webster, Harry, *Railway Motive Power*, 1952.
7. Heldt, P. M., *High-Speed Diesel Engines*, 1947.
8. Allen, O. F., *The Modern Diesel*, 1947.
9. Kettering, E. W., *History and Development of the 567 Series General Motors Locomotive Engine*, 1951.
10. Drake, P. E., 'Gasoline Electric Traction in Europe', *Electrical World*, 58, 1911.
11. 'The Sulzer Diesel Engine for Rail Traction', *Sulzer Technical Review*, 1947, no. 2, 42.
12. Franco, I., and Labeyn, P., *Internal Combustion Locomotives and Motor Coaches*, 1931.

FLUORESCENT LIGHTING

THE fluorescent lamp employs two scientific phenomena that have long been known: that certain materials are excited to fluorescence by ultra-violet radiation, and that an electric discharge through mercury under low pressure produces a high proportion of invisible ultra-violet radiation. Knowledge of fluorescent materials dates back to the sixteenth century; it was in 1852 that Sir George Stokes discovered that some of them are excited by ultra-violet rays. Becquerel constructed the first attempt at a fluorescent lamp in 1859, when he placed fluorescent materials inside a Geissler discharge tube, but this was a crude and inefficient device. Numbers of more recent attempts were made to build lamps in which fluorescent materials were excited by the rays from a vacuum tube, but they were also inefficient and are more closely related to the television tube than the fluorescent lamp.

The first low-pressure mercury discharge lamp was introduced at the beginning of the twentieth century by Peter Cooper-Hewitt, the American individual inventor. Sir Humphrey Davy had discovered the effect of a discharge through mercury early in the nineteenth century. Cooper-Hewitt's lamp was inefficient by modern standards, though better than contemporary incandescent lamps; it also produced the characteristic blue light of the mercury discharge lamp. It differed from the modern fluorescent lamp in being designed to produce visible radiation, not ultra-violet. In 1901 he used rhodamine dye, which fluoresces red, to improve the light's colour, but the rhodamine deteriorated too rapidly for this to be a success. The Moore and neon discharge lamps, introduced at much the same time as the Cooper-Hewitt, also contributed to the development of the fluorescent lamp: D. McFarlan Moore, an American individual inventor, was the first to apply the hot cathode used on it, and he also anticipated Wehnelt in constructing lasting electrodes. Georges Claude's introduction of the neon tube, and the desire to modify its colour, stimulated interest in fluorescent powders and in means of employing them with a lamp.

Professor J. Risler patented a method for applying powders to the outside of tubular discharge lamps in 1923; in 1933 the invention of a means of applying them to the inside of lamps led to the introduction of fluorescent lamps for floodlighting and advertising. The French Claude Company and some German companies, including Osram, seem to have been the pioneers in this work. The lamps used were either neon or low-pressure mercury tubes; they differed from the modern fluorescent lamp in operating at high voltage and having cold cathodes, which are larger and less efficient than hot cathodes. The fluorescent powders were also less efficient than those used later, serving to vary the colour of the light rather than to increase the amount produced. Improvements to the hot cathode designs of Moore and Wehnelt, which increased durability, were made during the 1920's. Dr. Albert W. Hull of the American General Electric Company made probably the greatest contribution to this work, his ideas being patented in 1927. Increased length of life of the cathodes and the discovery of improved fluorescent powders represent the chief recent inventions involved in the fluorescent lamp.

The general interest in fluorescence as a light source, which was stimulated among European lighting engineers and scientists by the introduction of the advertising and floodlighting fluorescent lamps about 1934, led most of the large lighting companies to commence work on a fluorescent lamp for general lighting. Late in 1934 the American General Electric Company's consultant, Dr. Arthur H. Compton, informed his firm of the European progress with cold-cathode fluorescent lights and it commenced to develop a model for general use. Dr. George Inman was in charge of General Electric's development programme; he received a patent on the appropriate combination of elements in a practical lamp. This involved a low-pressure mercury tubular lamp employing Hull's teachings as regards the hot cathode, together with Inman's discovery that to obtain the highest luminous efficiency it was necessary to ensure that the mercury-vapour discharge should produce a maximum of ultra-violet radiation and that fluorescent materials should be used (such as silicates, tungstates and borates) which were especially responsive to the radiation. The lamp was first marketed in America in 1938. A similar lamp had, unknown to General Electric, been invented in 1926 by three Germans, Friedrich Meyer, Hans Spanner and Edmund Germer, of the Rectron Company. This Company took no interest in the development of the lamp, and the patents on it were subsequently acquired by General Electric.

Similar work was being done in Europe alongside the development of high-voltage cold-cathode lamps in which mixtures of fluorescent powders were first used to give white light or desired colours. But in Europe there was less thought of early production; some, at least in the British industry, had doubts about the commercial value of the small American lamp. Because of the technical contacts and patent-sharing agreements between them, it is difficult to apportion credit to individual companies for the advances made. However, the General Electric Company in England was responsible for the chief advances in fluorescent powders, which were essential to increase the lamp's efficiency. It was not until the outbreak of war that the decision to market fluorescent lamps was made in England, when British Thomson-Houston pioneered a high-loading lamp particularly suitable for large-area lighting.

The fluorescent lamp is the result of development largely done by the big electrical companies and is based on scientific discoveries mostly made by the end of the nineteenth century. It could have been produced by a similar effort certainly at any time after Hull's work on cathodes in 1927, and probably after the invention of the Cooper-Hewitt lamp and the work on cathodes of Moore and Wehnelt in the 1900's. The immediate occasion for the starting of the development of the fluorescent lamp was the use of fluorescent powders in lamps in Europe, in some cases by companies other than the large electrical manufacturers.

REFERENCES

1. Bright, A. A., Jr., *The Electric-Lamp Industry*, 1949.

2. General Electric Co. *v.* Hygrade Sylvania Corp., 61 Fed. Supp. 476 (D.C.N.Y., 1944).

3. Claude, A., 'Lighting by Luminescence', *Light and Lighting*, June 3, 1939.

4. Randall, J. T., 'Luminescence and its Applications', *J. Royal Society of Arts*, Mar. 5, 1937.

5. Jenkins, H. G., 'Fluorescent Lighting', *J. Royal Society of Arts*, Apr. 3, 1942.

6. Davies, L. J., Ruff, H. R., and Scott, W. J., 'Fluorescent Lamps', *J. Institution of Electrical Engineers*, part II, no. 11, Oct. 1942.

7. Aldington, J. N., 'Fluorescent Light Sources and their Applications', *Transactions of the Illuminating Engineering Society*, June 1942.

GYRO-COMPASS

THE gyro-compass is essential to the navigation of the modern ship. Though the effects of iron construction on a magnetic compass can be corrected, those of electrical equipment cannot; warships and especially submarines are worst affected by this. The first serious study of the gyroscope was made by the French scientist, Foucault, who used it to demonstrate the rotation of the earth in 1852; it had been known for some twenty-five years before as the 'rotascope', a gyroscope with three degrees of freedom which could maintain a fixed direction. Foucault showed that the axis of a gyroscope in which one degree of its freedom was restricted would point to the geographical North Pole, and return to this direction if disturbed; the gyro-compass has been based on this discovery.

Foucault's gyroscopes were hand-driven and no practical use could be made of them until they were electrically driven. Trouvé apparently made an electrically driven gyro-compass in 1865 in France; Hopkins invented a means of driving gyroscopes electrically in 1878; in 1884 Lord Kelvin described a possible gyro-compass to the British Association, and the French navy tested a free gyro to check the deviation of a magnetic compass. No progress was made towards a practical gyro-compass by these experiments; it did not appear until the first decade of the twentieth century, as the result of the work of Dr. Anschütz-Kaempfe in Germany.

Anschütz was born in 1872, the son of a professor of mathematics and physics. He studied medicine but the Austrian art historian, Dr. Kaempfe, persuaded him to abandon it for the history of art. After his father's death, Anschütz was adopted by Dr. Kaempfe and took the name of Anschütz-Kaempfe. When Dr. Kaempfe died Anschütz-Kaempfe was left wealthy; he became interested in polar exploration and in 1897 conceived the idea of reaching the North Pole in a submarine. A magnetic compass would not be suitable for this, so he had to seek an alternative. After some preliminary experiments in 1900 he came across the gyroscope, and studied Foucault's experiments. He first worked with gyros having three degrees of freedom, which would act merely as course indicators;[1] a number of experimental models were built, the foreman of a firm of model makers in Munich named Keicher doing most of the construction work and supplying practical mechanical knowledge. Tests were made in ships on a lake in 1903 and on the sea in 1904, but the results were disappointing. The free gyro wandered too much to be useful, as has since been established to be inevitable with known manufacturing techniques, and so Anschütz decided in 1905 to abandon it for the gyro with one degree of its freedom limited which would seek north, and thus be a true substitute for the magnetic compass.

The German navy had already shown an interest in his work; Anschütz-Kaempfe and Keicher had moved to Kiel in 1904 to be in closer touch with the maritime services. They worked with a firm of precision instrument makers at first but, after starting work on the north-seeking gyro, Anschütz-Kaempfe decided to form his own establishment in 1905. Numbers of experimental north-seeking gyros, or gyro-compasses, were made over the next two years. Dr. Max Schuler joined him in 1906, and contributed much to the final design of this first Anschütz single-gyro compass, which was completed in September 1907. Sea trials in the battleship *Deutschland* followed in the spring of 1908; they established the value of the gyro-compass and led to an order for them from the German navy. Anschütz-Kaempfe and Schuler now devoted their efforts to improving the accuracy of the compass since it was liable to error when a ship was rolling on quadrantal courses; Schuler's invention of a three-gyro compass in 1911 represented a great advance. This compass was produced until it was replaced by Anschütz's two-gyro compass in 1925, which represented his final contribution to the art.

Other gyro-compasses followed these successes. The first and most widely used of these was the Sperry. E. A. Sperry was a successful American inventor, who had been experimenting with gyroscopes since the 1890's. About 1908 he became interested in the gyro-compass and asked his friend H. C. Ford to help with its design. They chose the single-gyro type; it was ready for sea trials by 1911, and their success led to its adoption by the American navy. It was later adopted by the British navy and became widely used. Though it has since been considerably modified, the basic single-gyro design has been retained. Another compass which is now widely used is the Brown, designed by S. G. Brown during the First World War. It is also of the single-gyro type, carefully designed to minimise friction; it was not made in large numbers until modifications by the Admiralty during the Second World War had improved its reliability, but since then many have been produced.

The gyro-compass was foreseen by Foucault; although it could not be made until technological skills had progressed, its history shows that more than this was required. Despite the need for a non-magnetic compass, its inventor was neither a scientist nor a sailor, but a young man with some scientific background whose chief interests were art and exploration. This shows from what unpredictable directions even a needed invention, based on known principles, can spring. The manufacturers of navigational equipment played no part in the invention of the gyro-compass.

[1] Anschütz-Kaempfe probably chose to work first on the free gyro because that with two degrees of freedom is useless as a compass in polar regions. Though the free gyro will rotate once every 24 hours at the Pole, this could be checked against a watch.

REFERENCES

1. Hitchins, H. L., and May, W. E., *From Lodestone to Gyro Compass.*
2. *The Admiralty Manual of the Gyro Compass*, 1931 and 1953 editions.
3. *Anschütz and Co. G.M.B.H. 1905–1955*, 1955.
4. Hunsaker, J. C., *Biographic Memoir of Elmer Ambrose Sperry*, National Academy of Sciences, 1954.
5. Rawlings, A. L., *The Gyroscopic Compass.*
6. Chaldecott, J. A., 'Léon Foucault', *Sperryscope*, Spring 1952.

HARDENING OF LIQUID FATS

ONE consequence of the rise in population and in the standard of living during the nineteenth century was that the demand for solid fats for soap and food tended to outrun the supply. Adequate supplies of liquid fats, such as whale-oil and seed-oils, were available; the invention of a process for hardening them in the first decade of this century has ensured plentiful and cheap supplies of solid fats to meet the ever-increasing demand.

It had long been known theoretically that a liquid fat could be solidified if hydrogen could be combined with it; but, despite many attempts, no means of doing so were discovered until the publication in 1900 of the classic researches of Sabatier and Senderens on catalysis showed that the use of nickel, or other metals, as a catalyst made it possible to combine hydrogen with an unsaturated, or hydrogen-deficient, gaseous substance. Sabatier failed to see that an oil might also be combined with hydrogen in this manner. But it was not long before an inventor discovered that oil could be hydrogenated, the first patent for a process to solidify fats being granted in 1902.

Dr. W. Normann, a German chemist, was the inventor and, although his patent was the subject of great controversy and was declared invalid by a British court in 1913, his method was proved practicable after a long period of development and is the foundation of all processes used today.

Normann was born in 1870 and studied chemistry at Wiesbaden and Freiburg Universities. In 1901, while working at the oil-mills of Leprince and Siveke, he discovered his fat-hardening process. He appears to have been experimenting to see if he could hydrogenate oil as Sabatier had hydrogenated gaseous substances by the use of a nickel catalyst, but to have been unable to repeat his successful experiment because he did not know what special feature of the materials used had been responsible for its success. This naturally aroused scepticism as to his claims and, although Leprince and Siveke took out patents on the process in 1902, manufacturers were unwilling to take it up.

In 1905, however, Crosfield's, the Warrington soap manufacturers, invited Normann to join them and later bought the British patent. One of their chemists, E. C. Kayser, was interested in the process and, with Normann's help, the firm set out to discover the secret themselves. They also began to buy up the patent rights in other countries to try to obtain a monopoly of hydrogenated oils. Crosfield's finally discovered the 'know-how' which made the process workable – the need for special preparation of the catalyst – although there remained difficulties with impurities and variations in the raw materials, especially in whale-oil. They began production of hydrogenated oil for soap in 1909 at the rate of 100 tons per week.

But the difficulties Crosfield's met with apparently discouraged them as to the chances of profit and they sold their German rights, and their rights in Britain for edible use only, to Jurgens'. Normann went to Jurgens' experimental station, his 'know-how' probably being essential to the use of the process, and for the next twenty years he acted as adviser to various firms dealing with oils and fats.

The practicability of Normann's process having been established, numbers of other patents were granted to inventors for various types of apparatus in which the oil, hydrogen and catalyst could be brought together by various means. These patents were acquired by the companies in the fats-using industries. The outcome was a prolonged and complicated legal battle over patent rights in which Brunner Mond, now the owners of Crosfield's, and Lever's were concerned. In the course of this, Normann's British patent was declared invalid on the grounds that the specification did not give the information needed to carry out the process. The patents situation remained confused (in 1913 Lever proposed a patent pool with van den Berghs, Jurgens and other firms but this scheme broke down); in fact, although the Normann process had not established itself legally it had done so technically, and it subsequently came into general use as the consumption of hydrogenated fats for soap and edible purposes rapidly increased after 1914.

The invention, therefore, of the process for hardening liquid fats by hydrogenation was made by a chemist working in the oil industry; he worked alone on the invention and it was he, rather than his firm, who made all the efforts to get the process adopted by the fats industry. In the circumstances, Normann may be classified as an individual inventor, but one who worked inside the industry which his invention was to revolutionise.

REFERENCES

1. Wilson, C., *The History of Unilever*, 1954.
2. Ellis, Carleton, 'The Hydrogenation of Organic Substances' (3rd edition of the *Hydrogenation of Oils*), 1930.
3. Ellis, Carleton, *The Hydrogenation of Oils*, 2nd edition, 1919.
4. Waterman, H. I., *The Hydrogenation of Fatty Oils*, 1951.
5. Obituary of Dr. W. Normann, *Fette und Seifen*, May 1939.

HELICOPTER

It is not yet certain what is to be the ultimate commercial value of the helicopter, but it clearly has possibilities and, for military purposes, it has already firmly established itself. The practicable helicopter was the product of a long period of evolution, but it was only in the twentieth century, with the appearance of light internal-combustion engines, that useful machines became feasible.

In any full account of the efforts to give stability and control to this type of flying-machine, very many ideas and very many names in numerous countries would call for mention. But the flapping rotor blade articulated to the hub, and the cyclic variation of rotor-blade pitch are perhaps the outstanding conceptions. The pioneers who appear to have made the most important contributions are: Renard, Crocco, Pescara and Cierva in the essential inventions; Cierva, Bréguet and Focke in the early crucial development work, and Sikorsky and Doblhoff in subsequent development work.

C. Renard first suggested the articulating of the rotor blade to the hub in 1904, but only with the intention of reducing the stresses on the blade. Cyclic pitch control seems to have been first suggested by the Italian, G. A. Crocco, in 1906. It was first used in a helicopter capable of flight by the Dane, Ellehammer, in 1912.

The Marquis de Pateras Pescara, an Argentinian, was the first to obtain horizontal propulsion from the lifting rotor, to demonstrate the full effectiveness of cyclic pitch control in flight and to provide for auto-rotation of the rotor in case of engine failure. His machines, built in Spain and France between 1919 and 1925, were thus the first to possess the features of the modern helicopter, except flapping rotor blades, but they had a short mechanical life and they lacked stability. Pescara spent some £100,000 on them, a part of the cost being advanced by the French Government.

Juan de la Cierva, who invented and developed the Autogiro, was a Spaniard who had become interested in aeroplanes when a boy and trained as an engineer. He did not consider the helicopter a practical proposition; his original object was to produce an aeroplane which could not stall and he realised that the use of the auto-rotating rotor instead of a wing was the answer. The Autogiro thus differs from the helicopter in that its rotor auto-rotates and the engine drives a normal propeller. Cierva began his work in 1920 and in 1922 he demonstrated conclusively that flapping rotor blades equalised their lift when advancing and retreating and made the aircraft stable. His first five machines, the first three of which were unsuccessful, were privately financed and the sixth was financed by the Spanish Government.

In 1925 Cierva was invited to England by the Air Ministry, which had decided to back his work. The Cierva Autogiro Company, financed by Lord Weir and his family, was founded in 1926 and the development of the Autogiro was carried on up to 1939. A number of Autogiros were also built by the Weir's company, G. & J. Weir Ltd., the marine engineers, during the 1930's. This work on the Autogiro, most of it Cierva's, meant that by the 1930's helicopter builders had available a body of knowledge on the attributes of rotating-wing aircraft which helped them design their machines. Cierva's great mathematical talent enabled him to analyse many of the theoretical implications of his discoveries in *The Theory of the Autogiro*, although he did not publish all his theoretical work before he was killed in an air-liner crash in December 1936.

In 1931 Louis Bréguet – who had been a pupil of Renard – and his company began work on helicopters again. Their design was based on the prolonged thought Bréguet had given to the helicopter, and in 1935–36 their machine, although still having difficulties with stability while hovering, reached a speed of 61 m.p.h. – faster than any earlier helicopter.

Dr. Heinrich Focke designed the first practicable helicopter, with two counter-rotating rotors arranged side by side on outriggers. Focke had been the chief designer of Focke-Wulf, the German Cierva licensees, and had begun work on the helicopter in 1932. Removed from his post by the Nazis as 'politically unreliable', he founded a small company in order to continue his work. After an intensive study of previous helicopter work, he proceeded systematically from first calculations to wind tunnel and full-scale experiment, flew a successful model in 1934, a full-scale machine in June 1936, which in 1937 reached a height of 8000 feet and a speed of 76 m.p.h. and proved both stable and controllable. Although his arrangement of two rotors side by side has since been abandoned, his demonstration that the problems of the helicopter could be solved encouraged many others to try to solve them, sometimes in different ways.

The next helicopter to follow the Focke was the Weir, which represented the culmination of the Weir's interest in Cierva's work, which had continued after his death. In 1937 the British Air Ministry asked G. & J. Weir Ltd. to try to obtain a duplicate of Focke's helicopter for test purposes. This proved impossible, but G. & J. Weir decided to build a machine to the design of C. G. Pullin, who was in charge of their Autogiro development. This machine flew successfully in June 1938, using the side-by-side rotor lay-out. The machine owed something to Focke's work, but it probably owed more to Pullin's experience with Autogiros.

Much work had meanwhile been going on in the United States; for example, the U.S. Autogiro Company, financed by Harold Pitcairn, had spent $4½ million on the development of the Autogiro, as Cierva's licensee. But success there came from Igor Sikorsky, who had been encouraged by Focke's success to build a helicopter. Sikorsky had in fact built two unsuccessful helicopters in Russia as far back as 1909. From 1929 onwards he had been considering a single rotor helicopter with an auxiliary rotor in the tail, a lay-out first used by the Dutchman, von Baumhauer, in 1924. In 1938 Sikorsky was able to persuade the company with which he was associated, the United Aircraft Corporation in the United States, to allow him to take up the work on it. The first helicopter to this design flew in October 1939; by 1941 it was flying well and production for military use could be considered.

Among outstanding later advances were those of an Austrian, Friedrich Doblhoff. He had no experience of helicopters before 1939 when he became chief engineer of Wiener Neustadt Flugzeugwerke. In his design a compressed mixture was fed to the rotor tips and burnt in combustion chambers. Doblhoff had few assistants in his work, which was financed directly by the German Air Ministry; it cost some 500,000 R.M. to build four experimental machines. A number of other types of jet-driven helicopters have been produced since the war using ram-jet or rocket motors on the rotor, but the Doblhoff system still enjoys some support; it was used on the Fairey Rotodyne.

The case of the helicopter is interesting on many grounds. Up to 1938 no large aircraft manufacturer, except Bréguet, had taken an active interest in the idea. Focke's firm was small and he was largely responsible for it as an individual, although he did have the use of wind tunnels and other facilities of German research establishments. The work of Bréguet arose mainly out of personal interest. For the rest the progress was made and the enthusiasm maintained by individual inventors and developers, usually with limited resources, working on their own or in small firms set up to exploit their ideas. In Great Britain the most active firm was a marine engineering company. It is, however, worthy of note that the British Air Ministry gave early and whole-hearted encouragement to a number of pioneers, notably Cierva. When, however, Sikorsky, who had long had an interest in helicopters and had been encouraged by Focke's success, persuaded the United Aircraft Corporation to allow him to take up the development of the helicopter, rapid success was achieved with a design which became outstanding.

REFERENCES

1. Brooks, P., 'Rotary Wing Pioneer', *The Aeroplane*, Dec. 9 and 16, 1955.

2. De la Cierva, J., 'New Developments of the Autogiro', *Journal of the Royal Aeronautical Society*, 1935.

3. Focke, H., 'The Focke Helicopter', *The Aeroplane*, May 25, 1938.

4. Francis, D., *The Story of the Helicopter*, New York, 1946.

5. Lamé, Lieut.-Colonel, *Le Vol vertical*, Paris, 1934.

6. Liptrot, R. N., 'A Historical View of Helicopter Development', *Bulletin of the Helicopter Society of Great Britain*, Mar. 1947, vol. 1.

7. Liptrot, R. N., 'Rotating Wing Activities in Germany during the Period 1939–45', *B.I.O.S. Overall Report*, no. 8, 1948.

8. Pullin, C. G., 'Helicopter Research and Development', *Bulletin of the Helicopter Association of Great Britain*, Mar. 1947, vol. 1.

9. Sikorsky, I., *The Story of the Winged S.*, New York, 1943 edition.

10. Sikorsky, I., 'Sikorsky Helicopter Development', *Journal of the Helicopter Society of Great Britain*, Oct., Nov., Dec. 1947.

11. Renard, C., 'Sur un nouveau mode de construction des hélices aériennes', Académie des Sciences, *Comptes-Rendus hebdomadaires*, 1904, Tome 139.

12. Bréguet, L., 'The Gyroplane', *Journal of the Royal Aeronautical Society*, 1937.

13. Pescara, R. P., 'Les Machines à pistons libres', *Journal de la Société des Ingénieurs de l'Automobile*, July-Sept. 1937.

INSULIN

INSULIN is secreted by the 'islet tissue' of the pancreas. In 1889 von Mehring and Minkowski had recognised that the functioning of the pancreas was related to the metabolism of sugar in the body, and by the first decade of this century it was generally realised that the islet tissue produced some sort of chemical messenger secreted into the blood which was responsible for regulating the blood-sugar level. But, before the work of Banting, all attempts to prepare active preparations of this substance had failed.

The crux of Banting's discovery was that insulin is rendered inactive by ferments in the pancreatic gland (in which the islet tissue is embedded) and that this had defeated previous efforts at extraction. He combined this path-breaking conception with an old observation that the tying-off of the duct connecting the main pancreatic gland to the intestine would produce atrophy of the main gland but leave the islet tissue largely intact. Extracts made from pancreas degenerated in this way were entirely successful, but the method had a serious drawback: the degeneration of the pancreas took three or four months and the amount of insulin extracted was minute. Banting then went on, in collaboration with Best, to search for direct chemical methods of separating insulin from whole pancreas. Success here was achieved by the use of low temperatures and moderate concentrations of ethyl and methyl alcohol, which slow up the ferments in the main pancreatic tissue and enable insulin to be extracted from fresh pancreas. This opened up the way to commercial production.

Frederick G. Banting (later Sir Frederick) was born on a farm in Ontario, Canada, in 1891, entered the medical school at the University of Toronto in 1912 and became particularly interested in orthopaedic surgery. After the war he started a practice in London, Ontario, and assisted Professor F. R. Miller, a well-known neurophysiologist at the University of Western Ontario Medical School, in an investigation of the results of cerebellar stimulations.

Banting first conceived his important ideas in connection with the extraction of insulin in 1920. At this point he knew little of the extensive literature in the field:

> 'Banting subsequently states that had he been familiar with all the complexities of the subject as revealed . . . in the voluminous literature, he might never have had the courage to attempt his own research.'[1]

He approached Professor J. J. R. McLeod, physiologist at the University of Toronto:

> 'McLeod was not impressed. He pointed out the vast amount of work which had already been done, the many attempts to isolate the internal secretion of the pancreas (if such a thing existed) and the handicap which Banting would experience from his lack of training in research.'[2]

Banting remained firm in his belief that his method deserved a trial, and he returned to McLeod some months later. This time McLeod agreed to let him work at the University with C. H. Best, a graduate in physiology, as an assistant, and together they prepared their first successful extracts from the pancreas of dogs and oxen. The discovery of chemical methods of extraction from fresh whole glands followed.

On January 11, 1922, the first human patients received insulin at the Toronto General Hospital. McLeod was sufficiently impressed with the experiments to set aside his anoxaemia work and to direct his staff to investigate the physiological activity of insulin. An important problem was the purification of the extract, which Dr. J. B. Collip successfully solved.

Production problems were placed in the hands of the Connaught Laboratories, which had been set up by the University of Toronto during the First World War to produce anti-toxins. Best worked there and, with the help of D. A. Scott and E. G. Noble, he perfected many of the techniques used in the factory production of insulin. At this stage, the American Eli Lilly Company co-operated with the Toronto group and, through the work and leadership of Dr. G. H. A. Clowes, they contributed a great deal.

One of the more significant later advances in the treatment of diabetes was the development of protamine insulin by Dr. Hagedorn of the Nordisk Insulinlaboratorium in Copenhagen during 1933–35. Protamine prolongs the effect of insulin, so that a whole day's supply can be given in one injection. Hagedorn's United States patents became vested in the Alien Property Custodian after Germany occupied Denmark and the patents were returned to the Nordisk Laboratorium in 1949.

The discovery of insulin and the methods for its extraction were due, therefore, to the insight and persistence of a doctor interested in research. He succeeded where better-trained and better-equipped experimenters had failed. Improvements on the original work came rapidly from individuals in university and industrial research laboratories.

REFERENCES

1. Nicholas, Henry O., 'Insulin in Discovery and Use', *Rice Institute Pamphlet*, no. 11, 1924.

[1] Lloyd, Stevenson, *Sir Frederick Banting*, 1947, p. 77. [2] *Ibid.* p. 72.

2. Report on the Supply of Insulin, British Monopolies and Restrictive Practices Commission, Oct. 14, 1952.
3. 'Protamine Insulin Patent Returned to Danish Firm', *Oil, Paint and Drug Reporter*, Apr. 25, 1949.
4. Banting, F. G., 'Early Work on Insulin', *Science*, June 25, 1937.
5. Stevenson, Lloyd, *Sir Frederick Banting*, 1947.
6. Harris, Seale, *Banting's Miracle: The Story of the Discoverer of Insulin*, 1946.
7. Banting, F. G., 'The History of Insulin', *Edinburgh Medical Journal*, Jan. 1929.

JET ENGINE

THE turbo-jet engine combines the principle of jet propulsion and the gas turbine. These ideas are quite old; as early as A.D. 150 Hero suggested the use of a jet of steam for propulsion, and the first patent on a gas turbine was granted to an Englishman, John Barber, in 1791. Many attempts were later made to produce an efficient gas turbine, particularly in the first quarter of this century, but it was not until the 1930's that these ideas were combined into a practical aircraft engine. One man in England and several in Germany conceived, more or less simultaneously, the idea of a propulsive jet provided by a gas turbine. The history of the invention of the English turbo-jet will be described first.

During the early 1920's A. A. Griffith, a scientist employed by the British Government at the Royal Aircraft Establishment, had worked on the design of a compressor of sufficient efficiency to make a gas turbine a practical engine for driving a propeller and, by 1929, he had constructed a single-stage compressor turbine unit with a fairly high efficiency. Unfortunately Griffith had to cut short these promising experiments because the government transferred him to the Air Ministry Laboratory at South Kensington where the facilities for research of the type involved were completely inadequate.

Frank Whittle (now Sir Frank) began thinking in terms of a jet-propulsion system for aircraft in the 1920's. He had entered the Royal Air Force in 1923 as an aircraft apprentice and had attended the R.A.F. College. In 1928 he wrote a thesis on developments in aircraft design in which the possibilities of rocket propulsion and a gas turbine driving a propeller[1] were mentioned, but not the linking of a gas turbine with a jet-propulsion system. In 1929 he continued his study of propulsion at the Central Flying School at Wittering, where he took a flying instructor's course. It was there that the idea of combining a gas turbine with jet propulsion occurred to him. He applied for a patent on this engine.[2]

Whittle found it impossible to gain the support of the British Air Ministry or aero-engineering firms, and became so discouraged that in 1935 he allowed his basic patent to lapse. In the meantime the Royal Air Force had arranged for him to go to Cambridge, where he studied for the Mechanical Sciences Tripos. Further progress with the engine depended upon the finding of a financial backer, and

[1] Whittle gave some consideration to a jet-propulsion system in which the propelling jet was generated by a low-pressure fan driven by a conventional piston engine. Dr. Harris of Esher had patented the basic concept in 1917. The Italians used this system in a Caproni Campini aircraft first flown in 1940. The system did not prove to be practical.

[2] An engine somewhat similar to Whittle's had been patented nine years earlier by a Frenchman, Guillaume. Whittle knew nothing of this prior work.

Sir Frank Whittle: jet engine

finally Whittle, through the assistance of some friends, was put into touch with the investment bankers O. T. Falk and Partners, who agreed to provide the relatively small sums necessary to carry on the work. Whittle had by this time filed improvement patents on special features of his engine; a company was formed, Power Jets Ltd., to which the patents were transferred; the British Thomson-Houston Company was given a contract to design an experimental engine according to Whittle's requirements; and a small Scottish firm, Laidlaw, Drew & Co., was entrusted with the development of the combustion system. The Air Ministry allowed Whittle to retain the rights in his patents and gave him special permission to work six hours a week for Power Jets Ltd.

The engine had its first test run in April 1937; it worked, but its performance was low, on several occasions it ran out of control and the problem remained of finding suitable materials for withstanding the high temperatures and pressures. The obscure technical prospects made it difficult to raise further capital; Power Jets Ltd. had no money to spend on replacements and Falk and Partners allowed their option to renew their agreement with Power Jets to lapse although they continued to provide loans. The British Government stepped in with some financial help in March 1938 but the staff at Power Jets still remained inadequate and the work went forward on a precarious hand-to-mouth basis.

Nevertheless, the technical obstacles were being surmounted. In 1937 the R.A.F. had placed Whittle on Special Duty List which enabled him to give full time to the engine. Late in 1937 he discovered that the commonly accepted design for turbine blades was unsound and, although he was an amateur in this field and he met strong resistance from the turbine blade specialists, he was able to show why.[1] In his efforts to find stronger materials to obviate the failure of the blades, he approached, among others, Firth-Vickers, who developed a new nickel-chrome alloy, Rex 78. Engine testing continued throughout 1938 and into 1939 and, although at this period Whittle seems to have come extremely close to the design which ultimately proved successful, government officials were lukewarm and moves were made to withdraw official support.

In June 1939, however, a dramatic change occurred in the attitude of the government; Dr. Pye (later Sir David), the Director of Scientific Research, saw the engine perform and expressed confidence in its future; the government undertook to cover the cost of all future work by Power Jets and the Air Ministry assigned to the Gloster Aircraft Company the task of designing and building an airframe for the new type of engine. In 1940 the fundamental combustion problems were solved when I. Lubbock of the Shell Petroleum Company brought to Whittle's notice a combustion chamber upon which Shell engineers had been experimenting. The first test flight of the Gloster jet fighter took place in May 1941.

[1] 'It may seem a very strange thing that specialists on turbine design had overlooked a phenomenon which I had more or less taken for granted. I heard somebody once define a practical man as "one who puts into practice the errors of his forefathers". This blade business was a good example of it and of how, if habits of thought become deeply rooted, errors may persist from generation to generation. Turbines had slowly evolved from the primitive form in which a few steam jets (often four only) spaced evenly, impinged on the blades or "buckets" of a single wheel. In such cases it had been reasonable to assume that the velocity and pressure in the blast from each jet was uniform and to design the blades accordingly. It had seemingly occurred to no one that as jets were made more numerous and placed closer together until "full peripheral" admission was achieved, there would be a fundamental change in the nature of the flow.' (Sir Frank Whittle, *Jet*, p. 73.)

The next hurdle, the development of an engine suitable for production in quantity, was a formidable one. The government, holding that Power Jets had not sufficient experience for development at this stage, gave the contract to the Rover Company, the automobile manufacturers. But the move was hardly a success and delays occurred, arising partly out of differences between Rover and Power Jets regarding their respective roles.[1] But by now the major aero-engine firms had become interested in gas turbine work. Rolls-Royce, in particular, had entered the picture in 1941 and, speedily pursuing their development of the W-2B engine, commenced its production, as the Welland engine, in October 1943. In 1944 the government nationalised Power Jets Ltd. on terms which seem harsh when it is considered that, by the end of the war, the British had only one type of jet engine ready for service, a Whittle-type centrifugal engine.[2]

The Germans were pursuing a parallel course of investigation on jet-propelled aircraft. The German work had originated from several sources. Hans von Ohain, a student of applied physics and aerodynamics at the University of Göttingen, had patented a centrifugal-type turbo-jet in 1934; Herbert Wagner of the Junkers Airplane Company had started a research programme on gas turbines early in 1936 with Max Mueller in charge of the work; it was later extended to include turbo-jets, one being designed in 1938. Helmut Schelp, when a research student at the German Institute of Aeronautics in Berlin in 1936–37, had become convinced that the turbo-jet was the only engine which could reach speeds of over 500 m.p.h. and that it was practical to build it then. As an official of the research staff of the Air Ministry from 1937 onwards, he was able to influence higher officials in favour of the turbo-jet. The fact that rockets, pulse-jets and ram-jets were all being worked on in Germany in the later 1930's and that the Air Ministry was financing the work may have made it easier for the turbo-jet to secure official support once it had been suggested.

Hans von Ohain was perhaps the most important single figure amongst these investigators; it was with an engine of his design that the world's first jet aircraft flew in August 1939. In its general lines it was similar to Whittle's although in detail it differed considerably. Von Ohain and Whittle were unaware of each other's inventions and development activity. One of von Ohain's professors, R. W. Pohl, happened to be friendly with Ernst Heinkel, the president of Ernst Heinkel, A.G., the aircraft manufacturers. Pohl spoke in glowing terms of von Ohain's invention and, although the firm had had no previous experience in this field, he persuaded Heinkel to take on von Ohain to develop his engine. The Heinkel engineers cast doubts upon the practicability of his ideas but the inventor was not deterred. With the aid of two assistants, he built a demonstration model in 1936 and the management was impressed enough with its performance to give an order for the development of a flight engine. The development proved to be far from simple; the combustion problems and material difficulties that bothered Whittle also troubled the Germans. None of the German industrial oil-burner builders would even attempt to solve the combustion problem, so that Heinkel's inexperienced staff had no alternative but to do this work themselves.

While the Heinkel Company carried on its experiments the Junkers Airplane

[1] Whittle argued that Rover was a poor choice for this sort of development and should not have been allowed to make any basic modifications affecting engine performance.

[2] An axial type designed by the Royal Aircraft Establishment in 1939 was not ready by the end of the war.

Company was also conducting research in the turbine field. Both Junkers and Heinkel kept their developments secret; neither knew that the other was even interested in these revolutionary engines. Herbert Wagner, a brilliant Junkers engineer, believed that conventional engines were unsuitable for the high speed of which aircraft were becoming potentially capable. As engine manufacturers were doing no research on alternative power units, he persuaded his directors to allow him to do so although Junkers had no previous experience with engines. At first the Junkers engineers, under the direction of Max Mueller, aimed at the development of a turbo-prop engine, in the belief that this offered most promise; but Mueller designed a turbo-jet with an axial compressor as contrasted with Whittle's and von Ohain's centrifugal compressors, and work was concentrated on it by 1938. This ambitious design was ready for testing in 1938, but it was 1942 before it could be run under its own power. When Junkers and Heinkel, acting independently, sought financial assistance from the Air Ministry they were turned down because the Air Ministry insisted that development of this kind could be satisfactorily completed only by the aircraft engine companies.

Another individual, Hans Mauch, head of rocket development at the Air Ministry, tried to convince the aircraft engine manufacturers in 1938 of the practicability of the jet-propelled aircraft. He had heard of Heinkel's and Junkers' turbo-jet experiments and took an immediate interest. Helmut Schelp was then working in the Research Division of the Air Ministry where he was in charge of the development of the Schmidt pulse-jet and the Walther ram-jet. Mauch met Schelp in August, and invited him to become his assistant in the Development Division. Mauch and Schelp visited the aircraft engine firms in an effort to enlist their support for turbo-jet development but the firms saw only the difficulties associated with such a project, and in general refused to touch it.[1] A few companies accepted study contracts: the BMW Engine Company began some investigation of these engines, and the Bramo Engine Company of Berlin-Spandau undertook some research mainly because it feared that, if it did not, the government would withdraw support from Bramo's reciprocating engine programme. This firm completed a design for an axial turbo-jet engine in 1939. The Junkers Engine Company agreed to start a limited development programme directed by Anselm Franz in the same year.

Work proceeded along two fronts; engine development and airframe development. As far as airframe development is concerned, Hans Antz of the Air Ministry Airframe Development Group secured the assistance of Messerschmitts in the design of an airframe for a jet fighter; this design (ME 262) was flown with a conventional engine in 1940 and with jet engines in 1942; but the refusal of the heads of the Air Ministry to realise the inferiority of German piston-engined aircraft delayed its production, and it was not ordered until June 1943.

Mauch strongly believed that the engine firms were the only organisations with enough experience to complete the engine development and planned that Junkers Engine should take over Junkers Airframe's axial turbo-jet, turbo-prop and piston-driven ducted-fan projects; Daimler-Benz should develop Heinkel's centrifugal turbo-jet; and BMW-Spandau (the Bramo plant which had been purchased by BMW) should work on a counter-rotating gas turbine, and a related turbo-jet suggested by Helmut Weinrich of Chemnitz. Heinkel, however, refused to transfer their work and continued for a while at their own expense. Although their staff was

[1] Several problems worried them. Among the important ones were (1) the efficiency of the compressor and turbine, (2) materials for the turbine blades.

small, they expanded their investigations to include studies of other types of engines. Mueller of Junkers Airframe Division disliked the idea of being transferred to the Engine Division where he would not have complete control over the work, so he moved to Heinkel with half his staff. After Mauch left the Air Ministry late in 1939, his successors provided Heinkel with government funds for the development.

The engine development programme suffered from a severe shortage of technically trained personnel which accounted in a great measure for Heinkel's failure to develop a production engine by the end of the war. BMW began production in 1944 but the only German turbo-jet produced in any quantity by the end of the war was the Junkers 004.[1] This model was a conservatively designed axial turbo-jet, and was used in the ME 262 fighter which went into service in the autumn of 1944.

The jet engine was thus invented and developed almost simultaneously in two countries by men who were either individual inventors quite unconnected with the aircraft industry, or by men who worked on the airframe side of the industry and were not specialists in aircraft engine design. That no suggestion for the invention should have come from aircraft engine manufacturers is one of the most significant facts in the story of this invention. The others are the reluctance of governments to support the development of an invention of military value, even at a time of heavy armament expenditure; and the rapidity and success of development work once the aircraft engine manufacturers entered the picture, whether they did so voluntarily or under government pressure.

REFERENCES

1. Whittle, Sir Frank, *Jet, The Story of a Pioneer* (1953).
2. Schlaifer, Robert, *The Development of Aircraft Engines* (1950).
3. Heinkel, Ernst, *He 1000* (1956).

KODACHROME

THE commercial Kodachrome process resulted from an attempt by two independent inventors to produce a colour film that, for the ordinary photographer, would be no more complicated to use than the conventional black and white. The film is a monopack composed of three separate emulsion layers; the top layer is sensitive to blue, the second to green and the bottom to red light, and thin coatings of clear gelatin separate the layers. The Kodachrome developing process is a subtractive three-colour dye-coupling one in which the couplers are contained in the developer solutions rather than in the emulsions.

Kodachrome's history can be traced back to the nineteenth century, although the early theoretical ideas of Kuhn, Schinzel and others were not commercially feasible mainly because of the primitive state of photography and dye technology. The work done by Dr. Rudolph Fischer and Siegrist in Berlin from 1910 to 1914 forms the

[1] The lack of nickel imposed considerable difficulties on the production of jet engines in Germany, as the heat-resisting powers of nickel-chrome alloys are generally considered essential for parts such as combustion chambers and turbine blades. The use of plain steel considerably reduced the life of the engine.

basis of all the modern dye-coupler colour processes (Kodachrome, Agfa, for example). Dr. Fischer laid down the basic principles for the dye-coupler processes[1] and described in detail, in a 1912 patent, a monopack composed of three emulsion layers. Between the green and the red sensitive layers he proposed the use of a colourless layer of gelatin. Fischer also developed the basic ideas for the colour processing of this monopack: he conceived the idea of employing the coupler, the chemical compounds that form the dye images in the emulsion layers, and he specified the kinds of ingredients that should be used for this purpose.[2] These same types of couplers were employed in the successful Kodachrome process. Fischer and Siegrist tried to exploit these ideas; they produced a self-toning paper which incorporated the coupling agents in the emulsion but, because the soluble couplers diffused from one layer to another, they were unable to obtain satisfactory colour separation in a tripack. Fischer had no means of controlling this diffusion, and even more serious was the diffusion between layers of the sensitising agents in the emulsions which made it impossible to separate the colours in the monopack.[3]

In 1921 Dr. L. T. Troland, a member of the research staff of Technicolor Motion Pictures, filed patent applications for the preparation and uses for monopacks. Troland had in mind a two-layer monopack, and the specifications do not indicate that he thought in terms of using 'the material to yield a reversed or direct three-colour transparency' as in the Kodachrome process. Eastman Kodak later acquired rights under these Troland patents.

It was, however, two independent inventors, Leo Godowsky Jr., and Leopold Mannes who achieved practical success. They were students of music. During their school-days in New York State photography fascinated them and together they made many experiments; when Mannes went to Harvard and Godowsky to the University of California their research was interrupted except during vacations. They first patented some attachments for improving projectors; then they began thinking in terms of a single film containing all the colours. They supplemented their slight knowledge of chemistry by private study; they earned a living by teaching music and giving concerts while experimenting intensively in Mannes's kitchen. At first they worked with a two-layer monopack, but later turned to the three-layer type, and by 1923 had produced a crude picture containing all the colours in a single film.

At this stage they had obtained the support of private financial backers, and Dr. Mees, of Eastman Kodak, was assisting them by providing coated plates. Even when on concert tours in Europe the inventors continued to experiment; they finally conceived a controlled diffusion process to develop the images which enabled them to confine the action of the development agents to single layers. In this respect they went beyond Fischer, and these ideas proved crucial to the success of

[1] After Eastman Kodak introduced Kodachrome, the I.G. Farben interests set to work to improve the original ideas of Fischer and Siegrist, and, from this research, produced Agfa.

[2] Joseph S. Friedman in his treatise, *The History of Color Photography* (1944), states that 'the disclosures of Fischer and Siegrist have described rather completely the chemical requisites for a substance to act as a coupler. So fully did they cover the field that, with the possible exception of pyrazolone and its derivatives, no new chemical configurations have since been disclosed, but merely substitutions made in the old that left the reactive portion of the molecule intact.'

[3] In the development of Fischer's ideas some twenty years later, I.G. Farben perfected methods for keeping the couplers in their respective layers.

Kodachrome. One difficulty which Fischer had had was the diffusion through the layers of the sensitising dyes, but later a number of new dyes were discovered which did not 'wander' easily. Dr. Mees then invited the two inventors to join the Kodak laboratories. They accepted the offer of substantial salaries, patent royalties, excellent laboratory facilities and technical assistance, and for ten years concentrated on the commercial development of their ideas. With the co-operation of a large number of workers in the Kodak Research Laboratories, Mannes and Godowsky were able to produce a practicable process which was placed on the market by the Eastman Kodak Company under the name of 'Kodachrome' in 1935.

Most of the basic principles, therefore, behind the Kodachrome process were anticipated by earlier investigators but they failed to produce a commercial process. Two novices entered the field and conceived of a highly successful one; a large photographic firm saw the possibilities of their work and helped them to develop their early ideas.

REFERENCES

1. Sipley, Louis W., *A Half Century of Color*, 1951.

2. Mannes, L. D., and Godowsky, L., Jr., 'The Kodachrome Process for Amateur Cinematography in Natural Colors', *Journal of Society of Motion Picture Engineers*, vol. 25, 1935.

3. Cornwell-Clyne, Major Adrian, *Colour Cinematography*, 1951.

4. Friedman, Joseph S., *History of Color Photography*, 1944.

5. Mees, C. E. K., 'Direct Processes for Making Photographic Prints in Color', *Journal of Franklin Institute*, Jan. 1942.

6. Mees, C. E. K., 'Modern Colour Photography', *Endeavour*, Oct. 1948.

7. Eastman, Max, 'Kodachrome, Its Colorful Beginnings', *Christian Science Monitor*, section 11, Mar. 13, 1951.

LONG-PLAYING RECORD

MUSIC lovers found in the long-playing record the answer to their desire to hear a work uninterrupted by record changing. The four fundamental features of the successful long-playing record system are: (1) the slow rotational speed of $33\frac{1}{3}$ r.p.m.; (2) finer grooves, usually from 224 to 300 per inch; (3) vinylite plastic record; and (4) lightweight pick-up. The industry had long been seeking a practicable high-quality long-playing record of the normal ten or twelve inch diameter size, but these four features had not been combined until the Columbia Company introduced their long-playing record in 1948.

Dr. Peter Goldmark was responsible for this innovation. He was a scientist in charge of the Columbia Broadcasting System's Research Laboratories where he had been responsible for inventions in colour television. Like most music enthusiasts, he disliked record changing, and he asked Columbia Records, a sister organisation of the Columbia Broadcasting System, if they were interested in the idea of a long-playing record. Columbia Records were enthusiastic and Goldmark started experiments in the C.B.S. laboratories just after the end of the Second World War. He decided to concentrate on a long-playing disc rather than on tape or wire because discs were cheaper and the public were accustomed to handling them. Working alone in the Columbia laboratory and in his home, he established the basic ideas for the narrow-groove vinyl plastic record. The use of this material reduced surface noise as well as allowing the use of finer grooves than were possible with shellac.

After Goldmark had worked out the basic ideas and experiments, he transferred his research to the Columbia Records laboratory where he was given at first three, and later fifteen, assistants to develop the equipment to produce the record.[1] Concurrently with the development of the long-playing record, Goldmark and his assistants had to perfect special reproducing equipment for the new record. With the existing 78 r.p.m. records, a heavy needle force had to be applied, but there had been some recent advances in recording, making it possible to dispense with heavy needle forces, bulky diaphragms and armatures for extracting the mechanical energy from the grooves. Goldmark's team developed lightweight pick-ups, unknown at that time, high-compliance cartridges and slow-speed, inexpensive, silent turn-tables.

After three years of intensive work, at a cost of about $250,000, Columbia announced the new type of record in 1948. The Radio Corporation of America attempted to counter Columbia's achievement by introducing a 45 r.p.m. record, but this was only successful for popular music. Since 1948 over 150 new companies have started to manufacture long-playing records.[2]

In this case the invention originated from a scientist working in a large organisation where his main interest had been in television. His idea for a long-playing record was at first a side-line but, since his firm was associated with a record company, he was encouraged to concentrate on this work, and later was given much assistance in the development.

REFERENCES

1. Aldous, D. W., 'American Microgroove Records', *Wireless World*, Apr. 1949.
2. Goldmark, P. C., Snepvangers, R., Bachman, W. S., 'The Columbia Long-Playing Microgroove Recording System', *Proceedings of Institute of Radio Engineers*, Aug. 1949.
3. Conly, J. M., 'Five Years of LP', *Atlantic Monthly*, Sept. 1953.
4. Correspondence with Dr. Peter Goldmark.
5. Gelatt, R., *The Fabulous Phonograph*, 1955.

MAGNETIC RECORDING

MAGNETIC recording is a revolutionary device by which sound is recorded on steel wire or plastic tape by means of magnetism. It is used for broadcasting, in high-speed electronic computers and other modern instruments and in the storage of all types of information.

Valdemar Poulsen, a Dane, was the inventor of magnetic recording. In 1893 he took a post with the Copenhagen Telephone Company, where he was employed as a 'trouble shooter'; fortunately, he had spare time to engage in his personal experiments.

[1] Several men contributed to its success, among whom were R. Snepvangers, W. S. Bachman, D. Doncaster, T. Broderick, B. Littlefield and E. Porterfield.

[2] Columbia could not get an exclusive patent on the micro-groove feature.

No one really knows how the idea of magnetic recording occurred to Poulsen. During his experiments he found that

> 'it would be possible to magnetize a wire to different degrees so close together that sound could be recorded on it by running the current from a microphone through an electromagnet and by either drawing the wire rapidly past the electromagnet or drawing the electromagnet rapidly past the wire.'[1]

This invention had several things to commend it: the wire or tape could be used over and over again by de-magnetising it and recordings could be played thousands of times without destroying the quality. Poulsen invented this Telegraphone in 1898; with it he won the Grand Prix at the Paris Exposition in 1900. He filed an application for a Danish patent in 1898 and within two years he had filed additional patent applications in the United States and most European countries. These early patents suggested, as recording media, steel wire and tapes and discs of material coated with magnetisable metallic dust, though he himself used only steel wire and tape in his machines. The basic principles enunciated by Poulsen are still applied in all types of modern magnetic recorders.

Poulsen and his associate, Peder O. Pedersen, had hopes of exploiting the invention commercially and they sought financial help in America. In 1903 the American Telegraphone Company produced dictating machines and telephone recorders and, although the machines performed fairly well, the difficulty of playback and the slow rewinding speed were drawbacks. These technical obstacles, and quarrels between the stockholders and the management, were responsible for putting the company into receivership.

After 1900 Poulsen focused his attention on radio but he still found time to make improvements in his magnetic recorder. In 1907 he and Pedersen received patents for the discovery of d.c. biasing which considerably improved the quality of magnetic recordings but still left much background noise during playback. This weakness was only overcome by the invention of a.c. biasing in the 1920's, for which credit is generally given to W. L. Carlson and G. W. Carpenter of the United States Naval Research Laboratory. The Navy had started this research because it was thought that magnetic recording might be used for transmitting telegraph signals at high speed. A form of a.c. bias was used for 'sensitising' a telegraphone to radio telegraph signals, but only twenty years later was it found that this improved the quality of recorded sound in addition to reducing background noise. The system is in general use today.

Powerful electronic amplifiers came into general use in the 1920's and thus corrected the most serious failing of the early recording machines. Yet manufacturers had still to be convinced of their importance. In the late 1920's Kurt Stille, with the support of German financiers, formed the Telegraphie-Patent-Syndikat in Germany for the purpose of selling licences to manufacture magnetic recording equipment. Blattner, a motion-picture promoter, acquired a licence and made several films in England with the sound tracks recorded on steel tape. He then sold his rights to the Marconi Company, which manufactured the Blattnerphone tape recorder as used by the B.B.C. Karl Bauer bought a licence and produced the Dailygraph, a combination dictating and telephone recording machine. Eventually the C. Lorenz Company, one of the largest German communications companies,

[1] S. J. Begun, *Magnetic Recording*, p. 3.

completely redesigned this machine and put it on the market; they also developed a steel-tape recorder.

The replacement of steel wire or tape by plastic tape coated with magnetic material helped to make magnetic recording suitable for general use. An Austrian private research worker, Dr. Pfleumer, was responsible for this advance; he had begun experiments in the late 1920's, first with paper tapes and subsequently with the more suitable plastic tapes. In 1937–38 Allgemeine Electrizitäts Gesellschaft decided to try these instead of steel tape in their Magnetophon machine. I.G. Farben undertook the development of the tape for A.E.G. and Dr. Pfleumer joined I.G. Farben to participate in the work. A cellulose acetate tape was put into production in 1939, but it had some weaknesses and was replaced by 'Luvitherm' tape of polyvinyl chloride. A major improvement in the quality of reproduction was made in 1940 when Dr. H. J. von Braunmühl, chief engineer of the German State Broadcasting Service, and Dr. Walter Weber of its staff, applied high-frequency a.c. biasing to the tape. Magnetophons incorporating these two inventions were widely used in Germany during the war and aroused much interest when Allied teams at the end of the war investigated German industry.

Magnetic recording equipment had not been made in the United States from the time of the disappearance of the American Telegraphone Company until 1937 when Acoustic Consultants, and later the Brush Development Company, produced the Soundmirror. Dr. S. J. Begun was instrumental in bringing about these developments, first with German machines and then with the Soundmirror and other recorders of the Brush Development Company. The Bell Telephone Laboratories designed a similar steel-tape recorder, the Mirrophone, which was used for announcing the weather. C. N. Hickman developed an excellent tape recorder for the Bell Company in 1937, employing a new vicalloy tape which had much more desirable magnetic properties than the earlier ones. With the exception of the Bell Laboratories, the large corporations in the radio, phonograph and electronic fields remained aloof.

The Armour Research Foundation, a non-profit research organisation in Chicago, entered the field through the work of a young engineer named Marvin Camras. While still a student at the Illinois Institute of Technology, Camras invented a much-improved wire recorder. Needing first-class material and technical assistance, Camras sought the aid of the Armour Research Foundation, where he was given a post in 1940. The results of his early work were incorporated in machines produced during the war by General Electric, Utah Radio Products Company and many other licensees. Camras introduced improvements in the use of high-frequency bias, coating materials for magnetic tapes, and the use of magnetic sound for motion pictures. He is one of the largest patent holders and one of the most important inventors in modern magnetic recording.

The Second World War provided the impetus for the development of better recording machines. The United States Alien Property Custodian held the patents on the German Magnetophon and its tape, and certain relevant patents were available for licensing, but the large companies remained cautious. Early in the war the German tapes were replaced by one perfected by the Armour Foundation which allowed a reduction in speed from 30 inches per second to the present standard of 7½ or 3¾ inches per second.

For several years after the war, high-quality recorders were manufactured mainly by small companies; a few larger firms, such as the Radio Corporation of America and Western Electric, became interested later, but it remains true today that the great percentage of recorders are manufactured by relatively small firms.

This is a clear case of invention by an individual. Many of the important improvements to Poulsen's unique device came from other individuals, notably Pfleumer and Camras. The large companies in the United States played only a small part in the early technical advancement, but some large German firms took up Dr. Pfleumer's invention and, aided by the ideas of engineers in the State Broadcasting Service, developed a tape recorder which represented a major advance.

REFERENCES

1. 'Sound Through Magnetism', *Fortune*, Sept. 1943.
2. 'The Development of the Magnetic Tape Recorder', *Engineer*, Mar. 18, 1949.
3. 'Magnetic Tape Recording', *Fortune*, Jan. 1951.
4. Begun, S. J., *Magnetic Recording*, 1949.
5. Begun, S. J., 'Magnetic Recording', *Scientific Monthly*, Sept. 1949.
6. F.I.A.T. Final Report, No. 705, *High Frequency Magnetophon Magnetic Sound Recorders.*
7. F.I.A.T. Final Report, No. 923, *Further Studies in Magnetophons and Tapes.*
8. B.I.O.S. Final Report, No. 1379, *Plastics in German Sound Recording Systems.*

METHYL METHACRYLATE POLYMERS: PERSPEX, ETC.

POLYMERISED methyl methacrylate, known as 'Perspex' in Britain and 'Plexiglas' and 'Lucite' in the U.S.A., is much the most successful transparent plastic yet produced. Its great value lies in its combination of transparency, freedom from discoloration and excellent light-conductivity, with strength, low weight and ease of moulding. Its weakness, which has prevented it from replacing glass at all generally, is its surface softness, which makes it easy to scratch.

Knowledge of methyl methacrylate polymers goes back to the nineteenth century, when Fittig seems to have been the first to observe in 1877 that methacrylic acid would polymerise into a hard, white mass. The ease with which the material polymerised made it difficult to purify it by distillation, and hence the polymerised product lacked the transparency of modern commercial products. In any case, it was only possible to produce the raw materials in small quantities in the laboratory.

The related methyl and ethyl acrylates were developed for commercial use before the methyl methacrylates. Their polymers form a much softer and more elastic product than those of the methacrylates, and can be used as protective coatings, to stiffen materials and as interlayers in safety glass. Dr. Otto Röhm of Darmstadt was responsible for their development. He worked on them from 1900 to 1902 under von Pechmann, and was apparently so impressed by their commercial possibilities that he continued to work on them from then on. He obtained the first patent in the field in 1912; in 1927 the firm of Röhm and Haas started production after Bauer, one of Röhm's assistants, had invented a process for preparing acrylic esters from chlorhydrin, which was freely available, being the raw material for mustard gas. In 1931 the Röhm and Haas firm of Philadelphia introduced the acrylates into the United States.

The successful employment of the methacrylates was closely bound up with the work of Dr. William Chalmers, who at the time was a graduate research student at McGill University, Montreal. Following up suggestions made to him, by Dr. G. S. Whitby in the winter of 1929–30, Chalmers found that methacrylic ethyl ester and methacrylic nitrile could be readily polymerised into a hard, transparent solid. The former he prepared from hydroxyisobutyric ethyl ester at Dr. Whitby's suggestion; the latter, on his own initiative, from acetone cyanohydrin. He applied for patents both in Canada and the United States, the first patent applications on the polymerisation of the methacrylates by themselves to form a glass-like substance. Early in 1931 he got into touch both with Röhm and Haas and with Imperial Chemical Industries; both these firms indicated their interest in Chalmers's work and subsequently recognised the importance of his contribution.

An I.C.I. research team under Dr. Rowland Hill concentrated its attention on methyl methacrylate, which had a considerably higher softening point than ethyl methacrylate and was stronger, harder and more rigid. In November 1931 Hill and I.C.I. were granted a patent on polymerised methyl methacrylate for the production of moulded articles. Commercial production depended on the development of an economic process of synthesising methacrylic acid; this was invented by J. W. C. Crawford of I.C.I. and patented in August 1932. Acetone cyanohydrin, as used by Chalmers to prepare methacrylic nitrile, seems to be the principal raw material in most commercial processes, with the addition of methyl alcohol and a number of other ingredients. I.C.I. granted du Pont a licence under their U.S. patent and du Pont started production of 'Pontalite', later renamed 'Lucite', in the autumn of 1936.

Röhm and Haas were also working on the methacrylates and they made an important contribution to the methods of its commercial production. Although methyl methacrylate is manufactured in the form of rods, tubes, commercial moulding powders and denture base materials, the plastic in the form of cast sheets is the most widely used. Dr. Röhm and Bauer, experimenting at the German firm, invented a highly efficient method of making these sheets, and 'Plexiglas', manufactured by the American Röhm and Haas firm, went into production in 1935.

To sum up: nineteenth-century scientists first observed that methacrylic acid would polymerise. The exploitation of the products derived from the acrylates can be attributed to Dr. Röhm and the firm of Röhm and Haas. In making use of the methacrylates, a post-graduate research worker, Dr. Chalmers, was the first to discover that they polymerised into a plastic glass. Imperial Chemical Industries, as the outcome of the work of Dr. Rowland Hill, were granted the first patent on the use of polymerised methyl methacrylate, while Röhm and Haas were the first to commence production of this plastic glass.

REFERENCES

1. Hill, R., 'Acrylic Resins' in *Synthetic Resins and Allied Plastics*, 3rd edition, Oxford, 1951.
2. Caress, A., 'Perspex', *Endeavour*, Oct. 1944.
3. Frederick, D. S., 'Acrylic Resins', *Modern Plastics*, Oct. 1938.
4. Fleck, H. R., *Plastics: Scientific and Technological*, 1951.
5. Wakeman, R. L., *The Chemistry of Commercial Plastics*, 1947.
6. Hearings before the Committee on Patents, United States Senate 77th Congress, 2nd Session, Part II (1942).

7. Communication from Dr. William Chalmers.
8. British Patent No. 395, 687.
9. Ellis, C., *The Chemistry of Synthetic Resins*, 1935.

NEOPRENE

NEOPRENE is a truly synthetic rubber with a very wide range of uses, in some of which it is superior to natural rubber. Julius A. Nieuwland was responsible for starting the chain that led to its discovery. A native of Hansboeke, Belgium, he graduated from the American University of Notre Dame in 1899 and he decided to devote himself to a theoretical study of acetylene. In 1903 he was ordained, a year later he received his doctorate and returned to Notre Dame to resume his experimental work. In 1906 he noticed that a reaction occurred when acetylene was passed into a solution of copper and alkali metal chlorides; although no violent reaction took place, a strange odour indicated the presence of a new product. During the next fourteen years he continued his efforts to increase the reaction rate so that the products could be obtained in large enough quantities to be isolated and studied. Finally, when he substituted ammonium chloride for the alkali metal chlorides, the reaction was hastened and a yellowish oil was formed along with the gas. He proved in 1921 that the oil was divinylacetylene, a polymer of acetylene formed by the chemical union of three molecules of acetylene. During his examination of divinylacetylene he treated it with sulphur dichloride, and produced an elastic material somewhat resembling natural rubber, but this material proved to be too soft for practical use.

In 1925 Dr. E. K. Bolton of the du Pont laboratories suggested that du Pont should begin an investigation of methods of synthesising rubber from acetylene. Very little progress had been made when Bolton heard Father Nieuwland lecture on divinylacetylene at the American Chemical Society; he realised the importance of Nieuwland's discoveries, told him of du Pont's interest in synthetic rubber and discussed with him the possibility of using divinylacetylene as a starting material. It was agreed that if any of the findings had commercial value, du Pont would undertake the development and, since Nieuwland had taken the vow of poverty, it was arranged that all royalties should be paid to his religious order. Du Pont's chemists visited Nieuwland to gain first-hand information. They first tried to prepare divinylacetylene without the hazard of explosion observed by Nieuwland. With this accomplished, the problem was to convert divinylacetylene into synthetic rubber, but the experiments were failures, since the resulting products possessed none of the properties demanded of a general-purpose rubber; they had little elasticity and were subject to rapid oxidation.

Du Pont's chemists then began investigating divinylacetylene with an entirely different purpose in mind. Arnold M. Collins set out to identify the impurities in divinylacetylene, and his study showed that two of these were monovinylacetylene and a compound containing chlorine. Wallace H. Carothers, who later discovered Nylon, suggested that Collins should investigate the reaction between hydrogen chloride and monovinylacetylene. This was a brilliant inspiration, for a new product, chloroprene, was produced by the reaction. Chloroprene polymerises spontaneously to form a material like vulcanised rubber, which is now known as Neoprene.

The gas which Nieuwland had first observed in 1906, but which he had been unable to isolate in quantities sufficient for study, was monovinylacetylene, and this compound was the important ingredient of a new type of synthetic rubber. The next task was to try to vary the conditions of Nieuwland's original catalytic reaction[1] in order to increase the proportion of monovinylacetylene and reduce that of divinylacetylene. This was found possible and Neoprene became an important commercial product after its introduction in 1932.

This is a case where a pure scientist, devoting his life to a study of acetylene, made observations which were actively taken up by a large chemical firm and which finally resulted, through the inspiration of a scientist in an industrial research laboratory, in an important new commercial product. Chance played its part in that it was the impurity and not the main product of the original reaction which produced the key to success.

REFERENCES

1. 'The Story of Neoprene', *du Pont News Bulletin*, Feb. 1937.
2. 'Synthetic Rubber', *Fortune*, Aug. 1940.
3. Barron, Harry, *Modern Synthetic Rubbers*, 3rd edition, 1949.
4. Carothers, W. H., Williams, I., Collins, A. M., and Kirby, J. E., 'Acetylene Polymers and their Derivatives: II, A New Synthetic Rubber: Chloroprene and its Polymers', *Journal of American Chemical Society*, Nov. 1931.
5. Nieuwland, J. A., Calcott, W. S., Downing, F. B., and Carter, A. S., 'Acetylene Polymers and their Derivatives: I, The Controlled Polymerisation of Acetylene', *Journal of American Chemical Society*, Nov. 1931.
6. Stine, Charles M. A., 'The Approach to Chemical Research based on a Specific Example', *Journal of Franklin Institute*, Oct. 1934.
7. Wolf, Ralph F., 'Rich Man, Poor Man', *Scientific Monthly*, Feb. 1952.
8. Nieuwland, J. A., 'Synthetic Rubber from a Gas', *Scientific American*, Nov. 1935.

NYLON AND PERLON

NYLON was the first of the truly synthetic fibres. Its inventor, Wallace H. Carothers, by his work on the synthesis and structure of high molecular weight polymers, added greatly to scientific knowledge and, through the elucidation of the theory of fibre structure, paved the way to the discovery of other commercially useful synthetic fibres.

The story of the discovery begins in 1927 when E. I. du Pont de Nemours & Company decided to support a programme of fundamental research. Dr. Charles M. A. Stine, head of du Pont's Chemical Department at the time, having persuaded the Executive Committee to set aside a yearly fund of about $250,000 for the purpose, sought for outstanding research chemists. Dr. Wallace H. Carothers, a brilliant young chemist, then 32 years of age, who had taught at the University of Illinois and at Harvard, was offered the direction of the group and he joined the firm in February 1928.

While at Harvard, Carothers had already given thought to polymerisation and

[1] The same catalyst that Nieuwland utilised in producing divinylacetylene is used in making monovinylacetylene. Nieuwland owned the patent on this catalyst and du Pont paid royalties to Notre Dame for its use.

the structure of substances of high molecular weight, and for du Pont he chose to study the structure and synthesis of long-chain molecules. His early work at du Pont revolved around the classic reaction between an organic acid and an alcohol, to form an ester. In this way it was possible to synthesise polyesters in which the acid and alcohol molecules are joined end to end somewhat like a chain of paper-clips.

At first this information was only of academic value, since Carothers and his associates were primarily interested in the phenomenon of polymerisation. But something occurred in 1930 during an experiment with a polyester made from ethylene glycol and sebacic acid which was destined to be of very great practical value. In an attempt to remove a sample of the heated polymer from the vessel in which it was prepared, Dr. Julian W. Hill, working under the direction of Carothers, noted that the molten polymer could be drawn out in the form of a long fibre. More important, he found that even after the fibre was cold it could be further drawn to several times its original length, and that such cold drawing greatly increase the fibre's strength and elasticity. This phenomenon had never before been observed with a compound of this type.

While fibres of this original superpolymer were not of practical use because they were easily softened by hot water, they nevertheless suggested that some related compound might produce fibres that would possess the characteristics desired for use in textiles. Further investigation was accordingly undertaken in an effort to create a material from which strong, elastic, and water-resistant fibres could be drawn or spun. Numerous new superpolymers were synthesised, but each was deficient in respect of one or more textile properties. Carothers actually gave up the project for several months in 1930 and eventually abandoned his study of the polyesters, reaching the conclusion that they had no commercial possibilities; the outlook at one time was so uncertain that du Pont seriously considered dropping the whole of the work.

It was at this time that Dr. E. K. Bolton, who had succeeded Dr. Stine as director of the Chemical Department, urged Carothers to review his findings in an effort to discover some basis for a superpolymer from which commercially useful fibres might be prepared. Carothers then turned his attention to the polyamides, made in somewhat the same way as the polyesters except that a bifunctional amine (nitrogen-containing compound) is used instead of a bi-functional alcohol. He found that several polyamides, when extruded through a spinneret improvised from a hypodermic needle, produced filaments from which fibres of the type now known as Nylon could be made. These new polyamides appeared so promising that extraordinary efforts were made to bring the development to commercial success.

After further experiments, Carothers and his associates developed a polyamide which they called '66' polymer, from which strong, tough, elastic, water-resistant fibre could be spun, capable of withstanding temperatures in excess of 500° F. As in the case of the polyesters, it was found that fibre spun from '66' polymer and related polyamides had to be cold drawn to some three or four times the original length in order to develop the desired strength and elasticity. During this stretching or cold-drawing operation, a highly interesting phenomenon takes place: the long, chain-like molecules which make up the undrawn fibre are arranged in a helter-skelter fashion but, on cold-drawing to several times its original length, they take on an orderly arrangement. They become parallel to the fibre axis and to one another, as in silk, and are also brought much closer together. Increased molecular proximity results in increased inter-molecular attraction, which in turn gives rise to greatly increased strength and elasticity.

Textile experts at du Pont selected '66' polymer, made by the reaction between hexamethylene diamine and adipic acid, for commercial development and, although several hundred different polyamides have been synthesised and evaluated in the course of du Pont's work on Nylon, '66' is still the type most widely used for textile purposes.

Du Pont threw very large resources into the task of development. Eleven years elapsed between the beginning of fundamental research on superpolymers and the production of Nylon on the first commercial unit, at a cost to du Pont of around $22 million – about $1 million for research and $21 million for plant investment.

After many trial runs, yarn for an experimental batch of stockings was ready in April 1937, and in July 1938 du Pont completed a pilot plant. The manufacture of tooth-brushes using Nylon monofilament as bristles commenced soon after, and large-scale production of Nylon got under way in December 1939. Du Pont remained the sole producer of Nylon in the United States until 1953, since when several other American companies have commenced production under du Pont licences. Imperial Chemical Industries secured a licence under the English patents in the 1940's and, in co-operation with Courtauld's Ltd., formed the British Nylon Spinners Ltd. Numerous other foreign producers of Nylon were also licensed under du Pont patents.

Brief reference may be made at this point to some closely related, although slightly later, research work in Germany, which ultimately resulted in the marketing on a large scale, mainly in Europe, of a similar fibre known as Perlon L. After the publication of the basic patents of W. H. Carothers in the United States, I.G. Farbenindustrie intensified their studies of linear condensation polymers and in 1938 they took out patents on results which emerged from the work of their research team directed by P. Schlack. The process for the making of their Perlon fibre did not conflict with the du Pont patents.

I.G. Farben had indeed taken out licences under the Nylon patents but, for various reasons, they decided to manufacture Perlon and not Nylon. Very rapid progress was made before the war in preparing for the large-scale manufacture of Perlon. The war seriously disturbed these plans, but after the currency reform in Germany large plants were put into production there and in a number of other European countries.

Here, then, is a case in which a large corporate organisation dominated the scene from the time research began until the commercial production came on to the market. Nylon's discovery can rightly be credited to du Pont, Dr. Carothers directing the research and supplying most of the crucial ideas. Another large firm, I.G. Farben, followed closely upon their track and produced a rival fibre which has also been highly successful commercially.

REFERENCES

1. 'Nylon', *Fortune*, July 1940.
2. Irvin, H. H., 'Nylon Polyamides – Their Chemical and Industrial Application', *Chemical and Metallurgical Engineering*, May 1945.
3. Hill, Rowland, 'Synthetic Fibres in Prospect and Retrospect', *Journal of Society of Dyers and Colourists*, May 1952.
4. *Encyclopedia of Chemical Technology*, vol. 10, 1953, p. 916.

5. Adams, Roger, *Biographical Memoir of Wallace Hume Carothers.*
6. Hunt, Jas. K., *Nylon: Development, Physical Properties and Present Status*, publication of E. I. du Pont de Nemours and Co., Inc.
7. Shor, Morton, 'Nylon', *Journal of Chemical Education*, Feb. 1944.
8. Urquhart, A. R., Hegan, H. J., Loasby, G., 'The Development of Some Man Made Fibres', Annual Conference of the Textile Institute, 1951.
9. Schlack, P., *The Historical Development of Polyamide Fibre Materials*, Textil Industrie, July 1954.

PENICILLIN

THE story of the discovery of penicillin by Alexander Fleming is by now so well known that nothing beyond the bare outlines need be recounted here. Fleming was a dour, reticent, patient, tireless and highly perceptive bacteriologist who, after qualifying in medicine, joined a research group under Sir Almroth Wright at St. Mary's Hospital, where new ideas in vaccine therapy were being enthusiastically pursued. His experiences as a doctor in the First World War had heightened his distrust of the value of the chemical antiseptics, at that time in common use, and had strengthened his belief in the power of the natural protective mechanisms of the human body. In 1922 he discovered that a substance, which he named lysozyme, found in numerous secretions and tissues of the body, had an extraordinary power of destroying some bacteria without injuring cell tissue: a classic instance of a discovery coming to a prepared mind. In 1928, when Fleming had become Professor of Bacteriology in the University of London, the acute observation of a chance occurrence led him to the discovery of penicillin. When the cover of one of the plates on which he was cultivating bacteria was momentarily removed, a spore of the mould *Penicillium notatum* settled on the plate and Fleming noticed that, around the mould, bacteria were being destroyed. His interest immediately excited, he went on to find that penicillin was neither toxic to animals nor to the white corpuscles. He recognised its vast potential value as an antiseptic; he drew up a list of those bacteria attacked by it and those which remained unaffected. His findings have not since been significantly altered. Although there are many experts who believe that Fleming's contribution to the final, successful use of penicillin has been exaggerated, his fame for the original and pregnant observation seems to be securely established.

Although Fleming continued to use the mould in his own researches, he got no further. The tasks of extracting penicillin from the growing mould, of rendering it stable and of increasing its yield, tasks which were a combination of invention and development, were taken up over a period of years by Professor Raistrick of the London School of Hygiene and Tropical Medicine, and later brought to a wholly successful outcome by the team at Oxford under the direction of Sir Howard Florey. A description of this later work and the development by which penicillin was put on to the market is given in Chapter I, pp. 31–2.*

REFERENCES

1. Ludovici, L. J., *Fleming – Discoverer of Penicillin*, 1952.
2. Sokoloff, Boris, *The Story of Penicillin*, 1945.

* [The story of the discovery of Semi-Synthetic Penicillins is told below pp. 351–4.]

W. H. Carothers: nylon

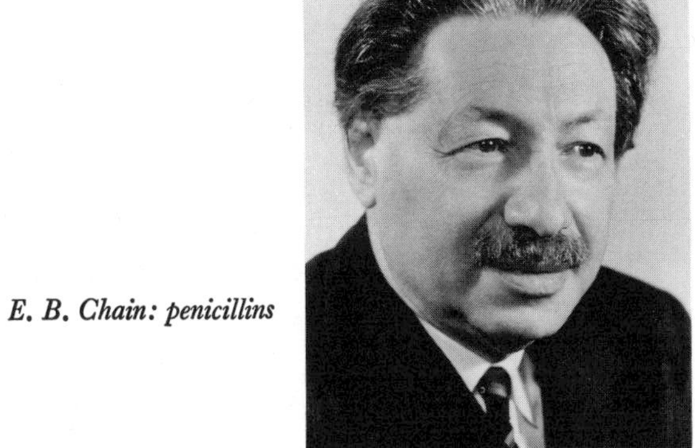

E. B. Chain: penicillins

K. Ziegler: low-pressure polyethylene

G. W. Jessup and F. W. Davis: power-steering

3. Ratcliff, J. D., *Yellow Magic – The Story of Penicillin*, 1945.
4. Masters, David, *Miracle Drug – The Inner History of Penicillin*, 1946.
5. Fox, Ruth, 'A Science Milestone – Sir Alexander Fleming and the Story of Penicillin', *Science Digest*, Sept. 1953.
6. Fleming, Alexander, *Penicillin – Its Practical Application*, 1946.

POLYETHYLENE

POLYETHYLENE, one of the most useful of modern plastics, is formed by the polymerisation of ethylene. In the original process, invented in the laboratories of Imperial Chemical Industries Ltd., this was carried out by the use of great pressures, but it has recently been discovered by Dr. Karl Ziegler in Germany and scientists of American companies that ethylene can be polymerised at normal pressures by the use of certain catalysts.

The history of the original discovery is not one of research workers proceeding deliberately towards a pre-defined goal. M. W. Perrin, one of the original workers in the use of high pressures, has written:

> 'The story of its initial discovery, however, provides an unusually clear-cut instance of the unexpected results that may come from research and of the importance of the rôle of chance in such work. At the present time, when so much thought is being given to the proper organisation of research in industry, it may be of value to record the full story of this particular discovery, which, as time passes, naturally tends to become idealized and presented in a way which suggests a steady and logical growth from the start of the general research programme to the discovery, recognition and development of the particular product.'[1]

In the 1920's I.C.I.'s Alkali Division was in touch with scientists in Holland. Professor A. Michels of Amsterdam had developed techniques for measuring the effects of pressure on the physical properties of matter, and two employees of I.C.I., R. O. Gibson and M. W. Perrin, were sent to work with him. Michels designed special equipment which was delivered to I.C.I. in 1931. By now interest was being directed to the possible chemical, as well as the physical, effects of high pressure. Encouraged by the discovery of P. W. Bridgman and J. B. Conant at Harvard that high pressures markedly affected polymerisation reactions, I.C.I. decided in January 1932 to authorise a study of the effects of high pressures on chemical reactions, as advocated by Perrin and J. C. Swallow.

The Dyestuffs Division of I.C.I. had meanwhile been interested in Michels's work and Professor Sir Robert Robinson, a member of its research committee, prepared a list of organic reactions to be tried. These were chosen in the hope that compounds not normally reactive might become so at high pressures or that reactions normally needing vigorous catalysts might occur without them. At first Gibson, with E. W. Fawcett, studied liquid phase reactions; later, the development of equipment for use with gases enabled the work to be extended. When a mixture of ethylene and carbon monoxide was tested a solid polymer of acrolein was found. In March 1933 an experiment involving ethylene and benzaldehyde yielded a white waxy solid which was presumed to be a polymer of ethylene; but the

[1] M. W. Perrin, 'The Story of Polythene', *Research*, Mar. 1953.

importance of this was not then recognised, and ethylene subjected alone to higher pressures caused explosions, so that the work was discontinued.

Late in 1933 I.C.I. set up a specially equipped laboratory where high pressures could be used in safety. Here Gibson and Perrin tried to determine the reason for any pressure effects. Fawcett meanwhile studied the polymerisation of linseed oil, in particular by pressure. In 1935 J. G. Paton found carbon monoxide reacted with aniline under pressure to form a solid; when it was decided to test the reaction between carbon monoxide and ethylene the 1933 results were remembered and ethylene was tested again, now in the hope that the pressure might polymerise it. In the experiment there was a defect in the apparatus, the pressure dropped and a small amount of white powder was discovered when the vessel was dismantled. This material had remarkable properties: it could be cold drawn like Nylon; moulded and formed into threads and film; it was chemically resistant and had excellent insulating qualities. I.C.I. filed applications for both British and United States patents.

The plastic and its useful properties had been discovered but the development tasks remained. Engineering of an extremely high order was required: high pressures and moderate temperatures are not normally associated; yet for the production of polyethylene both must be present. Sensitive methods of controlling the polymerisation reaction had to be found to prevent the violent decomposition of the ethylene.

Some twenty years later new processes for making polyethylene at normal temperatures and pressures were discovered by Dr. Ziegler and his team, E. Holzkamp, H. Breil and D. H. Martin, at the Max Planck Institute, Mülheim, which is largely financed by Ruhr companies chiefly interested in the study of the chemistry of coal. The discovery was made in the course of research on catalysts, the results of which have opened up prospects of revolutionary advances in polymerisation chemistry. Dr. Ziegler has himself explained that his discovery was not the direct outcome of attempts to solve a 'set' problem.

He in fact set out to follow a broad course of study in which 'my only guide was initially just the desire to do something which gave me pleasure'. This course threw up many interesting conclusions, some of them of highly practical value, and one of these led ultimately to a method of making polyethylene.

Dr. Ziegler had made his crucial discovery in 1950, and the patents describing it were published by 1954. American scientists, notably those of the Phillips Petroleum Company, had also discovered at much the same time that ethylene can be polymerised catalytically; but where Dr. Ziegler used an aluminium catalyst, Phillips used chromium. Both processes produce a denser and more rigid material than polyethylene made by the high-pressure process. Production began in Germany and the United States in 1956, though not until 1960 in Britain. The new polyethylene costs more than the older material, which still accounts for the bulk of British and American output. The application of Ziegler-type catalysts to polymerise propylene may have greater commercial importance; Dr. Natta of Milan Polytechnic Institute deserves most credit for this development, though American and German companies have also done much work on the process. Polypropylene is even harder and more rigid than high-density polyethylene; it may well be the first plastic to replace metal on a large scale, and it can also produce a textile fibre that rivals nylon in quality but costs less.

High-pressure polyethylene was discovered as a consequence of imaginative research and a stroke of good fortune in the laboratories of a large company. Low-pressure polyethylenes are the product of research by a small team in a non-profit-making research laboratory in Germany and by companies other than that which made the first important discovery.

REFERENCES

1. 'The Polyethylene Gamble', *Fortune*, Feb. 1954.
2. Perrin, M. W., 'The Story of Polythene', *Research*, Mar. 1953.
3. 'Polythene Grabs the Spotlight', *Modern Plastics*, Sept. 1955.
4. 'The New Polythenes', *Plastics*, Sept. 1955.
5. 'Low-pressure Polythene', *Plastics*, Nov. 1955.
6. 'Report on Germany', *British Plastics*, Nov. 1955.
7. British Patent No. 713,081, of Dr. Karl Ziegler.
8. Ziegler, K., Holzkamp, E., Breil, H., Martin, H., 'The Mülheim Normal Pressure Polyethylene Process', Paper delivered at a Meeting of the German Chemical Society, Sept. 1955.
9. Ziegler, K., 'The Indivisibility of Research', *Glückauf* 91, 1955.
10. Correspondence with Dr. K. Ziegler.

POWER STEERING

EXPERIMENTS were made with various forms of power assistance for the steering of vehicles in the nineteenth and early twentieth centuries, but it was not until the introduction of the heavy lorry and bus in the early 1920's that a real need arose to reduce the exertion involved in manual steering. Mechanical, vacuum, compressed-air and hydraulic systems were all tried in the late 1920's and early 1930's; hydraulic systems were those generally adopted.

Two Americans, Harry Vickers, who founded Vickers Inc., and Francis W. Davis, who worked privately, were mainly responsible for the invention and much of the development of hydraulic power-assisted steering systems. Vickers designed a hydraulic steering gear and a high-pressure vane pump for use on motor vehicles in 1925; Davis demonstrated his hydraulic power-assisted steering gear in October 1926 and in that year applied for patents on his design. Both men developed their designs during the later 1920's, producing numbers of experimental models, and both had produced systems which were ready for production soon after 1930. Vickers considered that booster units, which could be attached to existing vehicles with the minimum of alteration, would give the most practicable system; units of this type began to be produced in 1931 for fitting to heavy vehicles.

These early booster units lacked certain features to be found in modern power-steering systems. These are the preloaded spring and hydraulic reaction, invented by Davis, which act to give the driver with power steering the 'feel' of the road and return the wheels to the straight ahead position. Davis also invented the continuous oil-flowing 'open-centre valve' which is used on most modern power-steering gears, while he adopted the suggestion of another inventor, Aikman, for opening the valve quickly by the reaction of the steering column on the cam or sector.

Davis had been chief engineer of the truck department of the Pierce-Arrow Motor Car Company in the early 1920's, where he had become aware of the difficulty of steering heavy vehicles. After studying the literature on servo-mechanisms, he concluded that existing systems of power assistance were unsuitable and that a hydraulic system was most promising. Leaving the Pierce-Arrow Company in order to concentrate on this work, Davis rented a small engineering shop in Waltham, Massachusetts, engaged G. W. Jessup, a clever tool-maker and engineer, and together they developed a hydraulic power-steering system without the help of expensive equipment. The difficulty of producing a silent, cheap pump was overcome with the advice of a professor at the Massachusetts Institute of Technology. Pumps developed independently by Vickers Inc. and the Eaton Company are, however, widely used in modern hydraulic steering gears.

Davis's patents were issued in the years 1931–33. After the earlier demonstrations of his system, General Motors, on the basis of his patent applications, agreed to take a licence. But the depression caused General Motors to abandon in 1933 their plans to introduce power steering. Licences under the Davis patents for the pre-loaded spring and open-centre valve were secured by Vickers Inc. shortly after starting production of their own model in 1931, and other companies took licences in the later 1930's.

Davis found it difficult to convince manufacturers of the practicability of power steering for passenger cars; though General Motors had indeed planned to start production in 1941, the war intervened. After the war, the ease with which General Motors could then sell cars made them disinclined to introduce it.

The first company to introduce power steering for passenger cars was Chrysler in 1951, using the Gemmer Manufacturing Company design which incorporated Davis's 'open-centre' valve and hydraulic reaction. General Motors then made a licensing agreement with Davis in respect of certain improvements, and within two years their production of power-steering units reached one million a year. Davis said in 1954 that if he had known in 1926 what tremendous obstacles he would encounter before his invention reached the mass-production stage, he would never have attempted to invent the gear.

Work in Britain on hydraulic power steering systems was started later than that in America and designs have generally followed the lines of the American systems. One exception is the Lockheed differential-area system which has advantages for some applications.

The major contributions to the invention of hydraulic power steering for motor vehicles were thus made by Vickers, who was at the time the head of a small company, and by Davis, an individual inventor. The greater contribution was that of Davis, whose inventions made power steering practicable.

REFERENCES

1. 'Hydraulic Power Steering for Passenger Cars', *Lubrication*, Feb. 1953.

2. Davis, F. W., 'Power Steering for Automotive Vehicles', *Society of Automotive Engineers Journal*, Apr. 1945.

3. Heldt, P. M., 'Power Steering is Demanded by Heavy and High Speed Vehicles', *Automotive Industries*, Dec. 10, 1932.

4. Neil, Edmund B., 'Power Steering for Modern Vehicles', *Automotive Industries*, Dec. 1, 1951.

5. Geschelin, J., 'Gemmer's New Power Steering Gear, The Hydraguide', *Automotive Industries*, Feb. 1, 1951.

6. Personal interview with Francis W. Davis.

7. Taup, M. J., 'Twenty-Five Years of Hydraulic Power Steering', *Sperryscope*, Winter 1952.

8. Heacock, F. H., and Jeffery, H., 'The Application of Power Assistance to the Steering of Wheeled Vehicles', Institution of Mechanical Engineers – Automobile Division, *Proceedings*, 1953–54, p. 69.
9. 'Lockheed Power-Assisted Steering', *Automobile Engineer*, Apr. 1956.

RADAR

RADAR is a system of radiolocation which employs short-wave radio waves to detect objects. The waves are sent out in pulses, strike the objects, are reflected back to the receiving set and from the data gathered there, the position of the object is accurately determined. During the Second World War radar proved invaluable in the detection of approaching enemy aircraft and ships; in the years following the war, radar has been effectively applied to peace-time uses. The actual invention of the radar system can hardly be separated from the whole history of the science of radio, since most of the principles were first laid down by the scientists interested in radio waves.

In 1887 Heinrich Hertz showed that electromagnetic waves are reflected in a manner very similar to light rays, but it was not until the 1930's that radar equipment based on this discovery first became used. The possible utility of this property of electromagnetic waves had long been recognised by some individuals: in 1904 a German engineer, Hülsmeyer, had been granted patents on a radio-echo collision-prevention device. In 1922 Marconi said he had noticed the effects of reflection of electric waves, and suggested that apparatus could be designed to enable ships to discover the presence of other ships in fog by employing this phenomenon. In the same year A. H. Taylor and Leo Young, scientists in the United States government service, observed that ships passing a receiver and transmitter tuned to the same frequency produced a definite interference, and sketched out a radar apparatus which might be employed in convoying at sea.

The first use found for the reflecting properties of radio waves was in measuring the height of the Heaviside layer. This was done in Britain in 1924 by Sir Edward Appleton and M. F. Barnett, and in the U.S.A. in 1925 by Dr. Gregory Breit and Dr. Merle A. Tuve of the Carnegie Institute. Breit and Tuve were the first to apply the pulse principle.

By 1939 Germany, Great Britain, Holland and the U.S.A. all possessed military radar apparatus, while the first peaceful application had been made in France in 1935. In France and Germany the work was done by the scientists of radio companies: in Britain and the U.S.A. by scientists in government research stations.

Scientists of the Société Française Radioélectrique began to study the use of metric and decimetric radio waves to detect obstacles in 1934, first with a view to saving life at sea and then for military uses. An 'obstacle detector' working on decimetric waves and employing magnetrons and the pulse principle was fitted to the liner *Normandie* in 1935; it appears to have been successful, and equipment was installed to detect ships entering and leaving the harbour at Le Havre in 1936. Work was meanwhile done on equipment for detecting aircraft, though no radar warning system had been installed by 1939.

German work, begun before 1935, was carried on, under contracts from the Navy, by a new firm, G.E.M.A., and later by other firms. As a result, by 1939 a large number of radar sets were in operation for detecting aircraft. Though the German work began at least as early as that in other countries and had reached a very similar level of development by 1939, it lagged behind after the outbreak of war. German policy was based on the assumption that the war would be short, and consequently less effort was put into such basic work as radar than in Britain or the U.S.A.

Interest in the possibility of using radio waves to detect aircraft arose in Britain about 1934. The reflection of radio waves from aircraft had been observed in 1931 and 1933; on the latter occasion the possibilities for aircraft detection were carefully analysed. H. E. Wimperis, Director of Scientific Research at the Air Ministry and his assistant, A. P. Rowe, suggested that the country should increase its efforts to develop a method of detecting aircraft at a distance and an investigating committee of three scientists, Sir Henry Tizard, Professor A. V. Hill and Professor Patrick Blackett, was set up. It was soon realised that a radio beam would be the ideal alternative to the existing inadequate acoustic warning equipment.

Robert A. Watson-Watt (later Sir Robert), who was a lecturer in physics at University College, Dundee, before he began a research career in the government laboratories, played the major rôle in developing practical radar equipment in Britain. He was superintendent of the Radio Division of the National Physical Laboratories at the time the pressure for improved air defence reached its peak. He felt confident that radio waves could be employed to detect aircraft. His two memoranda of February 1935 described his suggested means for so using them; after a demonstration of the echoes produced by aircraft from the B.B.C. Daventry short-wave station, the Tizard Committee recommended that work on the lines suggested by Watson-Watt should be started. Working with six assistants, of whom A. F. Wilkins was the principal, Watson-Watt developed the first practical radar equipment for the detection of aircraft on the Suffolk coast in the summer of 1935. The main problems he solved were the construction of a high-power transmitter, the modulation of it with short pulses, the development of receivers to handle the pulses and of suitable transmitting and receiving aerials. The performance of the first equipment was considered promising enough for the Air Ministry to build a chain of five radar stations.

Development of radar equipment continued up to the outbreak of war, the Navy and Army finding that it was valuable for use on ships and with anti-aircraft guns, while the Air Ministry completed the chain of radar stations along the south and east coasts which were to play so important a part in winning the Battle of Britain in 1940. When the war began, an increased effort was put into radar research; one problem tackled was that of generating large powers at centimetre wave-lengths. The first attempts to use such short wave-lengths had been made in 1935 on the advocacy of Admiralty scientists, but suitable generators were not available.

In the autumn of 1939 the Admiralty asked Professor M. L. Oliphant and the physics department of the University of Birmingham to develop a high-power micro-wave transmitter. The majority of the scientists in the laboratory concentrated on the klystron, described by its inventors, R. H. and S. F. Varian of Stanford University, California, in 1939, which used for the first time closed resonators, described by W. W. Hansen, also of Stanford, in 1938, for the production of high-frequency power. J. T. Randall and H. A. H. Boot, struck by the difficulty of getting enough power from the klystron, considered instead applying the resonator principle to the magnetron, which had been invented by A. W. Hull of the American General Electric firm in 1921 but which, in its conventional form, lacked the properties they were seeking. The result was the cavity magnetron, which proved to be the needed generator, producing high powers on centimetre wave-lengths.

Great improvements in radar equipment were made possible by the use of centimetre wave-lengths; sets could be made more accurate and with greater range. Coming into service in 1942–43, centimetre-wave radar made possible improved blind-bombing techniques and naval search equipment.

The development of radar had meanwhile been proceeding independently in the United States. Military interest began after L. A. Hyland, an associate of A. H. Taylor, discovered accidentally in 1930 that aircraft cause interference in radio waves and Leo Young successfully applied the pulse apparatus to this. Despite the fact that radar looked so promising, the Navy was reluctant to spend any significant amount on it, but, through the persistent efforts of Harold G. Bowen, chief of the Naval Laboratory, $100,000 was allocated for radar research. Robert M. Page, head of the research section of the Naval Laboratory's radio division, developed some of the first modern radar equipment. In 1938, two years after successful laboratory demonstrations of the equipment, the American Navy finally fitted radar devices to some of its ships.

After 1940 Great Britain and the United States co-operated in radar development. In the United States the Office of Scientific Research and Development established a radar division to finance and co-ordinate radar research and engineering. Starting with a nucleus of twelve men, it organised the Radiation Laboratory in 1940 where activities expanded so rapidly that by 1945 the number of employees reached almost 4000. The staff included scientists from nearly every university in the United States. By the end of 1940 the bulk of the radar research was transferred from the Army and Navy to the Radiation Laboratory and several industrial laboratories. During the war the British Air Ministry organised a large research laboratory, the Telecommunications Research Establishment, where radar research was conducted.

Radar represents the use of an attribute of radio waves, the potentialities of which had been pointed out by many scientists and individual inventors for some thirty years before its first application. Though the first use of radar was a peaceful one, the incentive to develop it in most countries came from the military need to detect aircraft at a distance, and for this purpose it was first applied on a large scale. The actual devising and developing of radar apparatus was done more or less simultaneously in at least five countries, though possibly not independently in every case. Credit for this work goes to scientists working in government research establishments, radio companies and universities.

REFERENCES

1. Bright, Arthur A., Jr., and Exter, John, 'War, Radar, and the Radio Industry', *Harvard Business Review*, Jan. 1947.

2. Hightower, John M., 'Story of Radar', Senate Miscellaneous Documents, 78th Cong. 1st Session, Document No. 89, 1943.

3. Taylor, Denis, and Westcott, C. H., *Principles of Radar*, 1948.

4. Watson-Watt, Sir Robert, 'The Evolution of Radiolocation', *Journal of the Institution of Electrical Engineers*, Part 1, Sept. 1946.

5. 'The History of Radar', *Engineer*, Aug. 17, 24, 31 and Sept. 7, 1945.

6. Fink, Donald G., *Radar Engineering*, 1947.

7. 'Radar – The Technique' and 'Radar – The Industry', *Fortune*, Oct. 1945.

8. Wilkins, A. F., 'The Story of Radar', *Research*, Nov. 1953.

9. Wathen, Robert L., 'Genesis of a Generator – The Early History of the Magnetron', *Journal of the Franklin Institute*, Apr. 1953.

10. 'The Magnetron', *Fortune*, Oct. 1946.
11. Randall, J. T., 'The Cavity Magnetron', *Proceedings of the Physical Society*, May 1946.
12. 'Radar', *Encyclopaedia Britannica*, 1947 edition, vol. 18, p. 873.
13. 'Radar Pre-history', *Wireless World*, Dec. 1945.
14. Brenot, P., 'Réalisations d'un grand centre de recherche industriel pendant et malgré l'occupation', *L'Onde Électrique*, Sept. 1945.
15. Wimperis, H. E., 'Directing Research', *Engineer*, Sept. 14, 1956.
16. C.I.O.S. Reports, nos. I, 1; XXI, 1; XXV, 12; and XXXI, 38.
17. Smith-Rose, R. L., 'Radiolocation', *Wireless World*, Mar. 1945.

RADIO

THE modern radio system is not one invention; it is made up of the work of many inventors. Its origins lie in the work of nineteenth-century scientists, and especially Maxwell and Hertz. Radio telegraphy was its precursor; Marconi made the greatest single contribution to its practical realisation, though many other scientists and inventors, such as Lodge, Branly, Tesla and Stone, made inventions essential to its development.

The success of radio telegraphy soon aroused interest in the transmission of speech. Reginald Fessenden in America was one of the earliest pioneers of radio telephony; his consuming interest was in mathematics generally, and electricity in particular. After being educated at Bishops' College in Canada, he became one of Edison's chemical assistants, leaving him to become professor of electrical engineering at Purdue University. He later left teaching to help the United States Weather Bureau to develop apparatus for the wireless transmission of weather forecasts.

Radio telephony involves the use of continuous or undamped electromagnetic waves of high frequency; the chief difficulty in its early development was to make equipment producing such waves. Fessenden first transmitted speech in 1900, but the high-speed commutator he used for transmission was unsatisfactory. He then left the Weather Bureau and continued his work, financed by two Pittsburgh businessmen, concentrating on the improvement of the high-frequency alternator as a transmitter. Tesla had experimented with such alternators in the 1890's without success; but, as improved by Fessenden, they could be used to broadcast over a twenty-five mile range. After they had been further improved by Ernst Alexanderson of General Electric, who used iron instead of the wood Fessenden preferred for the armature, Fessenden was able to broadcast a hundred miles in 1907. Alexanderson continued to work on the alternator for the next three or four years, and raised it to a high degree of perfection. He later invented a magnetic amplifier, an electronic amplifier, a multiple-tuned antenna, which appreciably increased the efficiency wave of transmission, and, during the First World War, the Alexanderson-Beverage static eliminator.

Research along parallel lines to that of Fessenden and Alexanderson was being carried out simultaneously in Germany by Dr. Rudolph Goldschmidt, of the Allgemeine Electrizitäts-Gesellschaft, who developed excellent alternators in 1907. An alternative method of transmitting speech tried at this time was the Poulsen arc, invented by the Danish inventor Valdemar Poulsen in 1903. Several Europeans, including the German Telefunken Company, and H. P. Dwyer in America developed it; by 1910 Dwyer could transmit over five hundred miles.

Edwin H. Armstrong: radio

All the early detectors suffered from serious limitations and many inventors sought for more effective devices. The electrolytic detector was one of the earliest answers to this need; among its many forms those of Fessenden and Schlömilch seem to have been most widely used. The crystal detector dates back to Professor F. Braun's psilomelan detector used by Telefunken[1] in 1901. In the United States in 1906 H. H. Dunwoody, Vice-President of the De Forest Wireless Company, discovered the suitability of carborundum, and G. W. Pickard, an individual inventor, produced the cheap silicon detector. The crystal set first made radio popular and introduced the amateur to the science. In 1905 Fessenden applied for patents on his heterodyne system of detection, but it was not until the triode valve was included in the system that its value was fully appreciated.[2]

The triode valve, invented by Lee de Forest, eventually revolutionised radio. Sir Ambrose Fleming had applied the 'Edison effect' to produce his two-element valve in 1904. In February 1905 Fleming described his valve to the Royal Society and, it is said, de Forest, who at the time was running his own small radio firm, adopted it in his own experiments. The two-electrode valve was less sensitive as a detector than the electrolytic detector and, apparently out of a desire to increase its sensitivity, de Forest in December 1906 added a third electrode. This increased sensitivity, and the triode ultimately became a valuable instrument.

It does not seem to have been realised before about 1911 that the triode valve could be used to amplify current. This discovery seems to have been made independently in that year by von Lieben in Germany and Edwin Armstrong in America. This was the first of several inventions that Armstrong, who was a student under Pupin at Columbia at the time and was later a professor there, was to contribute to the development of radio. He worked independently of the companies throughout his career. The invention of the feed-back circuit, a means of multiplying the amplification of the valve, followed in 1912; Armstrong, Meissner, a scientist with Telefunken, de Forest, Irving Langmuir of General Electric and C. S. Franklin and H. J. Round of Marconi in England have all been credited with this invention. In 1913 another step in the evolution of the valve came with its use as a generator of high-frequency oscillations, which rendered wireless telephony practicable. Meissner in Germany, who had been working with von Lieben's valves, has been credited with this advance, as have Armstrong, Franklin and Round.

The introduction of the 'hard' valve, in which a high vacuum was obtained, was essential to its general use. Langmuir is generally given most credit for realising that valves which were entirely free from gas would operate more consistently, and for devising, in 1914–15, the means of producing valves with the necessary high vacua. However, some authorities state that Dr. H. D. Arnold of the American

[1] The Telefunken Company had been formed in Germany by Professor Slaby and Count Arco amalgamating with F. Braun and Siemens-Halske. It was the most active German company in the radio field during the pioneering period.

[2] In G. L. Archer's *History of Radio*, 1938, p. 89, there is a description of the heterodyne. Archer quotes Kintner's remarks about the invention. '"This was another bold stroke of Fessenden, in which he departed radically from methods practised by others. Like his other great inventions, it was made before he had suitable equipment with which to practise it. He required a source of local oscillations of adjustable frequency, and a high frequency alternator or oscillating arc was all that was available. These could be made to work, but with considerable inconvenience and a high degree of unreliability. The discovery of the oscillating tube provided the principal need of this great system."'

Telephone and Telegraph Company had realised this in 1912–13 and produced 'hard' valves.

Armstrong described the operating characteristics of the valve following a series of experiments made during 1914–15, using valves made with Langmuir's advice, and did much to direct attention to its potentialities. The need for radio equipment during the war gave a further impetus to the valve's development; its use for transmitters and receivers, in 'hard' form, was standard practice by 1918. Armstrong made another major invention, that of the superheterodyne circuit, while on service in France during the First World War. Westinghouse bought Armstrong's superheterodyne and feed-back patents for $530,000 after the war; later he received an even greater sum from R.C.A. for his super-regeneration invention.

One of the main developments in radio after 1918 was the discovery of the usefulness of the shorter wave-bands. It was generally considered that wave-lengths below 200 metres were useless except for short distance transmission, though cases were known of long ranges being obtained on short waves. These were regarded as freaks, however, and wave-lengths below 200 metres were, after 1918, allocated to amateurs who, encouraged by these 'freak' results, arranged trial broadcasts from America to England in December 1921 on 200 metres. Their success showed that short-wave low-power broadcasts could be heard over long distances.

Marconi, assisted by C. S. Franklin of the British company, had begun work on short waves in 1916. They had obtained promising results over short and medium distances; but they did not discover the long range possibilities of their transmissions because they failed to observe the phenomenon of the 'skip area'. In 1922 Marconi established an experimental station at Poldhu to determine the value of short waves for long distances, as demonstrated by the amateurs; tests in 1923–24 proved their effectiveness. C. S. Franklin developed directional aerials to 'beam' the short-wave signals during Marconi's experiments; these considerably reduced interference and have since been generally adopted.

Other companies followed up the successes of the amateurs. In 1924 R.C.A. set up a 100-metre transmitter on its South American service, following it by others and reducing the wave-length, but without using directional aerials. The American Navy and the Bell Telephone Company experimented with short waves at this time; Dr. Frank Conrad of Westinghouse was another pioneer of research on short waves, and the company set up a transmitter. The German Telefunken Company became interested in 1925, using short-wave transmissions to South America but, like the American companies, it did not use directional aerials until they were proved in use by British stations.

Frequency modulation is the greatest innovation in radio since the introduction of short waves, and it is another contribution of Edwin Armstrong. He became convinced in 1924, from his study of the problem of static, that frequency modulation would overcome it. The principle of frequency modulation was known; what was new was how to make it a practicable method of broadcasting. By 1933 Armstrong had perfected a system which overcame most static. R.C.A. refused to develop it and broadcasting companies opposed it, but Armstrong refused to abandon his idea. A friend, John Shepard, owner of the Yankee network, built a public station for the new type of broadcasting. Once the public heard FM, the demand for it grew and FM stations spread over the United States and the rest of the world, though reaching Great Britain only in 1955. W. R. Maclaurin has written:

'It is my own belief that the large receiving-set companies were not responsible for the pioneering work on FM and reacted to its introduction with varying degrees of stubbornness or apathy. The imagination of an independent inventor like Armstrong and a small but aggressive broadcasting concern like the Yankee Network were, I think, *essential*, both for the original research and the subsequent adoption of this important innovation.'[1]

The foundations of the radio industry were laid by German and English academic scientists. The majority of the basic inventions were produced by individual inventors who had no connection with established firms in the communications industry or by men who worked for new firms such as Marconi and Telefunken; large firms only entered the field after the potentialities of the invention had been established. Once the large firms started work, they made valuable contributions to radio technology, but it should be noted that a few individuals in the large industrial laboratories did the outstanding work. Nor have all the recent innovations come from these laboratories; an outside worker contributed the most radical of them. And amateurs were responsible for revealing, contrary to much expert opinion, the value of short waves for long-range transmission.

REFERENCES

1. Maclaurin, W. R., *Invention and Innovation in the Radio Industry.*
2. O'Dea, W. T., 'Radio Communication: Its History and Development', *Science Museum Handbook*, 1934.
3. Ladner, A. W., and Stoner, C. R., *Short Wave Wireless Communication*, 1943.
4. Archer, G. L., *History of Radio to 1926*, 1938.
5. Appleton, Sir Edward, 'Thermionic Devices from the Development of the Triode up to 1939', *Journal of the Institute of Electrical Engineers*, Mar. 1955.
6. Coursey, P., *Telephony Without Wires*, 1919.
7. Pierce, G. W., *Principles of Wireless Telegraphy*, 1910.

ROCKETS

WITHIN a period of twenty-five years the high altitude, long-range rocket has become a deadly weapon of war and a useful instrument for the study of conditions in the upper air.

The first scientific writing on the subject, published in 1903, was that of a Russian schoolmaster, K. E. Ziolkovsky, who worked out mathematically many of the requirements for space travel. Dr. R. H. Goddard, an American professor of physics, experimented on a small scale with rockets between the two World Wars, mainly with funds provided by private research foundations. Goddard was the first to make use of liquid propellants; his largest rocket, launched in 1935, reached a height of 7500 feet. He also worked on steering mechanisms and systems of engine cooling.

Hermann Oberth's book *The Rocket into Interplanetary Space*, published in 1923 while he was still a student, undoubtedly had great influence on modern rocket development. Ignorant of the work of Ziolkovsky and Goddard, his findings were based on careful theoretical calculations and he gave detailed designs. Although he

[1] W. R. Maclaurin, *Invention and Innovation in the Radio Industry*, 1949, p. 190.

failed at that time to interest scientists, his book was widely read by amateurs, and it inspired a number of young men in Germany to experiment with various types of rockets. Oberth spent many years as a mathematics teacher in his native Transylvania; in 1938 he accepted an invitation to undertake rocket research in Vienna and later in Dresden, but when he realised that he was being denied any information on the important German work then being done, he decided that his time was being wasted. When he tried to return to Romania the German authorities, fearing that he knew too much, refused to allow this and suggested that either he became a German citizen or retired to a concentration camp. He chose the former and was sent to Peenemünde, but it was then too late for him to have any direct influence on the building of the V2 rocket. However, he was described by Dornberger, the head of the Peenemünde establishment, as the 'originator of modern rocket theory', and after the first successful launching of the V2, Dornberger paid him this tribute: 'We all knew how much our work had derived, from the very start, from his pioneering spirit . . . the congratulations should go to him for showing us the way.'[1]

Amateur rocket societies were formed in various countries during the 1920's, the most active being the German VfR – Society for Space Travel. In fact, apart from Goddard, nearly all the practical work was carried out in Germany. In England, experiments with liquid-fuel rockets were discouraged by the authorities; an Under-Secretary of State wrote to the Chairman of the British Interplanetary Society soon after its foundation in 1933:

> 'We follow with interest any work that is being done in other countries on jet propulsion, but scientific investigation into the possibilities has given no indication that this method can be a serious competitor to the airscrew-engine combination. We do not consider we should be justified in spending any time or money on it ourselves.'[2]

Up to 1929 rocket research in Germany was carried out by individuals and the small group in the VfR, the most active members of which were Klaus Riedel, von Braun, Ley and Nebel. They worked with limited facilities and had little money. The VfR carried out numerous experiments over a period of six years; they had many failures but finally succeeded with their 'Repulsor' in reaching altitudes of over half a mile. The results of the society's experiments were freely published. When the Nazi Government came to power the society disintegrated and the active members dispersed to government and other research centres. Although their achievements had fallen far short of their hopes, the society had given an opportunity for enthusiasts to gather experience, and a number of them, in particular von Braun, were later key members of the Peenemünde group.

Advances in the development of rocket fuels and motors were made, among others, by Helmuth Walter at Kiel, and by Eugen Sänger of Vienna, best known for his theoretical work. In 1930 the Heylandt firm in Berlin became interested and employed W. J. H. Riedel and Max Valier on rocket research.

It is not impossible that the large guided missile would still be undeveloped if it had not been for an oversight on the part of the Versailles Treaty makers – rockets were not mentioned in the restrictions imposed on German armament. In 1929 the German Government decided to undertake research on rockets for military purposes; this decision ultimately resulted during the Second World War in a gigantic

[1] W. R. Dornberger, *V2*, p. 16. [2] P. E. Cleator, *Into Space*, p. 15.

development station erected at Peenemünde which employed 12,000 people and enlisted the assistance of many university and research institutes. At first public funds were made available to various individuals and groups but in 1932 the Army Weapons Dept., dissatisfied with the progress being made, set up its own experimental station under the direction of General Dornberger. His staff consisted of three assistants, von Braun, Walter Riedel and H. Grunow. They started to acquire knowledge of the principles:

> 'We were tired of imaginative projects for space travel. We wanted thrust-time curves of the performance of rocket motors . . . to know what fuel consumption per second . . . what fuel mixture would be best, how to deal with rising temperatures, what forms of injection, combustion chamber, and exhaust nozzle would yield the best performance.'[1]

Many individual inventors came to offer their ideas:

> 'It was our job to separate the wheat from the chaff, and that was no small task in a sphere of activity so beset with humbugs, charlatans and scientific cranks, and so sparsely populated with men of real ability.'[2]

With the A1 missile many set-backs were encountered but two of a new design, the A2, were successfully launched at the end of 1934; however, when bigger rockets were embarked upon it was discovered that experience derived from small-scale models could not necessarily be applied. The A3 was found to carry jet vanes too small to ensure stability. In 1939 repeated success was achieved with the A5; this model released a parachute and could be recovered, which enabled many different mechanisms to be tested. Before 1941 little thought had been given to the problem of guiding missiles. The guiding mechanism of the A5, for which valuable help was given by Boykow, a director of the Gyroscope Company, proved effective.

Plans for the A4, later known as the V2, had been ready for some years, but before going ahead with the development it was necessary to get endorsement from higher authorities. Hitler, however, was uninterested, and in the spring of 1940 rocket development was struck off the priority list; without this it was impossible to get the necessary materials and technical staff. Two years were spent in makeshift solutions and in desperate efforts to get support. Despite the frustrations, in October 1942 the V2, after ten years' work, was successfully launched: the missile covered a distance of 120 miles with a deflection of only 2½ miles from the target, reached a speed of over 3000 m.p.h. and a height of nearly 60 miles. 'The technical feasibility of a big guided rocket had been proven for the first time in history.'[3]

After this success further pressure was exerted to get priority for the production of the rocket. A decision was made that development should proceed, a launching site was to be built on the Channel coast, production plans were to be prepared. But in March 1943 a message came from headquarters: 'The Führer has dreamed that no A4 will ever reach England'. Hitler, however, relented in July 1943 and the highest priorities were given to the production of the rocket.

There was still much development work to do: the rocket had to be simplified

[1] W. R. Dornberger, *V2*, p. 20. [2] *Ibid.* p. 29.

[3] W. R. Dornberger, 'European Rocketry after World War I', *Journal of the British Interplanetary Society*, Sept. 1954.

if it were to be mass-produced; alterations had to be tested one at a time in order to isolate particular troubles. In the autumn of 1943 only between 10 and 20 per cent of the rockets were being successfully launched. The defects, however, were slowly eliminated; modifications continued until the middle of 1944 and it was not until the closing months of the war that launchings were completely successful.

To sum up: the early pioneers in rocket development were largely amateurs who experimented under hopelessly inadequate conditions in view of the magnitude of their ideas. Most of them were inspired by the classic book written by the mathematical student, Oberth. In a desperate war situation the German Army poured enormous resources into rocket development and, although at a crucial moment held up by one of Hitler's intuitions, this large team of scientists and technologists produced the first reliable and guided-rocket missile.

REFERENCES

1. Williams, B., and Epstein, S., *The Rocket Pioneers.*
2. Dornberger, W. R., *V2.*
3. Dornberger, W. R., 'European Rocketry After World War I', *Journal of the British Interplanetary Society*, Sept. 1954.
4. Gartmann, H., *The Men Behind the Space Rockets.*
5. Von Braun, W., 'Reminiscences of German Rocketry', *Journal of the British Interplanetary Society*, May-June 1956.

SAFETY RAZOR

KING GILLETTE conceived the idea of the modern safety razor at the end of the nineteenth century. He came from a family of inventors, but his formal technical training was slight. While employed as a travelling salesman he produced several minor inventions. He met William Painter, the inventor of the modern bottle-cap, and it was he who started Gillette on the search which resulted in the invention of the safety razor by suggesting that Gillette should try to invent something which the consumer can use, throw away, and buy again. According to Gillette's own account of his invention, the idea for a razor with a cheap disposable blade came in a 'flash' one morning in 1895 while he was shaving. He rushed out to purchase some 'pieces of brass, some steel ribbon used for clock springs, a small hand vise and some files' and built his first safety razor. He filed a patent application and the patent was issued in 1904.

The idea and a crude model did not ensure the commercial success of the invention. For six years Gillette sought for ways of making a cheap blade from sheet steel that would harden and temper suitably for taking a keen edge. He had almost no knowledge of steel, yet he felt confident that he would find a solution. The steel experts were not enthusiastic; from their experience they thought it would prove to be impossible. Gillette's friends thought the idea was a joke and refused to give him financial assistance. He remarked several years later:

> 'But I didn't know how to quit. If I had been technically trained I would have quit or probably would never have begun. I was a dreamer who believed in the "gold at the foot of the rainbow" promise, and continued in the path where

wise ones feared to tread, and that is the reason, and the only reason, why there is a Gillette razor today.'[1]

It took another ingenious inventor to solve the problem of making the cheap blade.

Gillette finally found men willing to risk capital. A business associate showed the razor to Henry Sachs, a Boston lamp manufacturer, who was impressed with it. In 1901 Sachs, his brother-in-law Jacob Heilborn, and an inventor William Nickerson, contributed $5000 and they organised the American Safety Razor Company. Sachs and Heilborn were fortunate enough to persuade Nickerson to study the mechanical problems. Although Nickerson had no previous knowledge or experience of sheet steel, he was a mechanical genius and had many inventions to his credit – such as the push-button mechanism that stops an elevator at the required floor, a device that prevents the premature opening of the elevator shaft door and a novel bulb-making process. But although a remarkable inventor, he was a poor businessman.

One of Nickerson's ideas was to make the razor handle heavy enough to facilitate accurate adjustment between the edge of the blade and the protecting guard. By 1902 he had determined the proper size, shape and thickness of the blade, found a process for hardening sheet steel, perfected a tempering process, set the dimensions for the blade cap and guard and designed the machinery to sharpen the processed steel. Nickerson, although a technician, used the 'cut and try' method of experimenting with outstanding success.

In the meantime, however, the manufacturing firm, now called the Gillette Safety Razor Company, had fallen badly into debt and the promoters grew impatient. Gillette persuaded an old friend, John J. Joyce, to give further backing to the enterprise. Gillette, a poor administrator, resigned as president of the company but returned later and remained president until 1910. Then he settled down in California where he devoted himself to the raising of fruit and to the devising of highly chimerical plans for the establishment of world economic unity.

The first blades were sold in 1904 when the company was still in financial straits; it showed its first profits in 1906 and after that sales increased at an enormous rate.

Individual inventors created the modern safety razor: King Gillette had the idea and built a crude model and Nickerson, by devices of a high order of ingenuity bordering closely on invention, converted it into the razor in use today.

REFERENCES

1. Baldwin, George B., 'The Invention of the Modern Safety Razor: A Case Study of Industrial Innovation', *Explorations in Entrepreneurial History*, vol. 4, no. 2.

2. *Gillette Blade*, June 1921.

SELF-WINDING WRIST-WATCH

THE idea of the self-winding watch can be traced back to the eighteenth century, if not earlier, when Abraham-Louis Perrelet, a Swiss, Abraham-Louis Bréguet, a Frenchman, and Louis Recordon, a Swiss settled in England, all produced pedometer pocket watches in which the mainspring was wound up by a small internal

[1] *Gillette Blade*, June 1921, p. 7.

weight swinging with the movement of the wearer. These watches remained curiosities; they were easily damaged, difficult to repair, bulky and expensive. Later, in the nineteenth century, a number of patents were taken out on self-winding watches, but these were still of the pedometer type and embodied no radical innovation.

The appearance of the now increasingly popular modern type of self-winding watch dates from the early twenties of this century. It was associated with the greater practicability of manufacturing small watches which in turn led to the practice of wearing the watch on the wrist where it is subject to more movement than when carried in the pocket. The crucial date is 1923 when John Harwood, an Englishman, took out on a self-winding movement a patent which appears to give him the right to be considered the real inventor in this field. Harwood is a watch-maker and repairer by profession; he completed his apprenticeship before the First World War. He conceived his original idea in 1922[1] without, apparently, knowing anything of the work on self-winding watches which had been done from the eighteenth century onwards. He was not attracted by the idea of perpetual motion or of saving the wearer the trouble of winding his watch, as some of the earlier inventors in this field had been. Harwood simply wished to produce a watch which had no openings where dirt might enter and which would be wound up regularly. As a repairer of watches he had been struck by the damage done by dirt, entering mainly through the external winding mechanism, and by the unsystematic habits of people in winding their watches. A self-contained watch would overcome these difficulties and would thus be more reliable and longer-lived.

Harwood made his first self-winding watch in 1922 and within a year had solved satisfactorily the problem of extracting enough power from an internal swinging weight to wind the main spring. At this point he applied for patents in Great Britain, Switzerland, France, Germany and America. The problem now was to get the idea accepted. No British firm could have made the watch, and Harwood, along with a Mr. Cutts, a business acquaintance whose name also appears on the original patent application, first paid a visit to Switzerland in the hope of persuading one or other of the Swiss watch manufacturers to produce the watch, presumably under licence. But at this time no Swiss firm was interested. Harwood and Cutts therefore, with British financial backers, formed the Harwood Self-Winding Watch Company which employed Swiss firms to make the watches and, by 1928, the Harwood watch was on sale in Great Britain. The 1930 industrial slump forced the business into liquidation but, between 1928 and 1931, about 30,000 watches were made. The patent was allowed to lapse in 1935.

Harwood's pioneering ideas, however, had set other minds to work. Some of the resulting efforts were commercially no more successful than Harwood's. But in 1930 the Swiss Rolex Watch Company produced its own automatic movement. The Harwood mechanism had embodied a weight swinging only through an arc; in the Rolex watch the weight could describe a complete circle – the so-called 'rotor' movement. With the lapsing of the Harwood patent, other Swiss manufacturers began to produce watches of this type. After 1939, however, the Rolex Company held the lead, for it possessed extensive manufacturing facilities, it continued to improve the design of its watch, and it further enhanced its efficacy by rendering it dust- and water-proof. Many other improvements in the self-winding watch

[1] Tradition has it that he invented the watch 'whilst in the trenches during the First World War'. This story is not correct.

have been made in the past fifteen years, but it remains true that a significant proportion of the self-winding watches now manufactured embody a movement of the Harwood type.

This is a case where an individual inventor, and one in a country where there was little or no watch-making, can claim priority. If there had been a vigorous watch-making industry in Great Britain, Harwood's invention might have proved a commercial success. As it was, the Swiss manufacturers were slow to see its possibilities. But when, after some years, the idea of the self-winding watch was taken up in Switzerland, the manufacturers there pushed forward its improvement with their customary energy and skill. So popular has the self-winding watch become that it now constitutes the larger part of the output of many Swiss factories.

REFERENCES

1. Chapuis, Alfred, and Jaquet, Eugène, *La Montre automatique ancienne* (*1770–1931*).
2. Harwood, John, 'The Birth of the Automatic Wrist Watch', *Journal Suisse d'Horlogerie et de Bijouterie*, May/June 1951.
3. *Revue Internationale de l'Horlogerie*, Apr. 1952.
4. Pipe, R. W., *The Automatic Watch.*
5. Interview with Mr. John Harwood.

SHELL MOULDING

THE invention of shell moulding is the first major change in foundry technology for a great many years. The shell mould is a simple device consisting of a thin mould of sand held together by phenolic resin. This type of moulding has at least two advantages over the conventional moulds: the castings have smoother surfaces and more accurate dimensions, which means reduced machining, and the moulds can be made on automatic machines without skilled labour. During the last few years there has been a tremendous growth in the industrial application of shell moulding, the automobile industry, for example, uses it extensively.[1] When the process was first introduced, its value was thought to be limited to moulding non-ferrous objects, but recent work has shown that ferrous metals can also be moulded in this way.

This revolutionary casting process was invented in 1941 by Johannes Croning, the proprietor of a Hamburg foundry. Like others in the foundry industry, he had for years been seeking a simple method of producing accurate castings; the well-known lost wax process was too expensive to be widely used. He discovered the

[1] The Ford Company was the first automobile company to appreciate the value of shell moulding. The story is an interesting one. Ford employed E. E. Ensign in 1939 in its Edison Institute. Ensign soon established himself as an expert on metallurgy. He sought for ways to use precision casting methods. He appears to have heard of Croning's invention even before the U.S. Government made their public announcement. Ensign informed the Valyes, two New York scientists, of this process and they immediately commenced independent experiments. The Ford Company did not keep their shell-moulding activity secret at this time because few shared Ensign's confidence in it. The Valyes claim to have turned out a practical mould before the Department of Commerce delivered their report on the Croning process in 1947. By late 1947 Ford produced feasible moulds; in 1948 they had a pilot line moulding 1000 exhaust valves a day.

answer to the problem, it would appear, by combining a modification of existing foundry practice with a process familiar in another industry. The use of liquid synthetic resins as the bonding agent in sand cores had begun some time before 1939 in Germany; for many years refractory and clay bricks and shapes had been produced by 'throwing' the clay into the mould. In the shell-moulding process, a powdered resin is mixed with the sand and the mixture 'dumped' into the heated metal pattern in a manner which is analogous to the 'throwing' of the clay; the mixture is, in fact, thrown on to the pattern. The mixture is next baked in an oven to harden it fully, the pattern is removed, the two halves of the mould joined together and it is ready to be filled with the molten metal.

Some authorities suggest that Croning started by using a liquid resin and pouring the resin-sand mixture on to a metal pattern, and that when he changed to using a powdered resin and a dry mixture thrown on to the pattern he obtained a quite unexpected improvement and thus established the basis of the process used today. It is probable that Croning received some assistance from the resin manufacturers over his use of powdered resins; he also received considerable assistance from the German Government, shell moulding being used to produce hand grenades and other metal objects before the end of the war.

When Allied technical teams were searching in Germany for developments made during the war, they discovered Croning's work and the U.S. Department of Commerce reported the findings in an F.I.A.T. Report in 1947. Litigation has developed over the patent rights, certain parties arguing that this government publication placed the invention in the public domain and free for all to use. Croning applied for patents in the United States and assigned his patent rights to Crown Casting Associates, a Boston organisation. The status of the patents has remained in doubt. In addition to Croning's patents, hundreds of improvement patents have been filed within the last five years. Both new and old firms in many countries are engaged in designing and developing machines for the shell-moulding process and it is now widely used, especially in the motor industry.

In this case a German foundry proprietor was responsible for a basic invention in a field in which many other firms and individuals had carried on research for a considerable period.

REFERENCES

1. Tindul, Roy W., 'Current Status of the Shell Molding Process', *The Foundry*, July 1952.
2. Dixon, M. C., and Bushnell, R. S., 'The "C" Process of Casting', *Foundry Trade Journal*, Mar. 26, 1953.
3. Du Mond, T. C., *Shell Molding and Shell Mold Castings*, 1954.
4. Bello, Francis, 'Plastics Remold the Foundry', *Fortune*, July 1952.
5. Ames, B. M., Donner, S. B., and Khan, N. A., 'Plastic Bonded Shell Molds', *The Foundry*, Aug. 1950.
6. Professor Piwowarsky, 'The "C" Process', *Foundry Trade Journal*, Feb. 19, 1948.
7. 'Synthetic Resins for Foundry Use', *Engineer*, July 18, 1952.

SILICONES

SILICONES are synthetic organic compounds of silicon, constructed of alternate atoms of silicon and oxygen, with organic groups such as methyl or phenyl attached to the silicon atoms. Silicones have unique qualities. They possess a relative constancy of properties over a far wider range of temperatures than any other organic material can withstand: consequently silicone oils, greases, resins and rubbers are used where high or low temperatures, or great variations in temperature, are experienced. Their good electrical properties enable them to be used in electrical insulation; combined with their heat-resistance, this gave them their first application, as a resin varnish binding fibre-glass insulation. The water-repellency and low surface tension of the fluids makes them ideally suited for use as water-repellent agents and mould release agents. All the variety of silicone materials are produced by small variation in the basic structure and the attached organic groups; consequently many more were discovered once the first industrial applications of silicones led to a general study of their nature.

Industrial silicones are the product of two lines of pure scientific research. The late Professor F. S. Kipping of Nottingham University made the major contribution to our knowledge of organo-silicon chemistry, devoting most of his career to its study, not because he saw any possibility of useful products being discovered, but because he wanted to discover how similar silicon was to carbon; when he had shown the similarity, he decided that he wanted to know more about the chemistry of silicon itself. This he considered could be best done by a study of the organic derivatives of silicon; the fifty-one papers he published on this subject between 1899 and 1941 provide the best source of knowledge of the chemistry of silicon. The commercial development of silicones, however, did not take place until knowledge of the fundamental mechanisms of polymerisation and the properties of macro-molecules was enlarged by the work of Professor H. Staudinger, Emil Fischer, W. H. Carothers and others between 1914 and the 1930's.

In the course of his work Kipping discovered in 1904 that organic groups could be attached to silicon by the Grignard reaction, which is still one of the two methods employed for the industrial production of silicones. At the time it eased his work and encouraged others to study the organic compounds of silicon thus produced. Among the products of some of his experiments were some glues and resins which he described as 'uninviting'. However he did make some investigation of them, which showed that they were large molecules formed by the union of small molecules, and brought out some fundamental facts on the nature of silicon compounds. But Kipping's main interest was in the study of pure compounds and not in those 'uninviting' glues and resins which were examples of the silicones now found so useful.

These commercially used silicones consist, as Kipping found in his glues, of large molecules. The work of Staudinger in the 1920's explained the chemistry of such substances, and he found out that materials of high molecular weight consisted of giant molecules or 'high polymers', which could be built up of smaller molecules. Carothers in America followed up Staudinger's work in the late 1920's by studying the principles underlying the preparation of high polymers and the laws governing the formation of various types. By the early 1930's there was, therefore, sufficient knowledge of the chemistry of high polymers and giant molecules to create certain materials of this type in the laboratory.

It was at this stage that interest in the commercial possibilities of silicones was aroused. The discovery of glass-clear plastics interested the research staff of the Corning Glass Works; Dr. E. C. Sullivan and Dr. W. C. Taylor wondered whether organo-silicon-oxygen derivatives might not also make glass-like plastics, or provide materials to extend the utility of glass. Dr. J. F. Hyde was engaged to investigate these 'silicones', several examples of which he was able to produce, including liquids and resins. He pointed out that some of these resins had a great resistance to heat. Their development for use in glass-fibre insulation was consequently begun. Dr. R. R. McGregor, the Corning Fellow of the Mellon Institute, Pittsburgh, also started work on the subject with his staff, investigating the chemistry of the materials and the problems of their production as well as their use for insulation.

At about the same time Hyde showed Dr. A. L. Marshall of the General Electric Company samples of silicone-bonded insulators. This resulted in General Electric itself starting research on silicones under Dr. W. I. Patnode and Dr. E. G. Rochow. The former discovered a silicone intermediate which made materials water-repellent, while the latter discovered a silicone insulation binder in 1938 and later a direct method of producing silicones which is simpler, safer and more economic than using the Grignard reaction, which involves the use of inflammable solvents. Most silicones are now produced by Rochow's direct process, but as it is unable to produce all types of silicone intermediates, some firms such as Dow Corning also use the Grignard process to make other intermediates.

The outbreak of war stimulated the development of silicones to the production phase. Not only were insulating materials wanted but the silicone fluids and greases were needed for uses, as in aircraft instruments, where their constant viscosity in high and low temperatures was valuable. By 1942 Corning was able to consider production and approached the Dow Chemical Company as having more experience of chemical production. The two companies formed the Dow-Corning Corporation the next year, and production was started as rapidly as possible to satisfy military needs. After the war, a wider variety of silicone products came on to the market as investigation showed how many types could be produced. In 1946 General Electric began production and other companies have since followed, including Midland Silicones and I.C.I. in Great Britain.

This is a case where pure scientific research has been found to have industrial applications. Kipping, Staudinger and Carothers provided much of the scientific knowledge. Corning Glass engaged Dr. Hyde to investigate the possibilities in a field in which no useful application has yet been found – the production of a plastic glass. But Hyde discovered that silicones could be used to make valuable fibreglass insulation material. Once a use for silicones had been found, other companies began to investigate them and other applications were rapidly discovered. Another discovery was that of Dr. Rochow at General Electric – the direct process for producing silicones. At no other point can an 'invention' be clearly discerned in this case. It is rather a case where pure scientific research has been applied to industrial uses by work more resembling development than invention.

REFERENCES

1. McGregor, R. R., *Silicones and their Uses*, 1954.

2. Letter from R. R. McGregor, Sept. 1955.

3. Bass, S. L., Hyde, J. F., Britton, E. C., and McGregor, R. R., 'Silicones – High Polymeric Substances', *Modern Plastics*, Nov. 1944.

4. Bass, S. L., 'Silicones – New Engineering Materials', *Chemistry and Industry*, Apr. 1947.

5. Emblem, H. G., and Sos, F. L., 'Silicon Organic Compounds', *Chemistry and Industry*, Dec. 1946.

6. Quail, F. J., 'Silicone Horizons', *Canadian Chemical Processing*, Sept. 1953.
7. Midland Silicones Ltd., *An Introduction to Silicones.*
8. Freeman, G. G., 'Silicones, An Introduction to their Chemistry and Applications', The Plastic Institute, Plastics Monograph no. C.9.
9. Rochow, E. G., *An Introduction to the Chemistry of Silicones*, 1951.

STAINLESS STEELS

STAINLESS steels are corrosion-resisting alloys composed chiefly of iron and chromium in certain crucial proportions with further elements added to produce desirable modifications. Alloys with 12–30 per cent chromium and 0·01–1·0 per cent carbon represent the basic stainless range, an addition of 7–35 per cent nickel covering a principal modification. They are usually divided into three groups, the distinction between the first two being less than that between the first two and the third.

(1) *martensitic*, characteristically containing from 12 to 17 per cent chromium and 0·10–1·0 per cent carbon, hardenable by heat and tough;
(2) *ferritic*, characteristically containing 15–30 per cent chromium with no nickel and a low carbon content, having excellent corrosion resistance but not hardenable; and
(3) *austenitic*, characterised by the indicated nickel content.

Alloys of iron and chromium had been produced by numerous investigators in the nineteenth century – Michael Faraday was responsible for several – but they were not true stainless steels, for their chromium content did not fall within the prescribed limits, and, indeed, the Englishman, Robert Hadfield, had reached the erroneous conclusion that chromium in the alloy impaired corrosion resistance. Few had seriously considered engineering alloys having chromium contents of 12–30 per cent because the high carbon content of their alloys obscured the engineering properties, and those who worked with chromium contents of less than 12 per cent had no opportunity of witnessing the 'stainless' quality which a critical amount of that metal confers. Late in the nineteenth century satisfactory methods for reducing carbon content were devised, but it was not until the early part of the present century that the importance of carbon content was recognised, when Carnot and Goutal showed that the corrosion resistance of the alloys increased when the carbon content was decreased.

The crucial discoveries originated in the twentieth century, but competent authorities differ greatly as to where the credit for them should properly lie. Harry Brearley, Elwood Haynes, B. Strauss and E. Maurer usually receive the main credit for stainless steel although at least one author has challenged their claims.[1]

In 1904 a French investigator, Léon Guillet, explored the mechanical and metallurgical characteristics of the low carbon iron-chromium alloys and published his findings in detail. He failed to notice the remarkable corrosion resistance of the alloys; nevertheless he produced and described the principal grades in both the martensitic and ferritic stainless steel groups. In 1909 another Frenchman, A. M. Portevin, published a study covering among others, the alloys investigated by

[1] Carl A. Zapffe, 'Who Discovered Stainless Steels?', *Iron Age*, Oct. 14, 1948, p. 120.

Guillet. Portevin also neglected to comment on their corrosion resistance, concentrating on their mechanical and metallurgical properties.[1] W. Giesen published an article in 1909 dealing with alloys, with the chromium and carbon proportions mentioned by Brearley in his United States patent of 1915.

Two Germans, P. Monnartz and W. Borchers, discovered and explained the anti-corrosion quality of the steels. Monnartz obtained a German patent on a stainless steel in 1911, and in the same year he wrote a paper on the corrosion resistance of carbon-free chromium steels.

Most authorities regard Harry Brearley in England and Elwood Haynes in the United States as the respective discoverers of the martensitic and ferritic alloys, although it might be more accurate to state that they were the first to visualise their uses and commercial value. Brearley, a self-taught metallurgist, was head of the research laboratory run jointly by the steel firms, John Brown & Co., Ltd., and Thomas Firth & Sons, Ltd., when he discovered his stainless alloy in 1912. Wishing to develop an anti-corrosive alloy for use in naval guns and predicting that iron-chromium alloys would serve, he made several casts. One made in the electric furnace contained 12·8 per cent chromium and 0·24 per cent carbon, and Brearley found that with a certain heat treatment the steel resisted corrosion. A complicated wrangle then arose between Brearley and Firth's. No interest had been shown in the steel by the military authorities for ordnance purposes. But Brearley had suggested its use for cutlery, and the company, apparently without consulting the inventor, had knives made from it by two cutlers, who reported it useless. Brearley, with steel obtained from Firth's, then had knives manufactured by Messrs. R. F. Mosley, and these proved successful. The production of the martensitic alloy thus began at Firth's in 1914 and at the Firth-Sterling mill near Pittsburgh in 1915; but their exploitation of the alloy was hampered because by this time Brearley had left them, carrying with him much of his 'know-how', to become works manager at Brown Bayley's steel works. In 1915 Brearley obtained an American patent, since Firth's had not considered this worth their while. In 1920 Brown Bayley's introduced commercially the ferritic alloys, a development of the martensitic.[2]

Elwood Haynes, also closely associated with the discovery of the martensitic steels, was an American individual inventor from Kokomo, Indiana. A pioneer inventor in the automobile industry, he had studied science and engineering at Johns Hopkins University. His interest in steels dated at least as far back as 1884 when he invented a process for producing tungsten chrome steel. While he was attempting to solve some of the metallurgical problems of the automobile industry, he discovered the 'stellite' alloys of cobalt, chromium and tungsten. These retained their hardness and toughness even under intense heat, and Haynes recognised their value as tool cutters. Additional experiments on tool-cutting alloys in 1911

[1] Portevin claimed that he noted and described the corrosion resistance of the stainless steels before 1911, worked out a way to make these steels for commercial use and filed a patent through the De Dion-Bouton Company, where he was chief engineer in the Chemical Department. A letter from Portevin is reported in *Iron Age*, June 30, 1949, p. 28.

[2] There was some very limited application of the low carbon variety around 1910. C. Dantsizen of the General Electric laboratories, in the period from 1910 to 1912, produced a low carbon chromium alloy which he used in incandescent bulb lead in wires. General Electric manufactured the wire and exploited it commercially in Christmas-tree bulbs. When the firm discovered a better wire, they dropped the alloy. Dr. Whitney, head of the General Electric Research Laboratory, continued the experiments and developed a chromium alloy for turbine blades.

and 1912 led him to the discovery of chromium-iron alloys which had remarkable corrosion resistance. He applied for a patent on one of them in 1912, but the Patent Office rejected his first application on the ground that these 'chromium-iron alloys are not new'. Haynes filed again, but shortly afterwards the Patent Office granted a patent to Brearley. In 1919 the Patent Office finally issued a patent to Haynes; his claims cover the 'production of a wrought metal article of manufacture having polished surfaces of the general character which is termed noble, in that such surfaces are incorrodible'. A syndicate composed of several large steelmakers in the United States bought the patents, and granted non-exclusive licences for the manufacture of the steel. The courts in the United States have upheld the Brearley and Haynes patents.

The third group of alloys, the austenitic, were discovered in Germany by Edward Maurer and Benno Strauss, of the research department of Krupps. After a long investigation of the chromium nickel steels, patents were applied for and Krupps produced such a steel in 1912.

Zapffe maintains that Guillet and Giesen produced the austenitic group but, although it is true that they described steels falling within the austenitic range, they made no mention of their corrosion resistance.

The story of stainless steel is, therefore, very complex. In the first decade of the century, metallurgists produced alloys in the stainless steel group but, until Monnartz and Borchers, they failed to recognise their outstanding property, the power of resisting corrosion. Two inventors, Brearley and Haynes, the one working in a research laboratory, though under conditions resembling those of an individual inventor, and the other an individual inventor, recognised the great virtues of stainless steel and were largely responsible for its general acceptance. Maurer and Strauss, working in Krupps, contributed much to the discovery and commercialisation of the austenitic group of steels.

REFERENCES

1. American Stainless Steel Co. *v.* Ludlum Steel Co., 290 Fed. 103 (C.A. 2d, 1923).
2. American Stainless Steel Co. *v.* Rustless Iron Corp., 2 F. Supp. 742 (D.C. Maryland, 1933).
3. American Stainless Steel Co. *v.* Rustless Iron Corp., 71 F. 2d 404 (C.A. 4th, 1934).
4. Zapffe, Carl A., *Stainless Steels*, 1949.
5. Zapffe, Carl A., 'Who Discovered Stainless Steels?', *Iron Age*, Oct. 14, 1948.
6. Monypenny, J. H. G., *Stainless Iron and Steel*, 2nd edition, 1931.
7. Thum, Ernest G., *The Book of Stainless Steel*, 1935.
8. 'Dr. Portevin Discusses Early Stainless History', *Iron Age*, June 30, 1949.
9. Brearley, H., *Knotted String: Autobiography of a Steelmaker*, 1941.

STREPTOMYCIN

STREPTOMYCIN, a useful antibiotic, is derived from a microbe found among a family of filamentous bacteria called actinomycetes, which have been shown to be extremely active against other organisms. Lieske demonstrated in 1921 that certain actinomycetes destroy some bacteria and inhibit the growth of others; Gratia, Dath and Rosenthal showed in 1925 that cultures of actinomycetes, which were designated as Streptothrix, can dissolve both living and dead bacterial cells. Nakhimovakaia and other Russian investigators examined the actinomycetes in the soil for their bacteria-killing properties in 1937.

The main credit for discovering streptomycin belongs to Dr. Selman A. Waksman, who migrated to the United States before the First World War and studied bacteriology and biochemistry at the New Jersey State Agricultural Experiment Station at Rutgers University, and later at the University of California. He returned to Rutgers in 1918, and gradually became recognised as an authority on micro-organisms of the soil. Before 1939 Waksman had not been particularly interested in the antimicrobial properties of the micro-organisms, but one of his students, R. Dubos, had isolated a powerful antibacterial agent, tyrothricin, from a soil organism. Waksman, struck by this finding, reasoned that the soil must contain numerous other organisms producing antimicrobial substances. His knowledge of the widespread distribution of the actinomycetes and their antagonism to other organisms prompted him to investigate them.

With the help of his students and the collaboration of the chemists and biologists from the near-by Merck & Company laboratories, Waksman started a systematic search, examining ten thousand soil cultures for their ability to inhibit bacterial growth and to produce antimicrobial substances, which he designated in 1942 'antibiotics'. The spectacular rise in the use of penicillin stimulated his drive to find an antibiotic effective against the diseases which penicillin did not control. In 1943 a strain of an actinomyces called *Streptomyces griseus* was discovered, and streptomycin was extracted from the culture of this organism. Waksman had, indeed, isolated the same organism from the soil in 1915 but had not recognised its remarkable antibiotic possibilities. By 1943, however, he had developed an extensive technique for the testing of antibiotics. Streptomycin was found to be active against numerous bacteria not sensitive to penicillin, the most important of which was the tubercle bacillus. The Merck Institute for Therapeutic Research immediately undertook additional tests which established its effectiveness against a variety of diseases.

Pilot plant production began in 1944 and the Merck Company developed a deep-vat fermentation process for the mass production of streptomycin. A large number of firms in many countries now produce this antibiotic by the submerged vat process.

Here the discovery was made by a university professor who was an authority on the micro-organisms of the soil. Although he received help from his assistants, from two philanthropic organisations and from an industrial laboratory, he controlled the experiments and takes the major credit for the initial discovery.

REFERENCES

1. Major, R. T., 'Cooperation of Science and Industry in the Development of the Antibiotics', *Chemical and Engineering News*, Oct. 25, 1948.

2. Waksman, Selman A., 'Streptomycin: Background, Isolation, Properties and Utilization', *Science*, Sept. 4, 1953.

3. *Antibiotics*, Oxford Medical Publication, 1949. H. W. Florey, Historical Introduction.

SULZER LOOM

THE orthodox type of loom for weaving cotton, wool and other fibres is a relatively primitive device which, despite many efforts, has long resisted fundamental improvements. Some of its highly stressed parts are made of wood, leather and other materials subject to rapid wear which results in the need for frequent adjustment and replacement of parts and difficulties in maintaining uniformity in the product. The vital part of the mechanism is the shuttle which travels across the loom inserting transversely threads one at a time (the weft) between two sets of longitudinal threads (the warp) held apart to allow the passage of the weft. In the orthodox machine the shuttle carries its own thread internally; it has therefore to be as large as possible but, even at its maximum practicable size, it can carry only a limited supply of weft. The shuttle, therefore, has to be replenished frequently; its movement is difficult to control, and if, as sometimes happens, it is ejected from the machine, it may be dangerous. Further, the shuttle has a discontinuous movement; it is hurled across the loom at speed and then brought to rest suddenly. This uses up energy and creates a great deal of noise.

Apart from other entirely different mechanical methods of enlacing fibres such as knitting, very many ingenious devices have been suggested for overcoming the serious drawbacks of the traditional loom.[1] No one of them can be said as yet to have established itself as a serious competitor to the orthodox machine; but by far the most promising invention, in a field where practicable new ideas have been rare, is that embodied in the Sulzer loom. Its essential novelty is that the shuttle is very much smaller and only about one-tenth the weight of the orthodox shuttle. It does not carry its own weft internally but picks it up at the beginning and drops it at the end of its run across the loom. At the end of its run the shuttle drops down and is conveyed back to its starting position. Each loom has a group of shuttles which shoot in turn across the loom, like a series of small metal projectiles, carrying a single thread. As a consequence it is claimed that the Sulzer loom can produce cloth more quickly, since the shuttle flies faster and its internal replenishment with weft is no longer required; that the cloth woven is much more uniform and that costs, particularly on mass production lines, are reduced for, although the new loom is more expensive than the older types, labour and other charges are smaller.

The early patents on this loom were taken out by Rudolf Rossman, who, though not originally a textile engineer, was employed up to 1928 by the Deutsche Wolle Company of Grünberg, Silesia, on development work connected with the conventional loom. He published a treatise in which he stated that the scope for the ordinary loom was narrow and that it would be necessary to look round for new methods of weaving. After the failure of his company he collaborated with a mill-owner in constructing a first trial machine. Rossman has also taken out patents in many other fields.

He appears to have tried unsuccessfully to interest various textile machinery manufacturers in his ideas about the new loom. In 1931 his patents were taken over by a financial consortium in Switzerland (Textil-Finanz A.G.), for which Sulzer Brothers of Winterthur carried out development. Rossman was in charge of the work at Sulzers from the end of 1934 until 1936, when he left the firm. The

[1] For an account of these see J. J. Vincent, 'The Advent of the Shuttleless Loom', *Times Review of Industry*, July 1953.

models built up to that time were not commercially practicable. Sulzer Bros., although one of the oldest and most distinguished Swiss industrial firms, with a high repute in the design and manufacture of steam and diesel engines, gas turbines, turbo compressors, centrifugal and heat pumps and other branches of engineering, had not previously engaged in the manufacture of textile machinery. They finally took the full financial responsibility for what was, for them, a new branch of development, partly because one of their outstanding engineers was greatly interested in the new machine; partly because the new loom, as sharply contrasted with the older types, clearly would have to be a precision product with many of its crucial parts involving accurate machine work of a kind in which Sulzer Bros. were adept; partly because they had already invested considerable sums in the new machine and success became almost a matter of prestige and partly, perhaps, because they were at the time hardly aware of all the difficulties which they would encounter.

The development, indeed, proved to be difficult and prolonged, involving the adoption of new principles and designs and leading to a considerable extension of the patents covering the machine. But, by 1945, progress was sufficient to enable Sulzers to grant a licence to an American Company, Warner and Swasey, which took up manufacture of the machines for cotton and wool weaving mills. In 1950 Sulzers equipped a factory for the series manufacture of the looms.

The Sulzer loom is a case where an original idea was conceived of by an individual inventor, where the firms in the textile machinery manufacturing industry took little interest in the invention and where the development, which proved to be prolonged, was undertaken by an enterprising firm outside the industry.

REFERENCES

1. Vincent, J. J., 'The Advent of the Shuttleless Loom', *Times Review of Industry*, July 1953.
2. Communications with Sulzer Bros. of Winterthur.

SYNTHETIC DETERGENTS

THE history of the appearance and the growing use of detergents can be summarised as follows. In the First World War the Germans produced some poor general substitutes for soap when soap was virtually unobtainable (the Nekals). Later, powdered fine-wash detergents only suitable for wool and silk were produced (as, for example, Dreft). Next, powdered or liquid dish-washing substances appeared, often based on Teepol in Great Britain. More recently heavy-duty general-purpose detergent powders based on alkylaryl sulphonates and complex phosphates have been marketed (Surf, Daz, etc.). Other types for other purposes are on the point of introduction or may be expected in the near future, such as enzyme-containing detergents, synthetic detergents for general toilet purposes, perhaps even in toilet bar form.

The synthetic soapless detergents have the merit that they do not form in hard water an insoluble scum which adheres to fabrics. This is of especial importance in extile finishing; the need in that industry for a cleansing and penetrating agent which did not suffer from the deficiencies of soap was so great that synthetic detergents were first used there, even when they were fairly costly. As the cost was reduced by improved methods of manufacture and by the discovery of new varieties obtainable from cheaper raw materials, and as their field of action was widened by the use of various 'builders', the use of synthetic detergents spread until today they are replacing soap in almost all its uses.

Since the early nineteenth century the structure of soaps has been known but it was not until the 1920's that it became commercially practicable to obtain more satisfactory detergents by modifying the structure of soaps or by the synthesis of compounds having related structures. The fact that many of these compounds can be prepared from raw materials other than the natural oils and fats has been of considerable importance in their subsequent development.

The first observations of the soap-like properties of non-soapy substances were made by the German academic chemist, Krafft, in 1886 and 1896. The American inventor, Twitchell, prepared other synthetic detergents as fat-splitting catalysts in 1898 and 1900. In 1913 a Belgian academic chemist, A. Reychler, found that the long-chain alkane sulphonates were good detergents and were more stable than soaps to acid conditions. In 1914 the British scientists, McBain and Martin, investigated the washing powers of alkane sulphonates; although these proved to be too expensive to be commercially produced, the work of these academic scientists showed that a molecule with a long hydrocarbon chain and ending in a solubilising group other than the carboxylic group, such as the sulphonate, would possess detergent properties but be free of soap's deficiencies.

The first attempt to market a synthetic detergent was made in Germany during the First World War when natural fats were almost unobtainable. This was Nekal, invented by Gunther and Hetzer of the Badische Anilin- und Soda-Fabrik in 1917. This substance, however, did not give satisfactory detergent properties but was later marketed by I.G. Farbenindustrie as a wetting agent for the textile industry. Much work was done in Germany during the 1920's by companies concerned with the supply of wetting agents to the textile industry. Their main object was to find a replacement for soap in textile processing. Their first successes were achieved by using the natural fatty acids as the raw material and modifying the carboxylic group to the sulphuric acid ester configuration, which could be done by converting the fatty acids to fatty alcohols and treating these with sulphuric acid. The fatty alcohols could be produced on a laboratory scale by the sodium reduction process invented in 1903 by Bouveault and Blanc; the invention of processes for the catalytic hydrogenation of the fatty acids into fatty alcohols in 1928–29 by Dr. W. Normann of H. Th. Boehme and by Dr. Walter Schrauth of Deutsche Hydrierwerke, cheapened these detergents and made their commercial introduction possible. Dr. H. Bertsch of H. Th. Boehme and Dr. Walter Schrauth were chiefly responsible for the development of the sulphated fatty alcohols. A somewhat different approach was taken by Daimler and Platz of I.G. Farbenindustrie, leading to their introduction of the 'Igepons' about 1930.

The sulphated fatty alcohols were good detergents for washing wool, but unsatisfactory for cotton. The 'Igepons' were quite good detergents and were supplied by I.G. Farbenindustrie to Unilever during the 1930's for certain uses. The American rights to the sulphated fatty alcohols were bought by the American Hyalsol Corporation which granted licences to du Pont, Proctor and Gamble and their jointly owned subsidiary, the Gardinol Corporation. They developed and marketed the

German Gardinol detergents and later introduced 'Dreft' for the washing of wool and silk.

The introduction of general purpose synthetic detergents, however, was dependent on the discovery that the addition of complex phosphates greatly increases the cleansing powers of detergent materials, makes them suitable for washing cotton as well as wool and reduces their costs. A Swiss, C. A. Agthe, had recognised and patented the use of certain phosphates as detergents and detergent-auxiliaries, though it is doubtful whether any use was made of them. The work on the use of complex phosphates in detergents was done by American, German and other European companies during the 1930's.

Another major development made in the 1930's was the production of detergents from petroleum-based raw materials and the bulk of modern synthetic detergents are now made from such materials. Although chemical companies both in Germany and the United States played a part in this work, the leading role in exploiting this class of detergent was taken by the oil companies in the late 1930's.

Another innovation which has become universally adopted since 1945 is the use of fluorescent brightening agents. These are substances which, when added to detergents in minute quantities, are taken up by clothes during the washing and make them appear brighter by converting the ultra-violet rays present in sunlight into visible blue rays. A German academic chemist, Krais, first noticed this phenomenon with natural fluorescent substances in 1929 and the first patents for the use of fluorescent materials in detergents were taken out by I.G. Farbenindustrie about 1941.

All the detergents so far mentioned belong to one chemical class, the anionic detergents, but there are other chemical classes of detergents less commonly used. One is the class of non-ionic detergents which possess good detergent properties but give little lather. The most important group of these was discovered by Schöller and Wittwer of I.G. Farbenindustrie in 1930 – in England 'Stergene' is an example of this class of detergent. Another class is that of the cationic detergents which are relatively poor washing agents but have special qualities; in particular many of them are germicidal. Krafft observed the soap-like qualities of such substances in 1896 and the first commercial production seems to have been by I.C.I. in 1933. In the U.S.A. at present some 79 per cent of production is of anionic detergents, 20 per cent of non-ionic and 1 per cent of cationic.

The invention of synthetic detergents is a case where academic research by German, Belgian and British chemists had shown that various classes of compounds having certain common structural features had useful detergent qualities, and industry then had to discover which compounds could be produced as a commercial proposition. The industrial work in the early stages was mainly done by German chemical companies experienced in the manufacture of synthetic dyestuffs and by companies producing wetting agents for the textile industry. The large soap manufacturers later took up the work.

REFERENCES

1. Couleru, M., 'Alcools gras et alcools gras sulfonés', Dix-Septième Congrès de Chimie Industrielle, Paris, Sept.–Oct. 1937. *Comptes-Rendus*, II.

2. Briscoe, M., 'The Fatty Alcohols and their Sulphonated Products', *The Industrial Chemist*, Feb. 1932.

3. Braybrook, F. H., 'The Development of Synthetic Detergents and Future Trends', *Chemistry and Industry*, June 26, 1948.

4. Kastens, M. L., and Peddicord, H., 'Alcohols by Sodium Reaction', *Industrial and Engineering Chemistry*, 41, 1949.

5. Reychler, A., 'L'Acide cétylsulfonique', Société Chimique de Belgique, *Bulletin*, année 27, 1913.

6. Reychler, A., 'Les Propriétés physico-chimiques de l'acide cétylsulfonique et du Cétylsulfonate de sodium', Société Chimique de Belgique, *Bulletin*, année 27, 1913.

7. Schwartz, A. M., and Perry, J. W., *Surface Active Agents*, 1949.

8. Lindner, K., *Textilhilfsmittel und Waschrohstoffe*, 1954.

TELEVISION

TELEVISION does not represent a single invention; discoveries basic to it appeared from several sources although a relatively small number of devoted men made it practicable.

It developed along two widely varying lines: the mechanical system and the electronic. Although at first mechanical television looked most promising, it ultimately proved to be a technical cul-de-sac. John L. Baird, the Scot, and C. F. Jenkins, the American, were the pioneers among the individual inventors, while Herbert Ives of the Bell Telephone Laboratories and Ernst Alexanderson of General Electric were the chief corporate inventors who made contributions to this system.

The roots of electronic television are found in scientific work on photo-electric cells and cathode-ray tubes. Most of the early television pioneers experimented with light-sensitive selenium cells which were unsatisfactory for dealing with moving objects because they reacted very slowly to variations in light; what was needed was a cell that would respond instantly to changes in light. The solution came from Germany in 1905 when two scientists, Julius Elster and Hans Geitel of the Gymnasium at Wolfenbüttel, perfected a much improved photo-electric cell based on an earlier discovery by the famous scientist, Hertz. Another important step was the development by Ferdinand Braun, of the University of Strasbourg, of the cathode-ray oscilloscope.[1] Cathode-ray tubes became available commercially, as a result, in 1897.

Boris Rosing, a professor at the St. Petersburg Technological Institute, first thought of using the Braun tube for the reception of images. In 1907 he proposed a system of remote electric vision in which a mechanical transmitter was combined with a cathode-ray tube as the receiver. At about the same time A. A. Campbell-Swinton, an English scientist, also proposed the use of cathode-ray tubes both for the transmission and reception of images. He further elaborated his ideas in 1911 and again after 1920, but they were never developed into a practical form.

The modern electronic system derives from the work in America of Vladimir Zworykin of the Radio Corporation of America and of Philo Farnsworth, an individual inventor. The systems of the two inventors were developed in the United States by R.C.A. and by Farnsworth independently, while Zworykin's system was also developed in Great Britain by Electrical and Musical Industries Ltd., who worked independently of R.C.A.

[1] The early work on cathode-ray tubes was done in the nineteenth century by Hittorf, Crookes and other scientists.

Zworykin studied electrical engineering in St. Petersburg and his interest in television began in 1910 when, as a pupil of Rosing, he worked on Rosing's scheme for a cathode-ray receiver. The work was dropped in 1912 when the limitations of the mechanical transmitting scanners were appreciated. It was already recognised that a complete cathode-ray system provided the answer, but no experimental work was started and the idea lay dormant until, after going to America in 1919, Zworykin joined the Westinghouse Company. His major difficulty centred on the electronic camera tube. He, like Campbell-Swinton, had conceived of the idea of charge storage by 1919. Zworykin lacked the necessary funds to carry his ideas forward into practical form at the time he conceived them, and it was several years before he could concentrate on their elaboration. Westinghouse took him on to their research staff but their laboratory devoted itself mainly to radio research, and, since Zworykin was given no freedom to pursue his ideas on television, he resigned to join a development company in Kansas. Returning to Westinghouse in 1923, he drew up an agreement whereby he retained the rights to the television inventions he had disclosed in 1919, while Westinghouse acquired the exclusive option to purchase the patents at a later date. Westinghouse now allowed him to start television experiments but, believing that television was a long-term gamble and wanting immediate results, the firm transferred him to research on photoelectric cells. In 1923 he had applied for a patent on the revolutionary Iconoscope, the device that transmits television images quickly and effectively; he thus removed a formidable obstacle from the development of commercial television. David Sarnoff, then the vice-president of R.C.A.,[1] became interested in Zworykin's progress in television, assigned four or five men to assist him, and steps were taken to improve his patent position. Zworykin recognised that a tremendous volume of development work remained to be done on circuits and synchronisation and he concentrated on these problems. When in 1930 research in the radio field was transferred from General Electric and Westinghouse to R.C.A., Zworykin moved to the R.C.A. Laboratories. By 1939 Zworykin and his group had substantially increased the clarity of the picture, developed a radio relay system for television and conducted extensive field tests. It is said that in this period R.C.A. spent over $2½ million on television research and advanced development. The United States Government authorised limited commercialisation of television in 1940.

Philo Farnsworth was essentially an individual inventor who, though fortunate enough to find substantial financial backing, always retained his autonomy in research. Largely self taught, he appears at an early age to have conceived of a completely electronic system. He was of the type which prefers to work on a small scale with relatively simple equipment. Working in laboratories in Los Angeles and later in San Francisco he was able to demonstrate a complete electronic system in 1927, when he filed his first patent application, including his image dissector tube, which constitutes his most important inventive contribution. After long-drawn-out

[1] R.C.A. was organised by General Electric in order to enter the international wireless field. General Electric purchased the American Marconi Company and the assets of this firm were transferred to the newly organised R.C.A. R.C.A. and A.T. and T. worked out a royalty-free patent cross-licensing agreement involving all current and future radio patents In 1921 Westinghouse worked out cross-licensing arrangements with R.C.A., and became a member of the radio group. These firms kept in close contact with respect to radio research developments. R.C.A. set up their own technical centre in New York in 1924. In 1930 General Electric and Westinghouse transferred most of their radio engineering development departments to R.C.A.

television

I. Schonberg

V. K. Zworykin

J. R. Whinfield: 'Terylene'

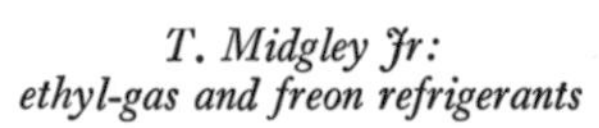

T. Midgley Jr:
ethyl-gas and freon refrigerants

patent interference proceedings, Farnsworth and Zworykin each received basic patents on their different systems of television transmission.

The Philco Corporation acquired an interest in the Farnsworth inventions, and for two years, until they became dubious about the prospects of the early commercialisation of television, they financed Farnsworth's research. By 1935 Farnsworth's system was showing much improved results but his financial supporters were becoming impatient for results (by 1938 over $1 million had been spent on development work) and, after seeking unsuccessfully to sell the patent rights to R.C.A. and Paramount, they formed the Farnsworth Television and Radio Corporation for the manufacture of television sets. A cross-licensing agreement on the Farnsworth patents had been made with the American Telephone and Telegraph Company in 1937; licensing agreements were made with the Baird television company in Britain, with the Fernseh A.G. in Germany in 1934, and with R.C.A. in 1939. Farnsworth, who preferred to carry on research rather than to assist in running a business, withdrew from active participation in the firm in 1940 and set up a laboratory in his home in Maine.

Although the basic inventions for the modern electronic system were made in the United States, it was first employed for general broadcasting in England in 1936. Baird's work with the mechanical system had indeed resulted in crude television broadcasts after 1929, but Electrical and Musical Industries Ltd. had realised the superiority of Zworykin's electronic system. This firm placed I. Shoenberg in charge of a television research team, which included A. D. Blumlein and P. W. Willans, who were to make important contributions. Working independently of R.C.A., this team produced a version of the Iconoscope, known as the Emitron, and other improved equipment, and E.M.I. was able in 1935 to offer to the B.B.C. a system with which regular broadcasting commenced shortly after. In 1936 an exchange of patent licences and technical information was made between E.M.I. and R.C.A.

Although many other inventions went into the first commercial television set, the important jumps in the field have been given here. The crucial inventions came from several directions. As for the ultimately abortive mechanical systems, individual inventors, such as Baird and Jenkins, fared as well as Ives and Alexanderson, of the large corporations. In the perfection of electronic television, individual inventors contributed a substantial number of the basic ideas. Rosing, early in the century, and Farnsworth later on, produced several ideas that are essential to the electronic television art. Farnsworth's almost single-handed battle to develop a practical television system before the R.C.A. may have driven the latter to increase the tempo of their research – just as the achievements of Baird with the mechanical system may have hastened the development of the electronic system and the beginning of television broadcasting in England. The progress made by R.C.A. is largely attributed to the genius of Zworykin. He inspired the original research and contributed most of the important ideas, some of which he conceived of even before he went to Westinghouse and R.C.A. The advanced development of the electronic system required many technicians, and here R.C.A. possessed a definite advantage over Farnsworth and his small staff. The British firm, E.M.I., deserve considerable credit for their work which led, in 1936, to the setting up in Great Britain of the first electronic television system in the world.

REFERENCES

1. Maclaurin, W. R., *Invention and Innovation in the Radio Industry*, 1949.
2. T.N.E.C. Hearings, Part III, 1939, Farnsworth testimony.

3. Moseley, Sydney, *John Baird.*

4. Garratt, G. R. M., and Mumford, A. H., 'The History of T.V.', in *Proceedings of Institution of Electrical Engineers*, Part III A, *Television*, vol. 99, 1952.

5. Everson, G., *The Story of Television – The Life of Philo T. Farnsworth.*

'TERYLENE' POLYESTER FIBRE

TERYLENE has already established itself as one of the great synthetic fibres possessing qualities such as toughness, resistance to abrasion and to sunlight, and resiliency (particularly in the form of staple fibre) which renders it the equal of, and in some ways superior to, Nylon. It was discovered in March 1941 by J. R. Whinfield and J. T. Dickson,[1] who, at the time, were research workers at the Calico Printers' Association. The crucial idea came to the inventors from a study of the work of W. H. Carothers, the inventor of Nylon. As explained elsewhere, the outstanding work of Carothers on condensation polymers opened up vast new fields to the polymer chemist. Carothers had at first devoted his efforts to the production of fibres from the polyesters but, finding them to possess an impracticably low melting point, he turned his attention to the polyamides and discovered Nylon, the first synthetic fibre. Whinfield and Dickson succeeded where Carothers had failed; they discovered a polyester with a high melting point, and, by good fortune, the fibre possessed other important qualities which could not have been predicted but were discovered later in the process of development.

Whinfield was a Cambridge graduate in chemistry who throughout his career had been much interested in fibres. On leaving the University he worked, to gain experience, in the small laboratory of C. F. Cross, a chemical consultant who himself is famous as the discoverer of the 'viscose reaction' for the production of Rayon. Subsequently Whinfield joined the research organisation of the Calico Printers' Association, a firm which had no direct interest in new fibres and in which Whinfield was for a number of years engaged in other branches of chemical research. His preoccupation with fibres, however, persisted. He followed with enthusiasm the early scientific papers and later practical achievements of Carothers, and, as early as 1935, was pressing upon his firm the case for research into the synthetic fibres. Finally in 1940 it became possible for him to devote some time, along with Dickson and a staff of two or three, to this branch of chemical research. Whinfield has himself described how an invention, formed slowly by long contemplation on the subject, finally led to success:

> 'Research is directed by knowledge, but it is enlivened by an element of unexpectedness. Knowledge suggested that polyethylene terephthalate would prove to be a fibre-forming polymer. Knowledge also suggested that in some respects it would differ from the aliphatic polyesters, but provided no real clue as to what the nature of this difference would be. This could only be settled by actual experiment.'[2]

The inventors were intrigued by the relation between crystallinity and molecular

[1] The names of W. K. Birtwhistle and G. G. Ritchie, who worked with Whinfield and Dickson, also merit mention.

[2] J. R. Whinfield, 'The Development of Terylene', *Textile Research Journal*, May 1953.

symmetry of condensation polymers and their chosen line of experimentation proved to be a happy one. By the first half of 1941 polyethylene terephthalate, the crude raw material, had been produced on a small scale and, using the simplest and cheapest techniques, as in the case of Nylon, crude fibres had been drawn from the polymer, fibres which were amenable to the subsequent process of 'cold-drawing'. It was also established, somewhat unexpectedly in view of their chemical structure, that these fibres were resistant to hydrolysis and had a high melting-point – qualities in the absence of which the fibre would have had no commercial potentialities.

Whinfield and Dickson had submitted a patent application in July 1941 but, owing to the war, the publication of the relevant specifications was delayed until June 1946 and the patent was only finally sealed in September 1946. Workers in the du Pont laboratories in the United States seem to have become aware, in 1944, of the success of the British scientists in the crucial foundation phase of the work and to have devoted attention to the development stage of the synthesis of the polymer. The priority of the British effort is, however, beyond dispute and in 1946 du Pont purchased from the Calico Printers' Association (to which company Whinfield and Dickson has assigned their rights) a licence for the development of Terylene for the American market. When the United States patent on Terylene was finally issued in 1949, the patent carried with it an assignment to du Pont.

The invention stage of Terylene, virtually complete by 1941, reveals four outstanding features. First, the idea was fundamentally a simple one but it came to inventors who had pondered long over the subject. Second, although by recourse to rough chemical analogies there seemed to be a reasonable presumption that fibres based upon the chosen chemical compound would have a minimum of the qualities essential for commercial purposes, even these qualities had to be established empirically and other favourable characteristics of the fibre, the scientific reasons for which are still not completely understood, came as unpredictable windfalls. Third, the invention stage was carried through with the simplest devices in a research laboratory of modest size of a firm with no direct interest in this branch of research by inventors who were able to devote only a limited amount of time to the task. And fourth, the inventors reached their results long before du Pont, the firm in which Nylon had been discovered and which had very much larger resources for research.

The principal stages in the subsequent development of the fibre may be briefly summarised. In the making of Terylene two raw materials were needed in quantity – ethylene glycol and terephthalic acid. The first was already in commercial production. The second was a mere laboratory curiosity since there had never before been a commercial demand for it. It was, moreover, needed in a particularly pure form. But there seemed to be good reasons to suppose that some tolerably inexpensive method could be devised to produce it on a large scale, and this proved to be the case. The problem at the next stage, the production of the polymer, was the type familiar in development: how to make a given compound with known chemical procedures and, although this involved hard and intensive work, the task was satisfactorily completed. At the third stage, the turning of the polymer into yarn and fibre, what was already known of the handling of Nylon contributed to a practicable solution.

The Calico Printers' Association decided that they were not in a position to undertake the tasks of developing Terylene and they sold their rights to Imperial

Chemical Industries and, so far as the American market was concerned, to du Pont. During the war Whinfield had joined the Ministry of Supply and he subsequently became an employee of I.C.I. The British Ministry of Supply took some interest in Terylene as a possible fibre for use in war time and preparation of Terylene on a small scale was carried out in the Chemical Research Laboratory of the Department of Scientific and Industrial Research. But the period of intensive development began after the war when I.C.I. and du Pont pursued this work independently. I.C.I., after running two pilot plants for several years, finally opened their large Wilton plant in 1955 with an annual capacity of 5000 tons. The costs have been substantial. Whereas the small textile laboratory in which Terylene was invented was costing less than £20,000 per annum, the cost of development and pilot plants has amounted to about £4,000,000 and the capital investment called for in the first major stage of commercial production to about £15,000,000. In the United States, du Pont put the fibre on the market first under the trade name of 'Fibre V' and later, 'Dacron'. The fibre first became available in the United States on an experimental basis in 1950 and the first full-scale plant began production in March 1953.

REFERENCES

1. Whinfield, J. R., 'The Development of Terylene', *Textile Research Journal*, May 1953.
2. Whinfield, J. R., 'Fibres from Aromatic Polyesters', *Endeavour*, 11, 29 (1952).
3. Whinfield, J. R., 'Chemistry of Terylene', *Nature*, 158, 930 (1946).
4. Osborne, W. F., 'Terylene is Here', *I.C.I. Magazine*, May 1955.
5. 'Polyester Fibres', *Encyclopedia of Chemical Technology*, vol. 13.
6. Izard, E. F., 'Scientific Success: Story of Polyethylene Terephthalate', *Chemical and Engineering News*, Sept. 1954.
7. Hill, Rowland, 'Polyesters and Terylene Fibre', *Journal of Royal Institute of Chemistry*, Jan. 1955.

TETRAETHYL LEAD

THE discovery of ethylised fuels provided a means of increasing the anti-knock rating of petrols and so of improving the operational efficiency of motor-car engines. The story of this discovery begins in 1912 when Charles Kettering, then of the Dayton Engineering Laboratories Company, became interested in the problem of 'knock'. Along with Thomas Midgley, a graduate mechanical engineer from Cornell, whom he had taken into his employment, Kettering set out in search of a cure. After eliminating the possibility that the source of the spark was the cause of the 'knock', the inventors had the idea that the remedy might be found by adding something to the fuel and that the colour of this additive might be important. They mixed iodine with the fuel because it was the only material with a bright colour available to them, and the knock disappeared. Still pursuing the idea of colour, they next experimented with aniline dyes but here the results were negative.

These investigations were stopped by the First World War but were resumed in 1919. In an effort to interest other scientists, Kettering published his results up to date and consulted large oil and chemical firms. Late in 1919 General Motors purchased Kettering's interests and thereafter he and Midgley worked in the research laboratories of that firm. Midgley, with a small staff, examined the merits of certain aniline compounds as fuel additives; but these compounds were available only in very small quantities and had a repulsive smell. At this stage, but for Kettering's pertinacity, the whole investigation might well have been dropped. He chanced to read of the work of Professor Victor Lenher, of the University of Wisconsin, who believed he had discovered a universal solvent, selenium oxychloride, and Kettering thought that this was one material that Midgley had not tried. Lenher helped Midgley to prepare and test the compound. The staff at General Motors[1] held out little hope of success for this material because it contained oxygen and chlorine, both of which were known to increase knock, but to their surprise the solvent greatly decreased it. They concluded that selenium, a rare metal, was responsible, and once again research was intensified. They made tests with an alkyl compound of a related material, tellurium, which compound proved to be twenty times more effective than aniline. Thus, through a chance observation, the direction of the research turned towards a study of the metals. As selenium and tellurium had disagreeable smells and penetrated the skin, Midgley looked for more suitable metals. Dr. Robert E. Wilson, then connected with the Massachusetts Institute of Technology, showed Midgley a chart of a periodic system he had drawn up in which he had arranged the elements according to their electropositive and electronegative properties. Midgley thought perhaps the anti-knock effect of an element could be predicted by its location on the periodic table. First he tried tetraethyl tin, and, although it reduced knock it also induced pre-ignition. It was, however, discovered that the heavier the metal the better it was for anti-knock purposes; lead was the heaviest but no one knew how to make a compound from lead that would be soluble in gasoline. Tetraethyl lead was known of, but it had never been made. Midgley and his assistants finally developed a zinc ethyl method which produced tetraethyl lead in quantities minute but sufficient for testing, and the compound performed much better than anything used previously. Tetraethyl lead was actually produced in December 1921, some nine months after intensive investigation of the metals had begun. But the crude laboratory process was unsatisfactory because ethyl is inflammable upon exposure to air.

The development of practical manufacturing processes for tetraethyl lead was then given priority at General Motors. Du Pont's services were called in; scientists at the Massachusetts Institute of Technology were employed and the Dow Chemical Company gave help in devising a practical method of extracting bromine from salt water. But, in the event, the best manufacturing process originated outside the industrial laboratories. Mr. Frank Howard of Standard Oil of New Jersey had been interested in Midgley's research for some years and he had initiated work along the same lines in his own firm but Standard Oil had never been so far forward as General Motors. Dr. Edward B. Peck, a research chemist in the Standard Oil laboratory, suggested that Dr. Charles A. Kraus of Clark University, who had made extensive studies of organometallic compounds and had recently perfected a process for obtaining high yields of tetramethyl tin, should be consulted. Standard Oil employed Kraus as a consultant although he remained at the university to do his research. He brought back to his laboratory a former graduate student, Dr. Conrad C. Callis, who had worked with him on tetramethyl tin. Some years later Kraus described the research that culminated in the discovery of the highly successful ethyl chloride process:

[1] Kettering and Midgley had a good deal of assistance at General Motors. Some of the other men on the 'knock' research staff were T. A. Boyd, C. A. Hochwalt, J. P. Andrew, C. P. Harding, and R. D. Wells.

'It seemed logical to us that if alkyl halides like methyl iodide and methyl chloride in reaction with sodium-tin alloy would produce satisfactory yields of tin halides, we would do well to attempt to make tetraethyl lead with some suitable member of the alkyl halide family.

'Inasmuch as bromine and iodine were both expensive, and my job was to find a process which would be commercially practical, it seemed to us that the reaction of sodium-lead alloy and ethyl chloride represented the only practical solution.

'Neither one of us was familiar with tetraethyl lead. We made it in experimental quantities by the established method of reacting sodium-lead alloy and ethyl iodide while we were getting together the necessary apparatus and materials. I might say that we had little to go on. We designed and built all our equipment ourselves. The first thing we made was an autoclave of about three-liter capacity from a section of heavy steel tubing, about fifteen inches long and about four inches in diameter, as I remember it. I did the welding myself.

'We attached a motor with a chain drive to the autoclave in order to revolve it while the reaction was taking place. Inside the autoclave we had steel balls about the size of large marbles so that the mixture could be agitated. The reaction was allowed to proceed to completion under pressure, which required about six hours. Tetraethyl lead was then separated from this reaction by steam distillation.'[1]

This Kraus-Callis process forms the basis for the method used today to obtain tetraethyl lead.

To sum up: this invention began with an alert, inquisitive, pertinacious individual, Kettering, who worked on the subject with a brilliant improviser, Midgley. Later these two inventors joined General Motors and, with the research facilities of the Corporation and drawing advice and suggestions from outside, they fell by good fortune upon a correct path. A university professor was responsible for the discovery of a practical process for producing tetraethyl lead.

REFERENCES

1. Story of Ethyl as gathered from the testimony of C. F. Kettering in the case of United States *v.* du Pont, Civil Action No. 49C 1071 Federal Dist. Ct. Ill.

2. Midgley, T., Jr., 'Problem+Research+Capital=Progress', *Industrial and Engineering Chemistry*, May 1939.

3. Schlaifer, R., and Heron, S. C., *Development of Aircraft Engines and Development of Aviation Fuels*, 1950.

4. 'The Story of Ethyl', *National Petroleum News*, Feb. 5, 1936.

5. Nickerson, S. P., 'Tetraethyl Lead: A Product of American Research', *Journal of Chemical Education*, Nov. 1954.

TITANIUM

DUCTILE titanium is a light, strong and extremely corrosion-resisting metal, the great potentialities of which have as yet barely been tapped. The metal has long been known but it is extremely difficult to obtain in the pure state; it is never

[1] S. P. Nickerson, 'Tetraethyl Lead: A Product of American Research', *Journal of Chemical Education*, Nov. 1954.

found free in nature and it contaminates very easily. These peculiar characteristics led most experts to regard titanium compounds as interesting for study, but offering little hope of the isolation of the metal in a pure form. In the 1880's, two Swedish chemists, Nilson and Petterson, obtained a metal which was 94 per cent pure by the use of high pressures but the method was not successful commercially. In the first decade of the present century, M. A. Hunter, experimenting in the General Electric research laboratory, achieved better results with an adaptation of the Nilson and Petterson process while trying to develop improved materials for filaments.

In 1914 Lely and Hamburger, in the laboratories of Phillips Metall-Glühlampenfabrik, A.G., in Holland improved Hunter's process; yet it was still quite unsuitable for large-scale commercial application.

Two Dutch scientists, Van Arkel and de Boer, adopted a somewhat better method in 1925 – the iodide dissociation process – which gave extremely pure material but was expensive. The principle involved here had been first proposed by a German, Weiss, in 1919 for the production of zirconium and tungsten. Van Arkel and de Boer were the first to produce a somewhat ductile titanium.

W. J. Kroll, a metallurgist, invented the first process suitable for large-scale production. It called for the reduction of titanium tetrachloride by magnesium under a noble gas at atmospheric pressure. Kroll, born in Luxembourg, studied iron metallurgy from 1910 to 1917 in the laboratories of a high school in Germany. When the First World War ended, the Metallgesellschaft in Frankfurt am Main sent him to their lead refinery where he spent one year. During the following three or four years he worked at several different metallurgical jobs.

He returned to Luxembourg in 1923, where he established a private laboratory. There he made substantial contributions to modern metallurgy. He discovered the age hardening of aluminium silver alloys and improved the production process for beryllium electrolysis. When he published this latter discovery, Siemens and Halske became interested in his work and the parties agreed to exchange information. Later he and Siemens and Halske entered into a contract for collaboration in the field of rare metals. Kroll acquired a knowledge of vacuum techniques which were to prove valuable to him in his later titanium work.

He first began experiments with titanium in 1930. He reduced titanium tetrachloride by the old Nilson and Petterson process. This involved the use of a steel bomb, and when some of these bombs exploded Kroll abandoned this approach as being too dangerous. He believed that a process that required a flash reaction and high pressure would never be a commercial success. Returning to his titanium experiments in 1935, he tried to reduce titanium dioxide with pure calcium obtained by vacuum sublimation. The products he obtained could be hot rolled into sheet but they were extremely brittle when cold. In June 1937 he reduced titanium tetrachloride with calcium under argon at atmospheric pressures and produced a metal which was malleable when cold. He later changed from calcium to magnesium as his reducing agent. At about the same time, another experimenter, Freudenberg, also working with a pressureless reaction, reduced titanium tetrachloride with sodium in a chloride flux under hydrogen but could not obtain cold ductile titanium with the process.[1]

[1] In 1921 Billy, a German, had also used a pressureless reduction process in which he reduced titanium tetrachloride with sodium and hydrogen, but Freudenberg operated his process on a much larger scale.

Kroll adapted some of the titanium equipment to the production of zirconium, and in subsequent research with zirconium he perfected a vacuum separation process for the salts produced which he could apply to titanium. In late 1938 he set out for the United States to sell his titanium process. He himself has described his visit:

> 'Six of the leading companies of the nonferrous and electrical industries of the U.S.A. were interviewed but the interest for titanium was nil, despite the fact that I stressed the availability of the ores, the good corrosion properties of the metal, the mechanical strength comparable with that of stainless steels. I left the United States in a sad state of mind, not having been able to interest anybody in my ideas.'[1]

In 1940 Kroll fled from Luxembourg to the United States where he served as a consultant for several years with the Union Carbide Research Laboratories. Anxious to learn the status of titanium research in the United States, he paid a visit in 1944 to the United States Bureau of Mines, where he suggested that he could produce zirconium for them within six months; he took over the direction of the Bureau's zirconium project in 1945.

Kroll did not play any important part in the large-scale development of his titanium process. In 1938 the American Government had become interested in the possible uses of domestic titanium ores. Dr. R. S. Dean outlined a plan for research and metallurgists at the Bureau of Mines' experimental station in Arizona began to seek for a process of refining titanium that could be adapted to large-scale production. The Bureau first considered the Kroll process late in 1940 after Kroll had published an article on titanium production in the Journal of the Electrochemical Society. They decided it showed the most promise, but the development posed serious problems.

> 'Dr. Kroll had produced a few pounds in his laboratory but there was no way of knowing whether his process could be adapted to produce the metal in large quantities. Also no one knew very much about titanium's engineering properties. How it would perform in structural use was still one of those questions that scientific investigation alone could answer.'[2]

This work was halted temporarily in 1941, but the Second World War gave added importance to metals possessing high-strength ratios, and it was resumed at the Bureau's Salt Lake City Station. F. S. Wartman, who directed the development, deserves a large measure of credit for its success. By 1944 a plant had been built in Boulder City, Nevada, capable of producing 100 pounds of consolidated metal per week. Production increased substantially in the following years and government scientists and engineers tested the metal for various applications.

The practicability of the Kroll process having been established, firms in more than one country adopted it, or some modification of it, for production. Research in the field has since expanded enormously. In 1951 Imperial Chemical Industries in Great Britain set up a small experimental Kroll plant. It was, however, finally decided that the use of sodium, instead of magnesium, to reduce the titanium tetrachloride offered prospects of lower production costs and the successful development

[1] W. J. Kroll, 'How Commercial Titanium and Zirconium Were Born', *Journal of the Franklin Institute*, Sept. 1955, p. 184.

[2] *Facts About Titanium*, U.S. Bureau of Mines, 1954, p. 8.

W. J. Kroll: titanium

C. F. Carlson: xerography

J. Bardeen, W. Shockley and W. H. Brattain: transistor

of the I.C.I. sodium process led to the building in 1955 of a plant capable of producing 1500 tons of metal per annum.

The invention of a commercially feasible titanium process, therefore, has its roots in the nineteenth century. But Kroll, a metallurgist working independently, conceived of the method which was developed first by an American Government research station and later improved and modified in important respects by research in industry.

REFERENCES

1. Kroll, W. J., 'How Commercial Titanium and Zirconium were Born', *Journal of the Franklin Institute*, Sept. 1955.
2. Kroll, W. J., 'Titanium', *Metal Industry*, July 22, 1955.
3. U.S. Bureau of Mines, *Facts About Titanium*, 1954.
4. Hunter, M. A., 'Early History of Titanium', *Journal of Metals*, Feb. 1953.
5. 'Titanium: A New Metal', *Scientific American*, Apr. 1949.
6. Barksdale, J., *Titanium*, 1949.
7. 'Titanium: The New Metal', *Fortune*, May 1949.
8. Mooney, R. B., and Gray, J. J., 'I.C.I.'s New Titanium Process', *I.C.I. Magazine*, Jan., Feb. 1956.

THE TRANSISTOR

FOR several decades the de Forest vacuum tube was the standard electronic amplifying device. Now the transistor, a tiny, rugged amplifying device conceived in the Bell Laboratories, is in a number of cases replacing vacuum tubes in radio sets, hearing-aids, telephone switching-systems, computers, bomb-sights and a variety of other electronic devices. The transistor, being only a small piece of crystal, is much less bulky than the vacuum tube; it does not require heating and so economises power and avoids the generation of unwanted heat. Although still in its infancy, it has already made its mark as a valuable creation.

The invention grew out of a study of a group of materials called semi-conductors; germanium and silicon belong to this group. Semi-conductors were utilised in communication systems in the first decade of the present century; crystal detectors, once so popular with amateur radio enthusiasts, depended upon these materials for their operation. Later the valuable electrical properties of the semi-conductors were harnessed to other uses: in lightning arresters on electric power lines and in regulating, measuring and modulating devices in wire telephone systems.

Before the Second World War, M. J. Kelly, the Director of Research at the Bell Telephone Laboratories, inaugurated a study of solid state physics with particular emphasis on semi-conductors. William Shockley and S. O. Morgan were chosen to lead the research. They speculated about the possibility of using a semi-conductor to control the flow of electrons in a solid, and Shockley

> 'predicted that it should be possible to control the meager supply of movable electrons inside a semi-conductor by influencing them with an electric field imposed from the outside without actually contacting the material'.[1]

The war interrupted plans for testing this theory. During the war silicon was

[1] Mervin Kelly, 'The First Five Years of the Transistor', *Bell Telephone Magazine*, Summer 1953, pp. 73, 74.

used as a detector of micro-waves in radar and, in the search for more satisfactory materials, knowledge was accumulated which contributed to the final discovery of the transistor. A group at Purdue University found that high purity germanium made an excellent detector, particularly at low frequencies, and other contributions were made by a number of laboratories, including the Bell Telephone Laboratories and the Radiation Laboratory of the Massachusetts Institute of Technology. It was, in large part, the discovery of suitable materials which made possible the working model of the transistor.

Immediately hostilities ceased Shockley organised a group for research on the physics of solids. On testing out experimentally Shockley's ideas, it was discovered that the projected amplifier did not function as Shockley had predicted; something prevented the electric field from penetrating into the interior of the semi-conductor. John Bardeen, a theoretical physicist, formulated a theory concerning the nature of the surface of a semi-conductor which accounted for this lack of penetration of field, and also led to other predictions concerning the electrical properties of semi-conductor surfaces. Experiments were carried out to test the predictions of the theory. In one of these, Walter H. Brattain and R. B. Gibney observed that an electric field would penetrate into the interior if the field was applied through an electrolyte in contact with the surface. Bardeen proposed using an electrolyte in a modified form of Shockley's amplifier in which a suitably prepared small block of silicon was used. He believed that current flowing to a diode contact to the silicon block could be controlled by a voltage applied to an electrolyte surrounding the contact. In the earlier experiments testing Shockley's ideas, thin films with inferior electrical characteristics had been employed. Brattain tried Bardeen's suggested arrangement, and found the amplification as Bardeen had predicted, but the operation was limited to very low frequencies because of the electrolyte. Similar experiments involving germanium were successful, but the sign of the effect was opposite to that predicted. Brattain and Bardeen then conducted experiments in which a rectifying metal contact replaced the electrolyte and discovered that voltage applied to this contact could be used to control, to a small extent, the current flowing to the diode contact. Here again, however, the sign of the effect was opposite to the predicted one. Analysis of these unexpected results by the two scientists led them to the invention of the point-contact transistor, which operates on a completely different principle from the one first proposed. Current flowing to one contact is controlled by current flowing from a second contact, rather than by an externally applied electric field. Brattain and Bardeen used extremely simple equipment, the most expensive piece of apparatus being an oscilloscope.

The Bell Telephone Laboratories announced the invention in June 1948, and, since then, development work has proceeded rapidly. The first point-contact transistor had several limitations: it was noisy, it could not control high amounts of power and it had a limited applicability. Shockley had meanwhile conceived the idea of the junction transistor which was free of many of these defects and most of the transistors now made are of the junction type. Many other improvements have been, and are being, made in the Bell Laboratories and elsewhere in the control of additives to give desired electrical properties; in the purification of materials; in replacing germanium by silicon, and in the design of new types of transistors for high frequency and high power.

The credit for the invention of the transistor belongs to the fundamental research group at the Bell Laboratories, and notably to Bardeen, Brattain and Shockley.

REFERENCES

1. Kelly, Mervin, 'The First Five Years of the Transistor', *Bell Telephone Magazine*, Summer 1953.
2. Interview with John Bardeen, formerly of Bell Laboratories and now Professor of Physics at the University of Illinois.
3. Bello, Francis, 'The Year of the Transistor', *Fortune*, Mar. 1953.
4. Brown, Ralph, 'The Transistor as an Industrial Research Episode', *Scientific Monthly*, Jan. 1955.

TUNGSTEN CARBIDE

CEMENTED tungsten carbide, developed in the second decade of this century in Germany, is a most versatile and useful new material. An extremely tough alloy, it is composed of tungsten carbide grains and a binder, usually of cobalt. It is used extensively for dies, for machine-tool cutting edges and for many sorts of wear and corrosion-resistant parts. Its discovery has aided the machine-tool industry, since tools equipped with cemented tungsten carbide cutting edges can cut metals at very high speeds.

Technologists at the turn of the century knew that tungsten carbide was a hard material, but in the pure form it was too brittle for industrial purposes. The French chemist, Henri Moissan, discovered it while attempting to synthesise diamonds in an electric furnace; he pulled out some pieces of this tungsten carbide from among the carbon crystals in his steel melt, but discarded them as 'worthless'.

During the twentieth century many efforts have been made to combine different metals with tungsten carbide in order to make it both hard and tough. Coles and Donaldson combined tungsten carbide and nickel and cast the product into sheets, which were used for burglar-proof safes, but they were far too brittle where high material strength and resistance to shock were required. In 1916 the German company, Voigtländer and Lohmann, received a patent on a mixture of tungsten carbide and molybdenum which they cold-pressed and sintered. They added molybdenum to prevent the crystallisation of the tungsten carbide but not for the purpose of binding the tungsten carbide particles, which was subsequently found necessary for the successful process. Two other inventors, Liebmann and Laise, suggested in a United States patent of 1920 that tungsten, carbon, and either iron or nickel in powder form be cold-pressed and then sintered to bind the materials together. However, they were vague as to the proper proportions of the materials and there is no real evidence that they created a practical material for drawing wire or cutting tools. The discovery of a practical cemented tungsten carbide, valuable as a die for drawing wire and as a cutting material, came about through investigations in the electric lamp industry.

During the First World War the Germans were anxious to find a substitute for diamonds in the drawing of tungsten wire and Karl Schroeter, of the Osram Gesellschaft,[1] was put on to this task. He turned first to tungsten carbide because he knew of its reputation as a very hard material:

[1] This firm, though an independent organisation, maintained connections with the German General Electric Company through an agreement for the exchange of patent rights.

> 'A vacuum arc-smelting furnace was constructed in quite a primitive manner, and in this small pieces of tungsten carbide were melted. Likewise, pieces of pressed powder were sintered just below the melting point in a carbon tube furnace. Very soon it was found that these products were very hard, but they were also extremely brittle.'[1]

Schroeter employed the tungsten carbide to draw tungsten wire, but the blocks always cracked. He observed that when he drew the wire unevenly the blocks cracked more easily and concluded that tungsten carbide in a pure form could not withstand the sudden and violent stresses imposed during the drawing process. In the course of subsequent experiments, he discovered that he could give the blocks a yielding tendency without decreasing their hardness by using a low-melting metallic matrix with tungsten carbide. The addition of certain metals, cobalt in particular, to the tungsten carbide gave it the necessary tenacity. Schroeter's method had other merits. Very high temperatures were required to produce satisfactory sintering results when pure tungsten carbide was employed, but by adding the cobalt matrix the temperatures for successful sintering were lowered and brought within the range where industry could easily handle the process. Schroeter finally developed a cemented tungsten carbide material (tungsten carbide and cobalt combination) which the Osram firm called Hartmetall. Although at this stage the carbide required further refinement for cutting material use, machine-tool engineers recognised its possibilities.

Baumhauer, also of the Osram firm, invented a somewhat different process for producing the material; the patents of Schroeter and Baumhauer seem to be basic, although in 1940 one United States court ruled that Schroeter's patents were invalid.[2]

From 1918 to 1923 Schroeter and his associates carried on systematic experiments with the material, testing its properties, structure and usefulness for special applications. The Krupp interests obtained the sales rights for the metallurgical application of this material and they themselves engaged in research, which resulted in a more homogeneous and somewhat tougher material called Widia, appearing in 1926.

The American General Electric Company in the same year became convinced of Widia's usefulness for cutting tool materials. They purchased the American rights to the Krupp patents and formed the Carboloy Company. Ludlum Steel and Firth-Sterling Steel Company also took licences under the German patents. Although the American firms were late in the field, they contributed much to the development and adaptation of cemented tungsten carbide for use in the machine tool industry.[3]

In this case, successful cemented tungsten carbide was discovered during a search for a substitute for diamonds. Other inventors had mixed various metals with tungsten carbide but Schroeter was the first to produce a hard, tough, practical

[1] Karl Schroeter, 'Inception and Development of Hard Metal Carbides', *Iron Age* Feb. 1, 1934, p. 27.

[2] General Electric Co. *v.* Willey's Carbide Tool Co., 33 F Supp. 969 (D.C., Mich., 1940). The court held that the work of Voigtländer and Lohmann, Liebmann and Laise, Coles and Donaldson, and Baumhauer clearly anticipated that of Schroeter.

[3] When Krupps first licensed their material, it was cut in 2-inch chunks and these could not be cut down. General Electric first tried to use the chunks as solid cutting tools but these soon fractured as the impact was too great. The company discovered that the material worked much more effectively as a tip on a steel-shanked tool.

material and to develop a commercial production process. The inventor, a research worker in an electric lamp firm, had no idea when he started his experiments of the potentialities of his discovery in the machine-tool industry. Krupps carried out development work on the material and commercialised it, and some American firms, principally General Electric, contributed to later technical and commercial development.

REFERENCES

1. Schroeter, Karl, 'Inception and Development of Hard Metal Carbides', *Iron Age*, Feb. 1 and 22, 1934.
2. Prosser, Roger D., 'Development and Application of Widia', *American Machinist*, Apr. 11, 1929.
3. Beardslee, K. R., 'Carboloy', *General Electric Review*, May 1953.
4. Jeffries, Zay, 'Hard Carbides', *Metals and Alloys*, Oct. 1939.
5. Comstock, G. J., 'Tungsten Carbide – The First Product of a New Metallurgy', *Iron Age*, Nov. 13, 1930.
6. 'Tungsten Carbide Cutting Tools', *Iron Age*, Feb. 7, 1929.
7. 'Powder, Pressure, and Heat', *Fortune*, Jan. 1942.
8. General Electric Co. *v.* Willey's Carbide Tool Co., 33 F. Supp. 969 (D.C. Mich., 1940).
9. Memorandum on Carboloy Co., Inc., Lewin Exhibit No. 11, Hearings Before the Committee on Patents, U.S. Senate, 77th Cong., 2nd Session, Part 1, p. 252 (1942).

XEROGRAPHY

AN individual inventor, Chester Carlson, conceived the idea of Xerography. This is a new photographic process which in a relatively short time has found numerous industrial applications. It is completely dry and is based entirely upon principles of photoconductivity and electrostatics. The process

> 'employs a plate which consists of a thin photoconductive coating on a metallic sheet. This coating can be electrically charged in the dark and will hold this charge until exposed to light. Thus an electrostatic image can be produced on the plate by exposing the plate to an optical image. When the plate is dusted with powder particles, the electrostatic image is transformed into a powder image which can be transferred to paper and fixed by fusing.'[1]

It has at least three advantages over conventional photography: the plates can be used repeatedly, the prints can be made on almost any type of paper, and the process is dry. Among its various uses are the making of master plates for offset lithographing, the enlargement of microfilm, and X-ray reproductions for the inspection of welds, castings and other products requiring non-destructive testing.

Carlson became attracted to the graphic arts during his schooldays and published a magazine for amateur chemists; he thereby acquired a working knowledge of the difficulties of setting type and of the drudgery involved in printing. Even in those days he had ideas about an improved duplicating process.

[1] R. M. Schaffert, 'Developments in Xerography', *The Penrose Annual*, 1954.

He worked his way through the California Institute of Technology where he specialised in physics and in 1930 he accepted a research position at the Bell Telephone Laboratories. Inventions and patents intrigued him, and he secured a transfer to Bell's patent department, subsequently studied law and, after receiving his degree, turned to the practice of patent law.

During the depression he was in financial difficulties: his comments on these early years reveal the influence of the economic motive on his inventive activity:

> 'I grew up in poverty and making and selling an invention struck me as one of the few possible avenues through which one could quickly change his economic status. I was greatly influenced by the success stories of Edison and others.'[1]

In 1934 he was employed as patent attorney for P. R. Mallory Y Company and he again noticed how laborious and costly were the conventional methods for producing copies of documents. He set out to examine the problem more critically.

> 'I purposely steered away from anything resembling silver halide photography or other known chemical processes because I felt that this ground had been already fully explored by Eastman Kodak and the other big companies in the field. An analysis of the requirements convinced me that the process would have to depend upon the effect of light in order to be universally useful, and by eliminating the chemical effects which were already in use I was led to investigate photoelectric phenomena. After several failures with experiments on methods which first occurred to me, I realised that I would have to go deeper into the subject in order to understand the subject better and, if possible, to discover some phenomena in the literature which had been overlooked by previous inventors.[2]

He spent a good deal of his spare time during the next three or four years perusing the technical literature in the New York Public Library. The use of an electrolytic effect occurred to him at first, but he concluded that this line was unsatisfactory because the currents required were too high. He reasoned that

> 'if the low voltages and high currents of electrolytic processes could be replaced by high voltages and low currents, the same amount of light could control a greater amount of energy. With the problem so sharply defined, the solution came almost as an intuitive flash. I was aware that a previous inventor had used powder to develop electrostatic images. By combining that with my concept of an electrostatic light-sensitive plate, the invention was complete and it only remained to search the literature to see whether a suitable photoconductive material was known.'[3]

He had developed the basic concept of Xerography by 1937 and immediately filed a patent application although he had yet to verify his ideas experimentally. In an effort to reduce his idea to practice, Carlson worked with sulphur-coated plates in the basement of his house. The lack of adequate facilities placed him at a disadvantage but he was handicapped to an even greater extent by his lack of skill as an experimenter. He then took on an unemployed physicist from Germany to aid him in the experimental work and together they successfully developed a small plate which produced a fairly good image. Carlson fitted up a method of demonstrating it to prospective purchasers and filed additional patent applications.

[1] Letter from Chester F. Carlson, July 26, 1954. [2] *Ibid.* [3] *Ibid.*

For several years he continued to improve his process and he tried to interest manufacturers in it, but without success. The war added further obstacles to his attempt to commercialise his invention.

His work at Mallory & Company carried him occasionally to the Battelle Memorial Institute, the world's largest non-profit research organisation, and on one of these visits he spoke of his invention. At this time Battelle wished to establish a research group in the graphic arts and they were looking for good ideas. The Battelle Development Corporation, a wholly-owned subsidiary of the Institute, had been organised several years earlier to finance and develop worth-while inventions. Carlson made an agreement with this Corporation whereby Battelle acquired the exclusive patent rights in return for which Carlson acquired the right to a substantial share in the proceeds from any future development.

The development of Xerography was turned over to Roland M. Schaffert, a Battelle research physicist with some previous experience in printing. For a year he worked alone but after the war Battelle assigned a few assistants to help him. By the latter part of 1946 two important developments were completed: a high-vacuum technique for coating plates with selenium; and a corona discharge wire, both for applying the original electrostatic charge to the plate and for transferring powder from the plate to the paper. The most significant contribution was the discovery of a method to keep the image background from being filled with stray powder. Thus Battelle improved Xerography to the point where industry became interested.

In 1946 the Haloid Corporation of New York, a firm producing photographic materials and photocopying equipment, began a search for a new product to add to their business and learned of Xerography through an article written in 1944. After investigating the process, Haloid accepted an exclusive sub-licence under the patents, despite the fact that the expense of development would be heavy for such a small firm, and in 1950 they announced the first commercial application of the process, the Xerox copying machine. The United States Signal Corps also took out a licence and has been conducting research on the application of the process to continuous tone photography. Several other firms now have licences for the development of particular uses for the process. In 1956 a joint organisation of the Haloid Corporation and the Rank Organisation Ltd. was set up for the development of the process outside North America.

An outsider was responsible for the basic invention in this case. Development work proceeded along many channels; a non-profit research organisation made the initial advances, then a small firm pushed the commercial development along rapidly, and, finally, the Army and large and small firms participated in the development.

REFERENCES

1. 'Printing with Powders', *Fortune*, June 1949.
2. Reid, W. T., *Xerography – From Fable to Fact.*
3. Schaffert, R. M., 'Developments in Xerography', *The Penrose Annual*, 1954.
4. Correspondence with the inventor, Chester F. Carlson.

ZIP FASTENER

A single individual fashioned the first zip fastener, but it was a later inventor who made it into a practical product. Whitcomb L. Judson, a mechanical engineer by profession, filed the first patent applications on this device in 1891 and 1894. The invention, originally designed for shoes, consisted of a series of separate fasteners, each having two interlocking parts which could be fastened either by hand or by a movable guide. Judson's invention was unique; there was nothing even remotely resembling it in the patent files. No one has discovered how Judson happened to invent this fastener, although the patent records reveal that this was by no means his first invention. It is interesting to note that from the time that Judson received his first patent on a slide fastener (1893) until the time he received his fifth (1905), no one else applied for patents on a similar contrivance.

Colonel Lewis Walker, a corporation lawyer, knew Judson through their mutual interest in the Judson Pneumatic Street Railway Company. The zip fastener aroused his interest and he organised the Universal Fastener Company of Chicago to manufacture the Judson fastener. For a while the Company manufactured the fasteners by hand and spent a considerable amount of money on developing a suitable machine. In 1902 the machine-building firm of Manville Brothers of Waterbury, Connecticut, constructed a machine for the Universal Company along the lines of one patented by Judson, but this proved to be too complicated and was abandoned. By 1905 Judson designed a new fastener which was easier to adapt to machine operations. Instead of linking the fastening elements together in a chain, he merely clamped them to the edge of a fabric tape. The Universal Company was reorganised as the Automatic Hook and Eye Company of Hoboken, New Jersey, and they sold, through house to house peddlers, the new fasteners under the name 'C-curity'. The garment manufacturers were not interested in the fastener, perhaps because the fasteners had defects such as the annoying habit of springing open at the most unexpected moments. Aronson, one of the firm's engineers, achieved some success in improving Judson's fastener.

The Company then took on Gideon Sundback, a Swedish electrical engineer who had settled in the United States in 1905 and had been employed by the Westinghouse Company. Sundback, after several years' work, filed a patent application and his model was sold as a novelty item, but it was also unsatisfactory in use. At this stage the Company lingered on the brink of bankruptcy but Sundback set out to develop a fastener in which the hooks would not disengage from the eye and force the whole fastener open. He invented a hookless fastener in 1912 that looked promising, but this model also proved impractical because the corded edges wore out quickly. Other inventors from various countries entered the field with improvement patents after 1905.

Sundback continued his search for a fastener which could be mass produced and for machinery to make it. On reviewing his previous work he decided that the new fastener should be flexible and he thought this could be achieved by reducing the size and reconstructing the shape of the interlocking parts; the tapes also required modification. Finally, in 1913, Sundback invented a fastener which, in all important essentials, is the modern zip fastener: the locking members were identical and interchangeable, their shape, size and construction constituting the most significant changes from the prior art. Sundback invented not only a practical fastener but also efficient machines for stamping out the parts and attaching them to the tape. Despite the virtues of the new fastener, the clothing manufacturers continued to oppose its introduction. However, in 1918, a contractor making flying suits for the Navy purchased 10,000 fasteners and a big jump forward in commercial application came in 1923 when the B. F. Goodrich Company put the fasteners in their galoshes. From that point on, the popularity of the zip was assured.

The origin of the zip fastener can be traced to an individual inventor, Judson. Sundback, after years of work, perfected the original ideas and devised a reliable commercial fastener.

REFERENCE

1. Federico, P. J., 'The Invention and Introduction of the Zipper', *Journal of Patent Office Society*, Dec. 1946.

PART III
SUMMARIES OF NEW CASE HISTORIES

Sir Christopher Cockerell: hovercraft

AIR CUSHION VEHICLES

THE air cushion vehicle, or hovercraft, has the unique ability of travelling with equal ease over water, land or ice at speeds of up to 80 m.p.h. Its military value is accepted and its commercial development is being energetically pursued. Large hovercraft started operating ferry services between Britain and France in the summer of 1968. Track-riding versions of the hovercraft are also being developed as replacements for the conventional railway; they promise speeds of up to 300 m.p.h.

Although since the eighteenth century methods have been suggested for supporting a vehicle on a cushion of air, only in the present century have workable machines been built. Three basic types of ACV have appeared: the fully amphibious, which use a flexible rubber skirt around the periphery of the craft to retain the air cushion that is produced by fans blowing air underneath it; the purely sea-going, which has sidewalls extending into the water to retain the similarly-produced cushion, plus flexible skirts at the bow and stern; and the ram wing, which rides on a cushion produced by its own movement. The tracked hovercraft, riding only a few inches above the surface on a cushion produced by fans, does not require such methods of retaining the cushion because air losses are reduced by the small clearance.

The first types of hovercraft to be built were those with sidewalls and the ram wing. From 1928 onwards an independent American inventor, Douglas K. Warner, experimented with sidewall type ACVs; but, after encountering troubles which included the failure of the air cushion in rough seas, he turned to the ram wing. A Finnish inventor, Toivo J. Kaario, also experimented with ram wing ACVs in the 1930's, he made several flights over snow in 1935–6 and persisted with his experiments into the 1950's.

The birth of the hovercraft industry has followed the work of Christopher Cockerell, a British electronics engineer. After studying engineering at Cambridge University, he decided to abandon a successful career in electronics and set up on his own as a boat-builder. By 1953 he had concluded that in order to increase the speed of a ship it would be necessary to reduce skin friction. This led him to try air lubrication – introducing a thin layer of air between the hull and the water. The early experiments were inconclusive but they did suggest that a thick layer of air was needed. Cockerell then moved nearer the true ACV, building a boat with deep sidewalls and hinged doors fore and aft, and blowing air into the central chamber. His next idea, towards the end of 1954, was to replace the doors with water curtains but he found it difficult to estimate the power required. He then thought of air curtains.

He fashioned an annular jet from two coffee tins, a hair dryer and tubing which he tested on the kitchen scales. These, and later tests on a two-dimensional system, proved that an air cushion could be created and maintained by blowing air inwards from a jet around the circumference of the machine, and so laid the foundations for the development of practical hovercraft. With the help of another boat-builder, Cockerell built a model hovercraft and in December 1955 applied for his first patent.

Support for his invention proved difficult to obtain and progress was slow.

But by the autumn of 1956 he had developed a closed vortex system that re-circulated the air continuously, and he demonstrated a model at the British Patent Office and Admiralty. The latter demonstration led to the first help for the hover-craft's development, though it also meant that the project was classified as being of military value. R. A. Shaw, then an assistant director in the Ministry of Supply responsible for aerodynamic research, had attended the demonstration at the Admiralty and concluded that the hovercraft deserved support. He therefore arranged for a contract worth £7,500 to be placed with Saunders-Roe, a small aircraft manufacturer with experience of flying-boats, to check the validity of Cockerell's results by tests on models. By the spring of 1958, Saunders-Roe had nearly finished their work and were showing promising results; but Shaw was unable to raise funds within the Ministry to carry on development, for military interest was waning. In April 1958 Cockerell therefore went to the National Research and Development Corporation, which had been formed with government funds in 1949 to finance the development of inventions, and it quickly agreed to support the project. The project was then declassified. In the autumn of 1958, the National Research and Development Corporation gave Saunders-Roe a contract to design and build a manned prototype.

This machine, the SRN1, cost only £120,000 for design, construction and early sesting, and was finished by the end of May 1959. This simple hovercraft used a tingle aero-engine for lift and propulsion, and followed Cockerell's basic design except that it used two instead of one peripheral jets; the single jet was found to make it unstable. Tests soon established the commercial and military potential of the hovercraft. The N.R.D.C. had, in January 1959, formed a subsidiary com-pany, Hovercraft Developments Ltd., in which Cockerell was a shareholder, to hold Cockerell's patents and develop the machine.

Work on ACVs had also been proceeding in the United States and Europe, but it has been of less practical significance than that inspired by Cockerell. An American doctor, W. B. Bertelsen, had achieved some results with small hovercraft primarily intended for overland use; a small research organisation, National Research Associates Inc., tried unsuccessfully to obtain support from government and industry for his ideas, finally securing the sponsorship of the Ordnance-Tank Automotive Command in 1957. The U.S. Navy also started a research programme on hovercraft in 1957, while the Marine Corps backed the work of Carl Weiland, a Swiss engineer who had invented a form of sea-going ACV with a 'labyrinth seal' for retaining the air cushion. The aero-engine manufacturer, Curtiss-Wright, built the prototype of an 'air car' in the summer of 1959, a simple ACV for overland use. Though this machine did not go into production, it was the first ACV to be fitted with a flexible rubber skirt around its circumference to help retain the air cushion. In this case the skirt was no more than a strip of rubber, but elaborations of the principle have transformed the efficiency and seaworthiness of the hovercraft without sacrificing its amphibious qualities.

The possibility of using some flexible material to fill the gap between the hull of a hovercraft and the surface seems to have been discussed among the engineers working on the British hovercraft project in the summer of 1959, but there were doubts about the feasibility of the concept. Cockerell had started thinking in 1957 about flexible structures for hovercraft, to let the machine conform to the rough surface of the water, and in September 1958 he applied for a patent that covered some features of flexible skirts, though not recognising the principle. A British consulting engineer, C. H. Latimer-Needham, had patented the principle of the flexible skirt a little earlier, but his ideas were not known to the hovercraft engineers at this time. The need to improve the seaworthiness of the SRN1 encouraged thought about ways of increasing its operating altitude. After reading a description of the Curtiss-Wright air car, R. Stanton-Jones, then chief designer of Saunders-Roe, tested flexible skirts. Experiments on models seem to have been made in the autumn of 1959 and on the SRN1 before the end of the year. The first skirt was six inches deep and lasted ten minutes, but development gradually improved durability and efficiency. By 1963 the SRN1 was fitted with skirts four feet deep. Hovercraft Developments followed a separate line of development which centred upon a patent applied for by one of its engineers, Denys Bliss, in 1962. This covered a 'finger' skirt, made up of segments of rubber held in position by cushion pressure, whereas the Saunders-Roe-Westland skirts were initially an inflated sausage-like bag. More recently Westland skirts have had fingers added beneath the bag, while H.D.L. have added a bag above the fingers, so that the two designs now have much in common.

In July 1959 Saunders-Roe was taken over by Westland, the helicopter manufacturers, and Westland built a series of larger hovercraft leading up to the 165 ton SRN4, completed in January 1968. Other companies had entered the hovercraft business in 1960 but they all dropped out except a small aircraft manufacturer, Britten-Norman, which concentrated on light hovercraft, and Vickers, which merged its hovercraft interests with Westland in 1966 to form the British Hovercraft Corporation, in which the N.R.D.C. holds ten per cent of the capital. At the same time the H.D.L. was made a division of the National Physical Laboratory. Cockerell resigned from the board of the H.D.L. in protest against these changes. Two more companies, Vosper-Thorneycroft and Hovermarine, were granted licences by the N.R.D.C. in 1967 and are developing purely sea-going hovercraft. Several American companies, including General Dynamics and Bell, are now developing ACVs; but the only machines to have been used, commercially or militarily, are Saunders-Roe designs built under licence by Bell.

Tracked hovercraft have been developed furthest in France, though the concept again has a long history. The Société Bertin have built an experimental machine running on an I-section concrete track, and in 1967 the French government authorised the construction of a longer track and more advanced vehicles as a step towards a proposed Paris-Orléans service. Small-scale experiments with tracked hovercraft had been made by H.D.L. while Cockerell directed its research, and in 1967 the N.R.D.C. formed a company, Tracked Hovercraft, to carry out full-scale experiments.

The key invention in the history of the ACV thus came from an electronics engineer turned boat-builder, while development has so far been most successfully pursued by a relatively small aircraft manufacturer. Most of the other interested firms in Europe are also small. Government finance played a crucial role in the early stages, but companies have provided the majority of the spending in Britain up to 1968: of the £17 million spent by the British Hovercraft Corporation and its constituent companies, the N.R.D.C. and the government have contributed about £3 million. The N.R.D.C. has spent about £4 million in all. Development spending has thus become substantial, if low by aircraft standards. It is also noticeable that development seems to have been pursued more energetically and more successfully in Britain than in the United States.

REFERENCES

1. 'Guide to the Hovercraft', *Engineering*, June 10, 1966.
2. 'Symposium on Ground Effect Phenomena', The Princeton University Conference and The Department of Aeronautical Engineering in co-operation with the U.S. Army (1959).
3. 'Release from the Wheel – The Hovercraft', C. S. Cockerell, *Flight*, Sept. 11, 1959.
4. 'G.E.M.'s', *Interavia*, June 1962.
5. *The Ground-Cushion Phenomenon*, Hearings before the Committee on Science and Astronautics, U.S. House of Representatives, 86th Congress, 1st Session, Apr. 13–15, 1959.
6. 'Research on New Transportation Methods – Ground Effect Machines', Hearings before the Committee on Science and Astronautics, U.S. House of Representatives, 87th Congress, 2nd Session, July 11–13, 1962.
7. L. H. Hayward, 'The History of Air Cushion Vehicles', *Hoveringcraft and Hydrofoil*, vol. 2, part 1 (Dec. 1962), p. 12; and part 2 (Jan. 1963), p. 12.
8. 'Hovercraft', N.R.D.C. Leaflet no. 9, 1968.
9. Erwin J. Bulban, 'Curtiss-Wright Tests Air Car Prototype', *Aviation Week*, July 8, 1959.
10. Interview with Mr. R. Stanton-Jones, Mar. 1968.

CHLORDANE, ALDRIN AND DIELDRIN

CHLORDANE, Aldrin and Dieldrin, three powerful new insecticides, were discovered in the middle forties and in each case Julius Hyman was intimately involved in the discovery. Hyman held degrees from the Universities of Chicago and Leipzig. In 1930 he gave up a post as research chemist in order to develop his own ideas for the manufacture of drying oils for paints, varnishes and other coatings from a waste product of petroleum refining. His cousin Joseph Regenstein, then President of Transo Envelope Company and the Arvey Corporation, agreed to give him financial support. The Varnoil Corporation, which later changed its name to the Velsicol Corporation, was organised. Hyman assigned his two relevant patent applications to this Corporation. As Vice-President, he acted as the firm's operating head, directing all research and manufacture.

By the summer of 1944, Velsicol had a laboratory staff of ten college-trained chemists and an equal number of assistants. The discovery of DDT in 1939 had led other chemists to search for new synthetic organic insecticides and Hyman organised experiments to determine whether any of the by-products of the polymers manufactured by Velsicol showed insecticidal activity. Several compounds did so, but nothing of commercial value followed from these experiments.

Early in 1944, the Rubber Reserve of the United States Government asked Hyman if he could make use of its cyclopentadiene, produced as an apparently useless by-product in the manufacture of synthetic rubber. In July 1944 Hyman came upon an unusual reaction discovered by Fritz Straus and his students in Germany in 1930. They had substituted six chlorine or bromine atoms for the six hydrogen atoms in cyclopentadiene and had produced hexahalocyclopentadiene ('hex'). On Hyman's instructions, S. H. Herzfeld, a research chemist in Velsicol, produced a sample of 'hex'. After distilling it, he discovered that a white solid remained. This unknown material ('237') was sent to Professor Clyde Kearns of the Entomology Department of the University of Illinois. Kearns reported that it showed marked insecticidal activity, though less than that of DDT. Its structure remained a puzzle until Hyman correctly guessed that '237' was a simple Diels-Alder adduct of cyclopentadiene and 'hex'. At Hyman's suggestion, '237' was chlorinated in order to increase its residual toxicity. The new compound, now known commercially as chlordane, proved to be more toxic than DDT. Velsicol then brought the laboratory process to the commercial stage, which took over a year and cost about $150,000.

As a result of a quarrel with Regenstein, Hyman refused to assign his chlordane patent to Velsicol, left the Company and set up one of his own, Julius Hyman and Company, which manufactured chlordane under the trade name OktaChlor.

Chlordane seems to have been discovered independently, and at about the same time, in Germany by R. Riemschneider and A. Kuhml. Riemschneider apparently worked at the Chemical Institute of the Free University of Berlin.

It is not altogether clear whether aldrin and dieldrin were discovered at Velsicol or later at Julius Hyman and Co., but it seems to be established that these discoveries emerged from the work of Hyman and his colleagues, some of whom had left Velsicol to join him when he set up his new company. A Colorado State Court, however, ruled that the inventions properly belonged to Velsicol. While still at Velsicol, Hyman became interested in the possibility of making bicycloheptadiene and, on his instructions, hexachlorocyclopentadiene was reacted with acetylene in a Diels-Alder manner. Tests of the resultant material indicated that this was an adduct of the reaction of the starting materials, but these results were not followed up while Hyman was at Velsicol. When he set up his own firm, his interest in bicycloheptadiene continued. One special difficulty was the danger of explosion, but this was overcome by the use of the 'flowing method'. In December 1947, E. Freireich, the firm's chemical engineer, began experiments involving the reaction of cyclopentadiene and ethylene and he isolated bicycloheptane, the product earlier isolated by Joshel and Butz of the United States Department of Agriculture. Freireich next substituted acetylene for ethylene and this reaction yielded bicycloheptadiene, a Diels-Alder product. Hyman remarked subsequently: 'We didn't even know whether it would exist up until the time we made it'.

When chemists at Julius Hyman and Co. reacted bicycloheptadiene with hexachlorocyclopentadiene, aldrin resulted. R. E. Lidov, a member of Hyman's staff, received the United States patent on aldrin. The oxidation of aldrin produced dieldrin. These two products showed strong insecticidal properties, contrary to all expectation. Hyman stated:

> 'In neither aldrin nor dieldrin had we any basis for belief that they would be the extremely powerful insecticides that they proved to be . . . . The hypotheses of insecticide activity current at that time were pretty well thrown into a cocked hat by these results.'

In the aldrin and dieldrin discoveries, the key roles were played by R. E. Lidov, A. Segel, E. Freireich, H. Bluestone and S. Soloway, together with Hyman.

The rights to the relevant patent applications covering chlordane, aldrin and dieldrin were the subject of extensive litigation between the Velsicol Corporation and Hyman, his company and employees. Velsicol won these cases and gained control over the valuable discoveries. The Shell Development Company subsequently purchased Julius Hyman and Co. and secured from Velsicol exclusive rights to aldrin and dieldrin.

This case has several important features. First, a small group of scientists in small firms were the innovators in an industry where very large firms predominate. Second, small firms brought these insecticides to the marketplace in a short time and at moderate cost. Third, chance played a part in all three discoveries; an unknown by-product of a particular reaction opened up the way for the discovery of chlordane; and aldrin and dieldrin which, according to predictions based upon a knowledge of their chemical structures, were not expected to show marked insecticidal activity, in fact did so.

REFERENCES

1. Julius Hyman and Co. *v.* Velsicol Corp., 123 Colo. 563, 233 P 2d, 977 (1951).
2. Hyman *v.* Velsicol Corp., 342 Ill. App. 489 (1951).
3. Velsicol Corp. *v.* Hyman, 103 F. Supp. 363 (D.C. Colo., 1952).
4. Velsicol Corp. *v.* Hyman, 405 Ill. 352 (1950), reversing 338 Ill. App. 52 (1949).
5. Transcript of testimony, Velsicol Corp. *v.* Lidov *et al.* District Court for City and County of Denver and State of Colorado. Civil Action No. A-65679-Division 1.
6. Lidov, R. E., Bluestone, H., Soloway, S. B., and Kearns, C. W., 'Alkali-Stable Polychloro Organic Insect Toxicants, Aldrin and Dieldrin', *Advances in Chemistry*, Series 1 (1950), p. 175.
7. Frear, D. E. H., *Chemistry of the Pesticides* (Van Nostrand, New York, 1955), 3rd edition.
8. Kearns, C., Ingle, L., and Metcalf, R., 'Chlordane', *Journal of Econ. Entomol.*, vol. 38 (1946), p. 661.
9. Kearns, C. W., Weinman, C. J., and Decker, G. C., 'Insecticidal Properties of some new chlorinated organic compounds', *Journal of Econ. Entomol.*, vol. 42 (1949), p. 127.

FLOAT GLASS

HISTORICALLY, there have been two basic methods of making flat glass: the window glass processes and the plate glass processes. Their development has been parallel but quite separate.

The window glass processes have all depended on forming a sheet by stretching a lump of molten glass either by blowing or pulling, but always in such a way that the natural fire-polished surface was not damaged. The disadvantage has been that any stretching process tends to cause distortion. Both process and product, however, have always been relatively inexpensive.

The plate processes have all consisted of casting a plate of glass, grinding the glass flat and then polishing it to make it transparent. Grinding and polishing were inherently lengthy and expensive processes. They also resulted in the destruction of the naturally hard surface of the glass, and the loss of up to twenty per cent of good glass that had to be ground away to produce the required flatness.

In the Float Glass process, a continuous ribbon of glass moves out of the melting furnace and floats along the surface of a bath of molten tin. Since the surface of the molten tin is dead flat, the glass surface produced is flat. The ribbon is then cooled down while still on the molten tin and passed through an annealing oven. Glass is thus made of uniform thickness and with bright fire-polished surfaces without mechanical grinding or polishing.

The advantages of the new process are great. The costs of the capital equipment and of the forming of a ribbon of glass from a stream of molten glass are much reduced; the loss of glass in grinding is avoided; a continuous process lending itself to long trouble-free runs replaces a non-continuous process; and variations in the width of the glass produced can be more quickly and cheaply made.

The invention and development of the new process occurred within Pilkington Bros. Ltd., a long-established private British company which, with its subsidiaries, was responsible in 1966 for some ninety per cent of raw flat glass and seventy-nine per cent of safety glass in the home market. The idea first occurred to Mr. Alastair Pilkington in October 1952, and he remained in charge of the subsequent successful development. A member of the family which controlled the Company, Mr. Pilkington was born in 1920, graduated in Mechanical Science at Cambridge, joined Pilkington Bros. Ltd. in 1947 and became Director in charge of plate glass production in 1955.

The earlier samples of the new type of glass were made on a 12″ wide pilot plant in 1953. There were several critical stages in the subsequent development. In 1954 it even appeared possible that the surface between glass and metal could never be brought to the standard of perfection required. With the backing of the Company, Pilkington persisted and the first production plant started in May 1957. For 14 months this plant continued to turn out nothing but unsatisfactory glass. In July 1958, however, the first square foot of good glass was produced. In 1959 the process was announced to the world and the plant continued to operate well for some months. The anxieties, however, were not at an end. As Pilkington has recounted it:

> 'As the set-up on the float bath was rather old and tattered . . . we renewed the worn out parts and expected to settle down to a long successful run. To our amazement we then made continuous cullet (bad glass) again and were back struggling and feeling more frustrated than ever before. It took us nearly three months of investigation to discover that when we first made good glass it was partly due to a fluke and that a vital part of our success had been due to a broken part of the plant. As soon as we understood the problem we made a set-up which reproduced similar conditions to those with the broken part. This made saleable glass immediately and we have never been in doubt again about the ability of the process to make consistently good glass.'

The team of developers enjoyed a second remarkable piece of good fortune. It turned out that, whatever thickness was rolled, the sheet of glass produced finally was approximately one-quarter of an inch thick. About one-half of the sales of plate glass by the Company was of this thickness. In consequence, the Company was able to sell glass profitably in large quantities while development work continued to devise techniques of producing glass both thicker and thinner than the original.

Construction of a £25,000 pilot plant in 1954 pushed the early costs to over £100,000 and, once the process had been scaled up to a production unit in 1957, costs began to soar. After fourteen months on the full-scale plant, the first foot of saleable glass was produced and development costs stood at £3,000,000. By the time the float process was developed to the stage where it could replace all plate glass, the total development cost had reached £7,000,000.

In the course of the application by Pilkingtons for patent protection, some facts of more than ordinary interest emerged. The British patent was applied for in

December 1952; complete specifications were submitted in 1954 and the patent was sealed in July 1957. The United States patent was applied for in December 1954 and issued in November 1959. But in the course of examining this latter application, the American Patent Office drew attention to much earlier thinking on this subject and to two earlier American patents. In September 1902, the American Patent Office had issued to a certain William E. Heal of Marion, Indiana, a patent for a process described as:

> 'The manufacture of sheet and plate glass of any desired thickness, and in continuous sheets by a new and improved method of flowing the molten glass from the melting tank into an adjacent receptacle, containing melted material of a greater specific gravity than glass, [preferably of tin or alloys of tin and copper] and causing the molten glass to float upon and spread into, a continuous sheet, and then drawing the sheet of glass therefrom, and causing it to pass into their lehr for annealing and by one continuous operation . . . to simplify, facilitate and cheapen the manufacture of sheet and plate glass to improve its quality.'

(Heal had applied for and been granted a patent on the same process in the United Kingdom in January 1903.)

In May 1905 H. K. Hitchcock of Walton, Pennsylvania, had been granted by the American Patent Office a patent which he described in these terms:

> 'The plastic glass . . . forced through a slot in the side wall of the furnace . . . enters a chamber in which it is hardened and annealed. The support for the sheet or plate of glass . . . consists of a practically continuous liquid bed, formed of molten metal . . . so that the sheet or plate will float on the surface of the molten metal.'

The emphasis placed by these two early inventors, and especially by Heal, upon the floating of molten glass onto a bath of molten tin, led the United States Patent Examiners to scrutinise the Pilkington application carefully and, indeed, to reject some of the claims made in it.[1] But it was clear that Heal and Hitchcock had not produced an operative process, had not carried out any large-scale experimental work and had certainly not produced satisfactory glass by employing their basic ideas. Indeed, the Float process is dependent for its success upon a continuous flow of glass from a furnace and no one seems to have produced such a flow until Pilkingtons themselves succeeded in doing this in the early 1920's.

The merits of the new method are well attested to by the speed with which it has been accepted almost universally. By the end of 1967 Pilkingtons had licensed their patents, together with important know-how in the formation of a ribbon of glass in a horizontal plane without the use of rollers, to large glass manufacturers in the United States, Belgium, France, Italy, Germany, Spain, Mexico, Czechoslovakia and the U.S.S.R.

Pilkingtons have continued to develop and exploit their process and in November 1967 announced new techniques making it possible, at no great extra cost and with little loss of time, to transform ribbons of clear glass into coloured glass with controlled physical characteristics. This new development involved an initial invest-

[1] In the United Kingdom it is the practice of patent examiners not to continue their searches back in time for more than fifty years. This is not the case in the United States.

ment of £25,000 on a small-scale pilot plant but, finally, it cost nearly £1 million and demanded the work of 30 chemists, physicists and engineers.

The Float Glass process is an outstanding case of successful invention and development in a large company which already had a dominating position in its home market but which recognised that large rewards awaited any discoverer of a method of producing high-quality flat glass without grinding and polishing. The basic idea of floating molten glass over molten tin was patented at the beginning of the century but remained unexploited. It seems certain that, in any case, its successful commercial development would have had to wait upon later technical advance in glass-making. Nevertheless, it is intriguing that the idea was not taken up much earlier by one or other of the large glass manufacturers in the world. It would seem that only a large company with ample resources could have succeeded in this. One such company had the imagination and the courage, and the unexpected good fortune, which enabled it to take the lead.

REFERENCES

1. Pilkington, L. A. B., 'Float Glass', *Advance*, Nov. 1966.
2. Earle, K. J. B., 'The Development of the Float Glass Process and the Future of the Glass Industry', *Chemistry and Industry*, July 15, 1967.
3. Patent Specification, London No. 769,692. Inventors: Lionel Alexander Bethune Pilkington and Kenneth Bickerstaff (1957).
4. Patent Specification, London No. 19,829. Inventor: William Ephraim Heal (1903).
5. 'Flat Glass', Report of the Monopolies Commission, Feb. 1968.
6. Cook, P. Lesley, 'The Effects of Mergers', *The Flat Glass Industry* (George Allen & Unwin, 1958), chap. IV.
7. Pilkington, L. A. B., 'The Development of Float Glass', *The Glass Industry*, Feb. 1963.
8. United States Patent Office File relating to United States Patent No. 2,911,759 issued to Lionel A. B. Pilkington *et al.*

THE MOULTON BICYCLE

THE bicycle must be one of the most useful and serviceable of all machines. There are probably more than a hundred million bicycles in use throughout the world and they continue to be produced by the million annually. The extraordinary fact about this machine, however, is that although it had long been manufactured by large firms, it had not changed in basic design in the seventy years from the introduction of the 'safety' model. It was left to an independent inventor outside the industry to take up the challenge and introduce a novel design which has been widely accepted.

The inventor was Alex Moulton. He came of an inventive and enterprising family. His great-grandfather, Stephen Moulton, after a visit to America, had brought back Goodyear's process for vulcanising rubber and had established in 1848 the rubber industry at Bradford-on-Avon in Wiltshire. Alex Moulton graduated in engineering at Cambridge and after working during the Second World War in the aero-engine industry, he set up a small research establishment on the family estate in Wiltshire. There he was successful in developing and selling to the British

Motor Corporation the rubber suspension system for their Mini car, and the hydrolastic system of suspension for larger cars. A parallel ambition at the time, however, was to produce a radically new type of bicycle. In 1956 he began studies of the evolution of this machine and of the many ideas, none of which have come to fruition, which had been advanced for the improvement of this form of propulsion. By 1959 the Mark I prototype of his new machine had been built and the first of a series of British patents had been granted to Moulton by October 1960.

The novelty of the Moulton bicycle lay in a new configuration, with an F-shaped frame and small wheels, together with a new method of suspension. The small wheel with a narrow section and high-pressure tyre reduced resistance; the method of suspension insured comfortable riding despite the narrow tyre. The new machine aroused popular interest. Patent protection was obtained in 1963 in the United States and in 1964 in the U.S.S.R. In 1965 Moulton was the recipient of a British Design Centre Award. By 1965 major cycle makers in seven countries had taken up licences for the manufacture of the Moulton bicycle.

From the first news of the Moulton model, large firms were interested, particularly so since the bicycle market seemed to be on the decline. It at first appeared that the Raleigh Company, the largest manufacturer of bicycles in the world, would manufacture the machine, under licence, but these earlier negotiations broke down. The Raleigh Company put on the market in 1965 a small wheeled bicycle of similar appearance which did not embody the Moulton patented elements. Moulton Bicycles Ltd. undertook the production and the marketing of their own model in 1963 which, by then, as a result of a series of improvements, had reached Mark IX. This had virtues, in ease of propulsion and comfort, which gave it an advantage over all imitations and it is, therefore, not surprising that subsequently the Raleigh Company acquired the rights to the Moulton bicycle.

Alex Moulton is a clear case of an independent inventor who, to quote his own words, believed that 'a new concept or innovation will always spring from the creative impulse of an individual, and to imagine that it can come from "group thinking" is folly'. Working with a small but carefully chosen team he has produced some most important designs and inventions. The Moulton bicycle is a striking case of innovation in a machine where further progress was generally regarded as unlikely.

REFERENCE

1. Discussion and correspondence with Mr. Alex Moulton.

OXYGEN STEEL-MAKING

PRIOR to the invention of oxygen steel-making after World War II, there had been no major change since the last century in the processes of converting iron-ore into steel. Each of the conventional methods, the Bessemer, Thomas, open hearth and electric furnace, had drawbacks. In the Bessemer, air is blown through tuyeres (pipes) at the bottom of an acid-lined converter into a bath of molten pig-iron. Because of the converter's acid lining, phosphorus and sulphur impurities in the ore could not be removed. In addition, air-blowing yielded steel with an undesirable nitrogen content. The Thomas process, by substituting a basic lining for

L. A. B. Pilkington: float glass

A. E. Moulton: Moulton bicycle

R. Durrer:
oxygen steel-making

J. von Neumann: computers

L. M. Moyroud: photo-typesetting

R. A. Higonnet: photo-typesetting

the acid, solved the problem of phosphorus and sulphur removal, but not that of high nitrogen content. The open hearth, where fuel oil or gas blown on the surface of the ore bath heats the pig-iron in a furnace, produces steel of excellent quality; however, it is slow and costly to install and operate. The electric furnace, relatively expensive to operate, is restricted primarily to specialty steels.

Today, most new steel-making facilities are oxygen units, which reduce both capital and labour requirements. In oxygen steel-making, a jet of pure oxygen is blown through a lance vertically from the top of a converter into the molten pig-iron below. To avoid refractory damage and a phosphorus residue in the steel, the jet is so regulated in distance from the bath and in pressure as to avoid a 'deep' penetration. The reaction generated by the oxygen jet yields steel of high quality.

Henry Bessemer mentioned the possible use of oxygen for steel-making in his fundamental British patent of 1856. However, large-scale production of pure oxygen became a reality only after the development in 1928 of the Linde–Fraenkl process. The first practical step in the steel industry was to add oxygen to the air blast of a Bessemer converter – as had in fact first been done in 1925 – and this became common practice in Europe after 1945. To achieve further reductions in the nitrogen content of the steel produced, nitrogen-free blasts of oxygen and steam or oxygen and carbon dioxide were also used; a Swede, T. Haglund, had suggested the latter mixture in a 1943 patent. Professor Bo Kalling in Sweden took up its development. Oxygen injection into open-hearth furnaces, to speed the reaction, also became common from 1950 onwards.

Professor Schwarz of the Technical University at Aachen applied for a German patent in 1939, which was issued in 1943 in the name of H. A. Brassert & Co., on a process of blowing pure oxygen from above deeply into a bath of molten pig-iron. While this patent revealed some elements of the ultimately successful method, it involved deep penetration, did not clearly set forth actual operating conditions, and was not developed commercially. In 1952 the German steel firm, Mannesmann A.G., licensed the Schwarz patent but never used it in their plant.

An English consulting engineer, John Miles, received patents in 1946 for a method of blowing a fluid containing from seventy-five per cent to ninety-eight per cent oxygen, preferably at an angle, from the top of the converter into the bath. He and some colleagues tested their ideas on a small scale, first at Messrs. Lloyds of Wednesbury and then at the La Boel works in Belgium. However, Miles failed to stimulate any interest among British steelmakers and his invention, like Schwarz's, remained dormant. During the forties, inventors blew oxygen into the converter from below, in the hope that this would remove excessive nitrogen from the steel; however, the idea failed because the heat from the pure oxygen blowing caused refractory damage. During this period of research, most of the experimenters knew of each other's work and often discussed common problems.

A Swiss, Dr. Robert Durrer, who from 1929 had advocated the use of pure oxygen in steel refining, provided the real breakthrough. He had received his Doctorate in Metallurgy from Aachen in 1915. Following a year of scientific work there, he took a position in the iron industry where he remained until 1929 when he accepted the chair of metallurgy of iron and steel at the Berlin Institute of Technology. Commencing in 1929, he performed small-scale experiments with oxygen in steel refining and smelting. The Germans, during World War II, adopted some of his ideas. In 1943 Durrer returned to Switzerland as a member of the management board at von Roll A.G. (formerly known as Gesellschaft der Ludw.

von Roll'schen Eisenwerke A.G.), the largest Swiss steel firm. Durrer became works manager at Gerlafingen, where he assumed the added responsibility of dealing with the firm's metallurgical problems. At this time the Swiss badly needed ways of transforming their low-grade ores into quality steel. Durrer purchased a two-and-one-half-ton converter in the United States, and brought Dr. Hellbruegge from Germany to assist him. Dr. Hellbruegge had served as Dr. Durrer's assistant for several years at the Berlin Institute of Technology and had worked with pure oxygen there as well as later while employed in the steel industry. Together they began small-scale tests in the spring of 1948, which continued into the next year, and proved Durrer's ideas of blowing oxygen from above the converter into the molten pig-iron bath below. Durrer described the beginning:

> 'On the first day of spring, our "oxygen man" Dr. Heinrich Hellbruegge carried out the initial tests and thereby for the first time in Switzerland hot metal was converted into steel by blowing with pure oxygen. . . . With this treatment of one ton of metal the first practical step on a thorny passage has been undertaken.'

Durrer encountered refractory damage and high phosphorus content in the steel, yet his results indicated he had conceived something new and important.

Dr. Theodor Eduard Suess, works manager of Vereinigte Oesterreichische Eisun- und Stahlwerke Aktiengesellschaft (VOEST), a government-owned Austrian steelmaker, contacted Durrer, whom he had known for many years. Like Switzerland, Austria was seeking ways to process their low grade ores. Neither the Bessemer nor the Thomas process was suitable for the Austrian ores of high manganese content. VOEST, faced with a rebuilding programme for their war damaged plant, wanted to replace, if possible, their open hearth with an improved method. The open hearth, upon which they had theretofore depended, used scrap for much of its raw material. However, as the Austrian steel firms employed more scrap, a shortage developed, and they were forced to rely on the more expensive hot metal. This threatened to make the open hearth economically unattractive.

In April 1949 Dr. Hellbruegge notified Dr. Herman Trenkler, plant manager at VOEST's Linz plant that 'we are now working for production'. The next month Trenkler visited Durrer and Hellbruegge at Gerlafingen. A meeting was held shortly thereafter with representatives of the following steel firms. Von Roll, VOEST, Osterreichisch-Alpine Montangesellschaft (Alpine), another government-owned Austrian firm, and Mannesmann, a German combine. The firms agreed to undertake a joint study of the use of pure oxygen in steel-making. Dr. Suess organised the experiments at VOEST's Linz plant, with Trenkler, Hubert Hauttmann and Rudolf Rinesch participating. They started testing with makeshift equipment in order to meet some of the problems Durrer had predicted for the process. Using a two-ton converter, they followed Durrer's instructions to blow vertically from above into the bath. At first their results were less satisfactory than those achieved by Durrer. Then an accident occurred which led them to new data indicating the process should be modified. They established certain essential operating conditions and got much improved results.

Simultaneously with Suess's group, Roesner and Kuehnelt of Alpine conducted a series of tests, employing five- and ten-ton converters. Both firms, confident of success, authorised pilot plant construction, and shortly thereafter, in 1950 and 1951, full scale production facilities. Commercial production commenced at

VOEST in 1952 and at Alpine the following year. Thus the revolutionary process moved from conception to production in three years.

VOEST acquired the Swiss firm, Brassert Oxygen Technik A.G. (BOT) and sold a fifty per cent interest to Alpine. Then they gave BOT the exclusive world-wide rights to exploit the oxygen steel-making inventions. Suess had applied for patents in his own name, claiming he was the sole inventor. After litigation with VOEST, Suess surrendered to VOEST the rights to the process. A small Canadian steel firm, Dominion Foundries and Steel, Ltd., was the first firm outside Austria to apply the process commercially. In the United States, Henry J. Kaiser Co. became the exclusive agent for BOT. McLouth Steel Corp., then a small U.S. steel producer, purchased from Kaiser for $100,000 the know-how and engineering services essential for operating the oxygen process. Interest soon spread to other countries, and inventors improved the process. Among the last to adopt the new techniques were the largest American steel firms and the steel industry in the U.S.S.R.

Thus, though the idea of using pure oxygen in steel-making can be traced to Henry Bessemer, the present commercial oxygen top blowing process stemmed from the experiments of Dr. Robert Durrer, who advocated employing pure oxygen in steel-making from 1929 when a Professor in Germany, and who tested his ideas in the late forties while associated with a Swiss steel firm. Dr. Suess and his staff at the Austrian steelmaker, VOEST, learned of Durrer's findings, and building on his foundation, perfected the necessary operating conditions.

In this case, individuals at small European steel firms, one a former Professor, all experimenting on a modest scale, were the prime movers in the invention and initial commercial development.

REFERENCES

1. Henry J. Kaiser Co. *et al.* v. McLouth Steel Corp., 257 Fed. Supp. 372 (Dist. Ct. E. D. Mich., S. Div., 1966), and the record in said case.

2. Staratt, F. W., 'L D in the Beginning', *Journal of Metals*, July 1960.

3. Cuscoleca, O., 'Development of Oxygen Steelmaking', *Journal of Metals*, July 1954.

4. Suess, T. E., 'Production of Steel by the Oxygen-Impingement Process', *Iron and Steel Engineer*, Mar. 1952.

5. Cuscoleca, O., and Rosner, K., 'The Development and Present State of the L D Process', *Iron and Steel Institute Journal*, June 1959.

6. Adams, W., and Dirlam, J. B., 'Big Steel, Invention and Innovation', *The Quarterly Journal of Economics*, May 1966.

7. Maddala, G. S., and Knight, P. T., 'International Diffusion of Technical Change: A Case Study of the Oxygen Steel-Making Process', *The Economic Journal*, Sept. 1967.

8. Allen, J. A., *Studies in Innovation in the Steel and Chemical Industries* (Manchester University Press, 1967).

ELECTRONIC DIGITAL COMPUTERS

WITHIN the past quarter of a century, the electronic computer has changed from a bulky primitive instrument employed in wartime ballistics research to a compact sophisticated invention of great economic value.

There are two major types of electronic computers: (1) the analogue, which represents numbers by measuring some physical quantity such as length or volume, and which is limited in accuracy by the precision with which the physical quantity may be measured; and (2) the digital, which counts things and whose accuracy is limited by the number of digits the machine can handle. The origin of the more widely used type, the digital, concerns us here.

The abacus, an ancient counting device still used today, is a form of digital computer. At the beginning of the seventeenth century, several men either described or built machines which were forerunners of the modern adding machine. The story is by now well documented of how Charles Babbage, a brilliant and eccentric Englishman, in the nineteenth century laid the foundations for the modern automatic computer in his 'Difference Engine' and his 'Analytical Engine'. But, sound as his theories were, he died in 1871 with his machines unfinished. They were designed for mechanical operation and required thousands of parts machined to a precision unattainable in that day. His ideas for the Analytical Engine lay dormant until well into the next century.

Other inventors, notably Schuetz and Wiberg of Sweden, and Grant of the United States, worked on less ambitious 'Difference Engines' to compute and print mathematical tables. Herman Hollerith, experimenting in the latter part of the nineteenth century, advanced the use of punched cards when he developed equipment for the entering, classifying, distributing and recording of data on these cards. His system was used to prepare the 1890 U.S. Census. Subsequently he formed a firm, which ultimately became IBM, to manufacture accounting equipment. Hollerith and others brought the calculating machine art to a high degree of perfection.

Theoretical studies of computable numbers and functions were published by mathematical logicians during the 1930's and helped to make the design of practical computers possible. The most influential of these logicians were A. M. Turing of King's College, Cambridge, and Emil Post of the College of the City of New York. Post's ideas date from 1920–2 but were not published until 1936, the year of Turing's fundamental paper (on the Entscheidungs problem) in the Proceedings of the London Mathematical Society. Concepts of computability, later shown to be equivalent to Turing's, are the theory of λ-definability of Alonzo Church (1932) and the Herbrand-Gödel-Kleene theory of general recursiveness published in 1934.

A mathematician at the Bell Telephone Laboratories, George R. Stibitz, designed a relay semi-automatic computer in 1937, which was built at Bell under the direction of Samuel B. Williams. Independently of Stibitz, but at about the same time, Howard H. Aiken, a candidate for a doctorate in physics at Harvard University, recognised the advantages for his theoretical research which a machine offered over manual means of computation. First, he conceived a computer to evaluate simple polynomials; then he proposed one to handle a variety of problems. IBM agreed to help with this work in 1939. Collaborating with IBM engineers J. W. Bryce, C. D. Lake, B. M. Durfee and F. E. Hamilton, and with financial backing from the U.S. Navy, Aiken built the Mark I; it was completed and operating at Harvard in 1944. An electro-mechanical device, it incorporated about three thousand telephone relays, and received instructions from punched paper tape. Though cumbersome, standing eight feet high and measuring fifty-one feet in width, slow and noisy in operation, and basically a mechanical device requiring external programming, the Mark I constituted an important advance.

A third investigator, Friedrich Züse, began computer experiments while a student at the Technical High School in Berlin in 1936. Experimenting on a small scale with a few colleagues, he developed several electro-mechanical computers with speeds comparable to that of the Mark I. The Germans used his Z4 to make aircraft design calculations in 1942. By this time he had received some help from two Technical High Schools, as well as from the German Experimental Aeronautical Institute. However, when he tried to develop a truly electronic computer with Dr. Schreyer of Charlottenburg, official support was withdrawn and his work stopped. Züse resumed the development of computers after the war; his firm was purchased by Brown Boveri in 1964.

Even before Aiken's Mark I reached completion, men were developing a truly automatic electronic computer. An abortive attempt, cut short by the war, was made between 1938 and 1942 at Iowa State University by Dr. John V. Atanasoff, a professor of physics and mathematics, with the assistance of Clifford E. Berry, a graduate student. In 1942, Dr. John Mauchly, a physicist at the Moore School of Electrical Engineering of the University of Pennsylvania, prepared a report on the possibilities of making an electronic computer. When his written report was lost, Mauchly and an electrical engineer at the University, Dr. J. Presper Eckert, were able to reconstruct it, using the shorthand notes of Mauchly's secretary. Eckert showed how to convert Mauchly's ideas into apparatus. At this point, the war speeded up development. Members of the Moore School, in conjunction with the U.S. Army's Aberdeen Proving Ground, were computing artillery firing tables for the Army. The computation was painfully slow; a Bush analogue computer and a corps of one hundred girls calculating by hand were the principal aids. Lieutenant Herman H. Goldstine, a former mathematics professor at the University of Michigan, who was in charge of the computations, recognised the potentialities of Mauchly's and Eckert's ideas and persuaded the Army to finance the construction of a machine. All the necessary components had been in existence for years. To prove the design for the ENIAC, Mauchly and Eckert tested five per cent of the machine; then with the help of ten engineers, they spent two and a half years constructing it. The ENIAC contained eighteen thousand vacuum tubes, and could complete five thousand additions a second using electronic pulses. While the ENIAC stored instructions internally, it required hours of manual rewiring to change from one programme to another. Invention of the stored programme was a fundamental advance for computers. With the stored programme, some of the operations are contained within the circuitry of the machine; they can be called into play by number, thus increasing the speed and flexibility of computation. Another important concept, 'conditional transfer', permits the computer to choose between alternative operational sequences once it obtains certain intermediate results.

John von Neumann, a renowned Hungarian-born mathematician, was a U.S. consultant on the atomic bomb project at Los Alamos during World War II. In his study of the interaction of shock waves, he, too, saw the need for replacing hand-computation with machines. In the summer of 1944, von Neumann, then also a consultant at the Aberdeen Proving Grounds, discussed the work of the Moore group with Herman Goldstine. In 1946 von Neumann, Goldstine and Arthur W. Burks wrote the landmark report 'Preliminary Discussion of the Logical Design of an Electronic Computing Instrument' in which the authors described the ideas of 'stored programme' and 'conditional transfer', concepts which were soon incorporated in computer hardware. The Moore group initiated a stored programme computer development, the EDVAC, in 1945, which was completed at the Aberdeen Proving Ground in 1950. The EDSAC, begun in 1946 and operational in 1949 at the mathematics laboratory of the University of Cambridge, was the first stored programme computer.

After the war, the original Aberdeen group split up; von Neumann and Goldstine went to Princeton where von Neumann supervised the construction of experimental computers of his design. Mauchly and Eckert decided to produce commercial computers. Their firm, Eckert-Mauchly Computer Corp., later became a division of Remington Rand, which in 1951 marketed the UNIVAC I, a stored programme machine installed at the Bureau of the Census. Remington also acquired Engineering Research Associates, a second important firm in electronic computers, thus making Remington foremost in the fledgling industry. Engineering Research Associates had been organised after the war by a group of Navy men who had worked on computers during the war. William Norris, now the head of Control Data, was one of the organisers. IBM, dominant in data processing and office machinery, remained cautious about the commercial potentialities of the computer, though they worked on experimental models following their participation in the Mark I development. But the loss of the Census Bureau business to Remington spurred IBM into action. Once committed, IBM took the lead within a few years. Their strong financial resources, established sales organisation and engineering force, and their determination to become the leader in the field contributed to their success. Other firms, such as General Electric, Radio Corporation of America, Burroughs, Minneapolis-Honeywell and National Cash Register entered the field. One specially interesting case is that of the American firm Control Data which, established only in 1957, by 1967 ranked in profits second only to IBM. Its success has been due largely to an outstanding technical director, Seymour Cray. Control Data produced the first solid-state computer in 1960 and became the leader in the industry for scientific computer systems.

British work commenced about 1946 after scientists and engineers visited the early American installations. Development programmes for the Admiralty began at Manchester, London and Cambridge Universities, and at Elliott Brothers Research Laboratory. Other projects included a co-operative development programme with the National Physical Laboratory and English Electric as participants; Manchester University's work with Ferranti's help; and the aforementioned Cambridge EDSAC computer programme under Dr. M. V. Wilkes's direction. The National Research and Development Corporation, a government supported institution, gave help to several projects beginning in 1950, and served as a patent-holding agency for key government and university patents. Among the patents held were those for the cathode-ray tube storage system pioneered by Professor F. C. Williams of Manchester University. Leo Computers, a firm established by J. Lyons, the caterers, developed the first computer for commercial use at a cost of about £129,000; this machine was based on the EDSAC. During the 1960's, the competitive pace and high development costs forced many of the British firms to merge their computer activity.

The invention, after the war, of the transistor and other semi-conductor devices; integrated circuits; the development of improved methods of memory storage and better means of feeding data in and out of the computer greatly increased its speed, reliability and compactness. A whole new industry has emerged to supply the computer business with components, peripheral equipment and services.

To sum up: Some of the fundamental ideas for the electronic digital computer go back to Babbage in the nineteenth century. They were actively revived in the present century by a Harvard student, Howard Aiken; a Bell Telephone Laboratories mathematician, Stibitz; and a German engineering student, Züse. They were advanced by the design and construction of the first electronic computer, the ENIAC, by Mauchly and Eckert of the Moore School of Engineering of the University of Pennsylvania with U.S. Army support, and were improved upon by the addition of stored programming and conditional transfer contributed by von Neumann and his associates at the U.S. Army Aberdeen Proving Ground. Small firms, one organised by Mauchly and Eckert, stimulated the creation of a computer industry. A larger firm, Remington Rand, entered the field through the acquisition of two of these small pioneering firms and the success of Remington forced IBM to move. IBM, once in the industry, soon became its leader, although several small firms has subsequently grew rapidly as a result of outstanding invention and development.

REFERENCES

1. Freeman, C., 'Research and Development in Electronic Capital Goods', *National Institute Economic Review*, no. 34, Nov. 1965.

2. Kendall, M. G., 'The Use of Computers in British Industry', *The Three Banks*, Mar. 1966.

3. von Neumann, J., *The Computer and the Brain* (Yale University Press, New Haven, 1958).

4. Serrell, R., Astrahan, M. M., Patterson, G. W., and Pyne, I. B., 'The Evolution of Computing Machines and Systems, *Proceedings of the I.R.E.*, May 1962.

5. Burck, G., 'The Boundless Age of the Computer', *Fortune*, Mar. 1964.

6. Burck, G., ' "The Assault" on Fortress I.B.M.', *Fortune*, June 1964.

7. Burks, A. W., Goldstine, H. H., and von Neumann, J., 'Preliminary Discussion of the Logical Design of an Electronic Computing Instrument', *Datamation*, (Sept. 1962); also p. 36 (Oct. 1962).

8. Stibitz, G. R., and Larrivee, J. A., *Mathematics and Computers* (McGraw-Hill, New York, 1957).

9. Bowden, B. V., Editor, *Faster than Thought* (Pitman, London, 1953).

10. Staff of the Harvard Computation Laboratory, 'A Manual of Operation for the Automatic Sequence Controlled Calculator', *The Annals of the Computation Laboratory* (1946).

11. Morrison, P., and Morrison, E., 'The Strange Life of Charles Babbage', *Scientific American*, Apr. 1952.

12. Bergstein, H., 'An Interview with Eckert and Mauchly', *Datamation*, Apr. 1962.

13. Thomas, S., *Computers* (Holt, Rinehart & Winston, New York, 1953).

14. Bernstein, J., *The Analytical Engine: Computers, Past, Present and Future* (Random House, New York, 1964).

PHOTO-TYPESETTING

PHOTOGRAPHIC methods of preparing the masters from which copies can be printed – somewhat illogically called photo-typesetting, for type is not in fact employed – are beginning to make a significant impact on the printing industry.

Several thousand machines are now in use and the rate of installation is accelerating.

With the conventional typesetting machine, the operator uses a keyboard to cast either an individual character (Monotype) or a line (Linotype or Intertype) in a soft, quicksetting metal. When a page has thus been produced and any necessary corrections made in the proof, a mould is taken from the type. This mould is used to cast the printing blocks which, fitted into a letterpress machine, produce the final printed page.

The photo-typesetting machine replaces metal type by a photographic image of the letters, under control from a keyboard or from a tape from a computer. The master copy of the page is, therefore, a piece of film instead of a block of metal. This film is usually employed to make a plate for offset-lithographic printing but it could also be used to produce a plate for letterpress printing from a photosensitive material.

As against hot metal setting, the advantages of photo-typesetting are that it can be much faster, that difficult matter can be set more easily and that the capital locked up in stocks is much reduced. The disadvantages of photo-typesetting are the difficulties in inserting corrections and the higher first cost of the machine. But cheaper and faster photo-typesetting machines and the fuller utilisation of their great capacity should make them substantially more economical than the conventional types.

The photo-typesetting machines so far marketed fall into three classes: first came adaptations of hot metal machines capable of setting three or four characters a second; then electro-mechanical machines capable of 30 to 250 characters a second and more recently electronic machines which use cathode ray tubes to form letters (and are known as CRT machines) capable of 100 to 8,000 characters a second.

Although many ideas were patented in earlier years, it was not until 1936 that a successful adaptation of a linecasting machine was patented by a Swiss, C. E. Scheffer. Under his invention, the linecasting mechanism was replaced with matrices which could be assembled to produce the line wanted and which were then photographed character by character. Scheffer's invention was taken up by the American Intertype Corporation which, after nearly a decade of development under the direction of H. R. Freund, produced the Intertype Fotosetter, the first machine being installed in the United States Government Printing Office in 1946. In 1954 the Monotype Corporation marketed the 'Monophoto', the second hot-metal machine to be successfully transformed into a photosetting machine.

The first wholly novel and successful photosetting machine was the Photon (known in Europe as the Lumitype) invented by two French telephone engineers, René Higonnet and Louis Moyroud. In their own homes, they started work on the design of a photographic type-composing machine in the summer of 1944 and by April 1946 they had completed their first prototype at a cost of under $1,000. This crude machine was good enough to prove the basic design. The two inventors demonstrated the machine to the American Lithomat Corporation (later to become Photon Inc.) and they were given an advance of $10,000 to develop a pre-production machine. At this time Moyroud gave up his ordinary job to work full-time on the machine, which was completed by the spring of 1948. The inventors returned to the United States with the machine in July 1948 to continue its development. Lithomat, however, could not raise further funds for this purpose. The Graphic

Arts Research Foundation was then formed, with the aid of Dr. Vannevar Bush, to raise funds from the printing and publishing industries in order to finance the development of the machine, which by 1953 had cost $720,000. Development continued throughout the 1950's. Sales were at first low but by 1967, 545 machines had been sold in a number of countries and by 1968 Photon had won a significant share of the United States market.

The introduction of the Photon encouraged other firms to develop similar types of machine; the first of these to appear was the Linofilm from Mergenthaler-Linotype. The most revolutionary development, a photosetting machine on which the image is formed on a cathode ray tube, dates back to the patent applied for in 1948 by E. Dinga and assigned to Mergenthaler-Linotype. Commercial developments of such machines did not start until about 1960 and the first installations were made in 1966. Interest in large and fast machines of this type was stimulated in the United States by the Government, which in 1963 issued a requirement for a machine to set such massive documents as the spare parts lists of the U.S. Air Force. The Government Printing Office awarded a development contract in 1964 to Mergenthaler-Linotype which, in collaboration with the CBS Laboratories, produced its Linotron machine. On this an entire page is composed at a time on the cathode ray tube and 1,000 to 2,000 characters can be set per second. Other CRT photosetters have been developed in the United States with private funds. Alphanumeric, a small company, has developed a machine, entirely electronic, capable of setting 8,000 characters a second. The cost of developing and constructing the first of these machines from 1963–7 was $518,000. In 1968 this company was bought up by IBM, which will use the photosetter in conjunction with its computers.

In two other cases CRT machines developed in Europe by small organisations have been taken up by much larger companies. The Digiset was developed by Dr. Rudolf Hell in his small electronic company in Germany. The machine was introduced in 1966 and in 1968 the Radio Corporation of America, which had been working on a CRT photosetter of its own design, announced that it would be selling the Digiset in the United States as the R.C.A. Videocomp. The second machine was the Paul-PM. This was invented in England by P. Purdy, a former photographer, and R. Mackintosh, a journalist, who ran a small printing company. K. S. Paul and Associates, importers of printing machinery, supported the development for a time but later assistance had to be obtained from the National Research and Development Corporation. When the development was completed at a cost of about £150,000 in 1967, the Mergenthaler-Linotype Corporation bought control of K. S. Paul and Associates and it is now marketing the machine as the Linotron 505.

Photosetting machines have thus been, in large measure, the creation of individual inventors and small companies, whose innovations have been taken over by larger organisations. The most significant contribution to come from the established producers of printing machinery was Mergenthaler-Linotype, with its early patent on CRT machines and the Linotron.

REFERENCES

1. *The Penrose Annual* (1920–67).
2. Interviews with Mr. H. W. Larkem, Mr. Peter Purdy and Mr. Ronald Mackintosh.

3. Correspondence with Mr. Louis Moyroud.
4. B.P. 496,886 and file report.
5. Annual Report, Alphanumeric Inc. (for 1966).
6. Balzer, F., 'Survey of Filmsetting Installations', from *Print in Britain*, Sept. 1967.
7. Duncan, C. J., 'Computers and their Impact on the Publishing Process', *Printing Trades Journal*, (Feb. 1968).
8. U.S.P. 2,624,798.

THE PREVENTION OF RHESUS HAEMOLYTIC DISEASE

THE first recorded description of haemolytic disease of the newborn was in 1609 but until the discovery of the human blood groups during this century, and in particular the description of the group known as the Rhesus (Rh.) factor in 1939, no treatment was possible. The American physician and research scientist, Dr. P. Levine, discovered and described the hitherto unknown antibody (later named the Rhesus factor by Drs. Wiener and Landsteiner, then at Wisconsin University) when he established that the disease occurs when a Rh. negative mother produces a Rh. positive baby, who inherits this blood type from its father. Since the baby's cells usually, although not always, invade the mother at delivery the first baby is not normally affected; such foreign foetal cells entering her circulation may cause her to react, as the body does to any foreign material, by producing antibodies to destroy them. The presence of these antibodies becomes a permanent factor in the mother's blood stream and, as the level of antibodies may rise after each birth, subsequent Rh. positive children's cells are attacked with increasing severity. Although about eight per cent of all pregnancies may give rise to antibodies, only about 1 in 180 babies in western countries are affected because the child may not be Rh. positive, may be a first child, or may be protected by one or other of several factors. In the milder form of the disease anaemia and jaundice develop; increasing severity leads to stillbirth, to infant death or to brain damage of the surviving children. After the discovery of the cause of the disease, the only remedy available was the drastic one of immediate blood transfusion of the newborn to clear the antibodies; later a technique was developed of changing the blood of unborn babies repeatedly until labour could be safely induced, but this also involves dangers to the mother and child.

An entirely new method, that of prevention rather than treatment, arose almost simultaneously in England and the United States. In England it had an unusually fascinating beginning. During the 1950's, Dr. Philip Sheppard, an Oxford geneticist, was working on the behavioural genetics of the butterfly *Papilio machaon* and advertised in the Amateur Entomological Society Journal for some living specimens. Dr. C. A. Clarke, a consultant physician in Liverpool, who took an amateur interest in butterflies, was able to supply some hybrids he had bred. These surprised Sheppard as, scientifically, he would have expected them to look intermediate between the two species; instead they all resembled just one of their parents. So in 1953 they began to investigate the butterflies genetically. Inheritance of genetic variability occurs essentially in the same way in all animals – in human populations hair and eye colour and blood groups are examples of inherited variants.

In 1956 Dr. Sheppard went to Liverpool in order to work more closely with Dr. Clarke, taking with him a small grant of £2,000–£3,000 which he had received from the Nuffield Foundation for his research on butterflies. Dr. Clarke suggested they should study similar systems of inheritance in humans and a year later, together with Dr. R. B. McConnell they set up an heredity clinic. The Rh. complex with its system of closely linked genes particularly attracted their attention.

Dr. Levine had first observed in 1943 that another factor influenced the incidence of Rh. haemolytic disease. He noticed that in the majority of families with a history of the disease the mother belonged to the same ABO group as the father and that when the opposite was the case Rh. antibodies were not usually formed. Further supporting evidence came in 1956 from the work of H. R. Nevanlinna and T. Vainio. The explanation of why ABO incompatibility gives protection is that when foreign foetal cells enter the mother's circulation they are immediately eliminated and consequently she is not stimulated to form anti-Rh. antibodies. In 1957 Sheppard and Clarke and two other doctors at Liverpool, Dr. McConnell and Dr. R. Finn became greatly interested in the interaction between the rhesus and ABO blood systems. A large family study was carried out and an article describing the results was published in 1958. These and further studies have confirmed that ABO incompatibility reduces the risk of Rh. haemolytic disease by about one-fifth.

Early one morning Mrs. C. A. Clarke woke up her husband to suggest he might use the antibody formed against the Rhesus factor to prevent the occurrence of rhesus haemolytic disease; this idea was discussed in the Department and at a Liverpool symposium in 1960 Dr. Finn tentatively put forward the view that it might be possible to destroy foetal blood cells in the maternal circulation following delivery by means of suitable antibody, thus mimicking the natural protection afforded by ABO incompatibility.

During the next four years the Liverpool workers carried out experiments. Male volunteer blood donors with Rh. negative blood, who had previously received injections of Rh. positive red blood cells, were given injections of anti-Rh. antibody. After some initial failures, improvements in the type and dose of antibody led to the effective clearance of foreign blood cells before active antibodies could be formed against them. The trials were carried out in co-operation with the Liverpool Regional Blood Transfusion Service under the direction of Dr. D. Lehane. This organisation approached the Lister Institute, a non-profit-making body, which agreed to produce the antibody serum. The results of the first experiments of the Liverpool workers were published in an article in the British Medical Journal in May 1961. In 1964 they started strictly controlled clinical trials in Liverpool's five maternity units; teams in Sheffield, Leeds, Bradford and in Baltimore (where Dr. Finn went for a year) also ran similar trials. Another group experimented at Freiburg.

Meanwhile Dr. J. G. Gorman of Columbia University, an Australian who had emigrated to the United States in 1955, was assigned to the blood bank at Columbia Presbyterian Hospital, New York and there came into contact with Dr. V. J. Freda, an obstetrician. In 1959, Gorman and Freda, together with Dr. W. Pollack, an English immunologist of the Ortho Research Foundation in the United States, made an approach to the study of this disease which was direct, as contrasted with that of the English scientists. Gorman first learnt of the principle of passive antibody mediated immuno-suppression from Florey's textbook on General Pathology which he 'happened to read only because I was presented with a free copy by the publisher'. Of the many research schemes they discussed, this seemed the most promising avenue and there was no theoretical reason why it should not work if applied to Rh. negative mothers. They had mapped out a project almost identical with that of the Liverpool workers, although based on a slightly different line of reasoning, when the B.M.J. article, together with a hopeful editorial on the subject, appeared in May 1961. Although their first reaction was 'dismay at being scooped for publication by Liverpool, the effect in New York was to give a tremendous boost to our own plans. We were now suddenly working on a feasible and reputable research project whereas before we had been largely talking to deaf ears.' But even before the article appeared, they had been successful in receiving a grant from the Health Research Council of New York City and afterwards the Ortho Pharmaceutical Foundation contributed considerable funds for the research. The Company was responsible for producing, as early as 1961, the gamma globulin which, as an antibody, was superior to the raw serum because of its purity and freedom from hepatitis. Ortho Pharmaceuticals is the only American producer of this gamma globulin.

The American research workers carried out trials for three years on male volunteers at Sing Sing Prison. There was free exchange of information between them, the Liverpool workers and others in the field, in particular when some trials revealed there was an enhancement rather than the desired suppression of immunity with the passive antibody. It was discovered that the danger was dose-related: small doses might increase the risk of immunisation, large doses were safe and completely immunosuppressive. Clinical trials were started on American women at risk in 1964 and by 1967 some 3,000 mothers had been treated.

An unusually interesting case of individual initiative was revealed in 1967 when an American doctor, Eugene Hamilton of St. Louis, stood up at one of Dr. Gorman's lectures and announced that he had started clinical trials in April 1962. He had read the B.M.J. article of May 1961 giving the first results of the Liverpool workers, and had been impressed with the logic of the approach. He carefully prepared his own plasma from Rh. negative mothers who were severely sensitised and had delivered stillborn foetuses but had no history of hepatitis or other communicable disease. Unknown to others working in the field, he had for five years given protection with successful results to three times as many mothers as any other of the clinical trials, thus doubling the world figures of proven effectiveness of the treatment.

In Australia, since August 1967, about fifty per cent of women at risk have been given the treatment through the agency of the Red Cross; in Germany, due to the work of Schneider and Preisler, who started trials on male volunteers in 1962, the anti-Rh. gamma globulin is available and is being used. The United States and British health authorities have been cautious in approving the treatment; in the former it was passed for general use in June 1968; in Britain it has been available for selected, high-risk cases from the beginning of 1968.

This is a case where the important discoveries were made by scientists in Britain and the United States who became acquainted with each others' work through the free publication of results. In England the work began with the study of butterflies and a method of thought originating in one branch of science was successfully applied to another, apparently unconnected, area of study. A practising doctor in America, operating independently, carried out early clinical trials with extraordinary success. One pharmaceutical company in the United States was interested in the work from the start and made substantial grants for research. But in the early stages of this remarkable discovery, especially in England, the funds available or needed were comparatively small.

REFERENCES

1. Clarke, C. A., Finn, R., McConnell, R. B., and Sheppard, P. M., 'Intern. Arch. Allergy', *Applied Immunology*, 13 (1958), pp. 5–6.
2. Finn, R., *Lancet*, I (1960), pp. 526–7.
3. Finn, R. *et al.*, *British Medical Journal* (1961), pp. 1468–90.
4. Finn, R. *et al.*, *Nature*, 190 (1961), pp. 922–3.
5. Freda, V. J., and Gorman, J. G., *Bulletin of the Sloane Hospital for Women*, 8 (1962), pp. 147–58.
6. Clarke, C. A. *et al.*, *British Medical Journal*, I (1963), pp. 979–84; and (1965) pp. 279–83.
7. Clarke, C. A., and Sheppard, P. M., *Lancet* (Aug. 14, 1965), p. 343.
8. Clarke, C. A., *Vox Sang*, II (1966), pp. 641–55.
9. Freda, V. J., Gorman, J. G., and Pollack, W., *Science*, 151 (1966), p. 828.
10. Gorman, J. G., Freda, V. J., Pollack, W. J., and Robertson, J. G., *Bulletin of the New York Academy of Medicine* (June 1966).
11. McConnell, R. B., *Annual Review of Medicine*, 17 (1966), pp. 291–306.
12. Smith, C. H., *Blood Diseases of Infancy and Childhood*, 2nd edition (The C. V. Mosby Co., St. Louis, 1966), pp. 121–65.
13. Hamilton, E. G., *Rh. Isoimmunization: A Simple Method of Prevention using Anti-D Antibody* (unpublished paper).
14. Correspondence and interview with Professor P. M. Sheppard.
15. Correspondence with Dr. J. G. Gorman.

SEMI-SYNTHETIC PENICILLINS

THE early penicillins, despite their great value, had serious limitations. Penicillin G, to be effective, had to be injected and had virtually no effect on gram-negative bacilli. The later Penicillin V could be given orally but constituted no advance in controlling a wider range of infections. In the meantime, a penicillin-resistant organism, *Staphylococcus aureus*, was causing widespread infection, especially in the confined quarters of hospitals. Efforts to overcome these problems were made in many directions. Some scientists, especially J. C. Sheehan in the United States, worked on the synthesis of penicillin.[1] Others, such as E. B. Chain, believing that many details of the mechanism of biosynthesis of the penicillin molecule remained unsolved, continued their studies of penicillin fermentation. The more important of the American firms producing antibiotics, with their large research laboratories, had largely abandoned research on penicillin fermentation in favour of a search for new antibiotics.

It was the Beecham Group which made the important breakthrough. In their laboratories in 1957 the penicillin nucleus (6-amino penicillanic acid: 6-APA) was isolated. Once 6-APA became available it was possible to add side chains to it by chemical reactions (hence the term semi-synthetic penicillins) and thus produce

[1] In 1959, two years after the fundamental discovery in the Beecham Laboratories, J. C. Sheehan and K. R. Henery-Logan announced the successful synthesis of 6-APA. Of this Chain has written:

> 'This method cannot compete economically with the biological method of production but, in the absence of the latter, apart from its scientific interest, it would have had great practical importance as a way to obtaining penicillins modified in their side chain.'

large numbers of new penicillins which could then be tested to determine whether they were improvements upon the older penicillins.

Beecham was a long established British company which had much earlier become known all over the world for its proprietary medical products. After 1928 the Company expanded by the purchase of other firms selling proprietary medicines. It later acquired interests in the food, drink and toiletry industries. Although the pharmaceutical side of the Beecham Group was gradually becoming more important, even in 1965 sales in other products accounted for two-thirds of the total.

Up to 1946 Beecham had conducted little or no research. In that year the Beecham Research Laboratories were set up at Brockham Park. The Company apparently believed, as it proved wrongly, that with the introduction of the National Health Service, the demand for their popular proprietary medicines would quickly decline and that the future would lie mainly in the sale of ethical drugs. The decision to devote a large annual budget to pure research was largely due to the influence of Mr. H. G. L. Lazell, who up to 1951 was a Director of the Beecham Group, became Managing Director in that year and Chairman in 1958 and was responsible for the later energetic development and commercial furtherance of the Beecham discoveries.

The Beecham Group had appointed Sir Charles Dodds as its Chief Scientific Consultant. At his suggestion in 1954 an approach was made to Dr. (now Professor) E. B. Chain to advise the Company on possible ways of entering the antibiotic field. E. B. Chain was, of course, intimately associated with the early discoveries of penicillin in Oxford by the team working with (the then) Professor Florey. In 1948 Chain had gone to Rome to become the Scientific Director of the International Research Centre for Chemical Microbiology, Istituto Superiore di Sanità. He strongly recommended the Company not to try to discover new antibiotics through screening techniques but to intensify the work in which he was still engaged, the attempt to modify the penicillin molecule, particularly with the aim of obtaining penicillins resistant to the action of the enzyme penicillinase which was produced by the staphylococcal strains then encountered with alarming frequency in hospital wards.

His advice was accepted. The penicillin work of the Beecham Research Laboratories started in 1956 under the direction of the late J. Farquharson and with a team including F. P. Doyle, J. H. C. Naylor, G. N. Rolinson and F. R. Batchelor. The last two scientists mentioned went for a time to Rome to work with Chain and his colleagues, making use of the pilot fermentation plant there while a similar plant was being built at the Beecham Research Laboratories. During this period, at Chain's suggestion, a new penicillin, para-amino benzyl-penicillin, was made by fermentation. In this penicillin the 'tail' was highly reactive, thus lending itself to modification by chemical treatment for the purpose of producing new penicillins.

But a startling new turn was given to the Beecham research programme when in 1957 the existence was first predicated, and then established, of the penicillin nucleus itself. This discovery had in it an element of chance in the sense that it derived immediately from the reconsideration of an observation which had been made many times in the past but had been set to one side and its significance thereby missed.

During 1957 the scientists at the Beecham Research Laboratories noticed what had been previously observed: that the chemical and microbiological assays of the penicillin content of experimental broths gave very different results (the microbiological method measures the degree to which the penicillin present kills or stops the growth of bacteria; the chemical method measures the total quantity of penicillin in the broth). It was further noticed that the difference proved greatest when no precursor was used and vice-versa. In particular, when the precursor favouring the formation of p-amino benzylpenicillin was omitted, the chemical test showed more penicillin present than the biological test could account for in terms of dealing with bacteria.

This led to a hypothesis by the Beecham scientists that this might be the elusive molecular core of penicillin. The hypothesis was tested by adding to the brew the appropriate chemical to make a known penicillin such as G or V. By this means, within three days, the presence of 6-APA was proved and was confirmed in the succeeding month by chromatographic techniques.

A patent application was taken out in 1957, and granted in 1959, for the isolation of 6-APA under the names of F. P. Doyle and J. H. C. Naylor of the Beecham Research Laboratories. The scientific results were published in January 1959 in *Nature* in an article by Batchelor, Doyle, Naylor and Rolinson.

The next stage was to improve the techniques for the extraction of 6-APA in its pure form. This had its own particular difficulties but the flow of new compounds and their screening for effectiveness had begun and by February 1959 two very promising compounds had emerged. In October 1959 an oral penicillin, 'Broxil', was put on the market. In September 1960 'Celbenin', which resisted staphylococcal penicillinase, was made available. In July 1961 'Penbritin' was introduced; it was highly effective in a broad range of diseases caused by bacteria.

The cost of the research and development by Beecham has been considerable. Between 1947 and 1957, when the existence of the penicillin nucleus was predicated, investment in research was about £2½ million. After the nucleus was isolated as a stable solid, annual investment on R and D was increased and by 1966 the cumulative total was about £12 million. The research and development staff of about 50 (18 graduates) in the Company in 1952 rose to 650 (175 graduates) in 1968.

The details of the manufacturing and commercial tasks confronting the Company in the full exploitation of their new products do not fall within this story. But it may be mentioned that Beecham originally had no experience of the intricacies involved in building an antibiotic plant and an American firm, Bristol Myers, in return for certain concessions, gave valuable assistance. The collaboration of the research teams of the two firms and also their independent approaches widened and speeded up the introduction of the new penicillins. Other firms in the United States, Germany and Belgium had also gone ahead after the first published report of the Beecham scientists; some of them purchased licences for Beecham products, others applied for patents on their own semi-synthetic penicillins.

The commercial success of this discovery can be judged by the facts that Beecham sell their brand-named penicillins in seventy-five countries; that overseas sales rose from less than £200,000 in 1959–60 to nearly £8¾ million in 1966–7, and that the sales by Beecham licensees are of the order of £35 million.

This is a case of an important breakthrough brought about by a large company. Whereas the early penicillins were the discoveries of scientists working in universities, these later products came from a large industrial research laboratory, although the latter enjoyed the advice and co-operation of eminent outside scientists. The Company was comparatively new to the field of large-scale industrial research; it deliberately and courageously chose a narrowly defined objective; it achieved its success along a route which could not have been predicted beforehand, and was indeed regarded in some quarters as unpromising both scientifically and commercially. This success was 'science-based'; but although the Company has spent, and continues to spend, large sums on research and development, it is still predominantly engaged in the profitable manufacture and sale of proprietary medicines, food and toiletries.

REFERENCES

1. Beecham Group, documents, correspondence and discussion.
2. E. B. Chain, correspondence and discussion.
3. H. Raistrick, correspondence.
4. Batchelor, F. R., Doyle, F. P., Naylor, J. H. C., and Rolinson, G. N., *Nature*, 183 (1959).
5. Ballio, A., Chain, E. B., Dentrice di Accadia, F., Rolinson, G. N., and Batchelor, F. R., *Nature*, 183 (1959); 187 (1960); and 195 (1962).
6. Reports on the Progress of Applied Chemistry, *Society of Chemical Industry*, 1959–65.
7. *Proceedings Royal Society*, Series B, 1961.
8. Chain, E. B., 'The New Penicillins' from *New Perspectives in Biology*, B.B.A. Library, vol. 4 (Elsevier Publishing Co., 1964).

THE WANKEL ROTARY PISTON ENGINE

THE Wankel is the first of many rotary piston designs to have challenged the dominance of the reciprocating piston engine for small power units. Compared with the reciprocating internal-combustion engine, it offers reduced size, weight, vibration, noise and production costs, combined with comparable thermal efficiency. Commercial production was started by N.S.U. Motorenwerke in Germany and Toyo Kogyo in Japan in 1967 as a petrol engine for cars; a joint N.S.U.–Citroën car will follow later. The engine is also suitable for industrial, marine or aeronautical uses, and there is much interest in possible diesel and multi-fuel versions. Sixteen European, American and Japanese companies held licences to build the engine by January 1968.

The history of the rotary piston engine begins with two rotary pumps invented by an Italian, Ramelli, in 1588. Most of the designs – and all of the applications – of rotary piston machines until the present were pumps or compressors, but the idea of a rotary piston engine has fascinated inventors since the steam engine was invented. The difficulties which caused early designs to fail were greater with the internal-combustion engine than with the steam engine. In both engines effective seals have to be provided between the rotating piston and the casing: but in the internal-combustion engine these seals have to separate the four cycles of inlet, compression, expansion and exhaust which take place in the chambers formed between the piston and the casing; and there are also few geometrical forms which allow the four-stroke cycle to be performed within a single casing, with adequate compression of the charge, and without the use of valves. Wankel's contribution has been to design effective seals, and to discover the geometrical form of an engine that can perform the four-stroke cycle in one chamber without valves, that gives a usefully high compression ratio and that provides a simple form of sealing.

The thoughts which led Wankel to this success began in 1924, when he was twenty-two. Financial troubles caused by the post-war inflation had prevented him from attending a university, and he had established a small car repair workshop. He was attracted to the rotary piston engine because 'I considered the shaking and pounding of the reciprocating piston engine unaesthetic as compared with the running of a turbine or an electric motor'.[1]

He started designing rotary piston engines in 1926, but in 1929 he turned to experiments on rotary valves for reciprocating engines, which raise the same sealing problems as the rotary piston engine itself. Wankel worked on rotary valves until 1945, though his real interest remained the rotary piston engine. His work on rotary valves was backed by B.M.W. from 1934 to 1936, and by the German Air Ministry from 1936 to 1945. In 1945 the research institute which was then working for the Air Ministry was closed; it was not until 1951 that he was able to establish a research institute of his own, financed by consulting work, to continue the development of rotary valves and studies of rotary piston engines. N.S.U. supported his work on rotary valves, but he worked on engines at his own risk until 1953.

The ideas which led Wankel to his successful engine covered two basic types of rotary piston machines: those with a stationary outer casing, and those on which the outer casing revolves as well as, and even faster than, the rotary piston itself. Many engines of similar basic shape can be converted from one type to the other, and many variations of each type have been invented over the past century.

Wankel began in 1945 with a design that had originated as a pump, invented by the Italian scholar Cavalieri in 1647, and that was the idea on which the Powerplus and Zoller compressors of the 1920's were based: a vane, mounted on a rotor, rotates within a stationary casing. In 1946 Wankel realised that the vane and the rotor that carried it could be replaced by a lenticular-shaped rotor; unknowingly, Wankel had thus reinvented a design invented by the British engineer, Galloway, as a steam engine in 1834. This design was, however, only suitable as a compressor or steam engine; the four-stroke cycle could not be performed within it. Wankel next turned his attention to machines with three-sided or X-shaped rotors; but these designs were only able to perform the two-stroke cycle, and effective sealing was difficult. This work was not entirely wasted, however, for these designs helped Wankel to improve his sealing system so that it was virtually ready for his final design.

Wankel's next step towards this design was taken in 1953, when he reverted to his 1946 design and realised that it could also operate with a rotating outer housing, and with the housing rotating at twice the speed of the internal rotor. (The outer housing then becomes the piston rotor from which power is taken.) Again this design could not perform the four-stroke cycle within one unit, but Wankel proposed the design to Borsig A.G. as a compressor and to N.S.U. as an engine comprising a compressor and a separate expansion unit. Both companies agreed to support further research. Before any development work was done on this design, however, Wankel began to investigate whether one could also use the speed relationship of this machine – which is unusual in that the big outer rotor runs faster than the small inner rotor – with ratios other than the 1:2 it used. In March 1954 he found that a 2:3 ratio was possible, which allowed the use of a three-sided rotor that formed three chambers within the outer rotor or casing; on April 13th he found that this engine could perform the four-stroke cycle without poppet or rotary valves, and the present Wankel engine had been invented. It was only when

[1] Letter from Herr Felix Wankel, Mar. 5th, 1968.

he applied for a patent that Wankel found that many features of his engine had been anticipated over the previous thirty years by the Swedes Wallinder and Skoog and Fixen, the Frenchman Sensaud de Lavaud, and the Swiss Maillard; but none of these engines could perform the four-stroke cycle with a high enough compression ratio for efficiency. Wankel doubts whether his work would have been speeded by knowledge of these previous designs; he might have wasted time in vainly trying to eliminate their drawbacks.

It was obvious from Wankel's experience with the 1:2 ratio engine that the engine with a rotating casing could be converted into one with a stationary casing, but the latter places greater stresses on the seals. Wankel therefore insisted that the early development of the engine at N.S.U. should be concentrated on the rotating-casing design; this design was also used by Borsig as a compressor and by N.S.U. as a supercharger. The first engine ran successfully at N.S.U. in February 1957. Despite Wankel's doubts, N.S.U.'s director of research, Dr. W. Froede, and his chief designer, H. D. Paschke, went ahead with the design of an engine with stationary casing, because it seemed more suitable for production. N.S.U. first ran this type of engine in July 1958, and development has since been concentrated on it. Production of small engines began in 1962, and limited production for cars started in 1963; large-scale production for cars followed in 1967. Development by licensees of Wankel and N.S.U. is now world-wide.

The invention of the successful rotary piston engine was thus the work of a consultant with his own small research institute, who obtained the support of manufacturers for his final ideas. Its early development was carried out by one of the smaller German motor companies, N.S.U., which was then switching back from motor-cycle to car production; it may have felt that a new engine would help it break into the car market.

Felix Wankel: rotary piston engine

REFERENCES

1. Wankel, F., *Rotary Piston Engines* (Iliffe Books Ltd., 1965).
2. Correspondence with Herr Felix Wankel.
3. *Development of Rotary Combustion Engine* (N.S.U. Motorenwerke A.G., 1965).
4. Froede, Dr. Ing. W., 'Recent Developments in the N.S.U. Wankel Engine', Automobile Division, Institution of Mechanical Engineers (1966).
5. Ansdale, R. F., 'Rotary Combustion Engines', *Automobile Engineer* (Dec. 1963; Jan. and Feb. 1964).
6. Information supplied by N.S.U. Motorenwerke A.G.

INDEX

Abacus, 342
Ackroyd-Stuart, H., 57
Acoustic Consultants Co., 271
Acrylic fibres (Orlon, etc.), 66, 71, 75, 166
Adams, W., 204, 341
Aerial navigation, 58
Aero-engines, 208, 211; fuels for, 64
Aerofin, 101
Aerofoils, 59
Aeronautical Society, 59
Aeronautics in the nineteenth century, 11, 58
Agriculture and farming, 132, 200
Agthe, C. A., 306
Aiken, H. H., 342, 343, 345
Air-conditioning, 66, 73, 84, 92, 101, 161, 166
Air cushion vehicles, 11, 66, 73, 74; case history, 329–32
Aircraft industry and the aeroplane, 58, 118, 122, 123, 146, 174, 196, 200, 211; development costs, 215; research expenditure on, 199
Airship, 174
Akers, Sir W., 114
Aldrin, *see* Chlordane
Alexanderson, E., 286, 307, 309
Alizarin, synthetic, 63
Allegheny Ludlum Steel Co., 240
Allen, J. A. 341
Allen, W., 63
Allgemeine Elektrizitäts Gesellschaft, 271, 286
Allis-Chambers Co., 243
Alpine & McLouth Steel Co., 207
Aluminium industry, 54, 131, 207
Aluminum Company of America, 239–40
American Association for the Advancement of Science, 62
American Intertype Corporation, 346
American Journal of Science, 61
American Locomotive Co., 251
American Philosophical Society, *Transactions*, 61
American Rolling Mill Co. (Armco), 242
American Sheet and Tinplate Co., 242
American Smelting and Refining Co., 240
American Telegraph and Telephone Co., 218, 288, 309
American Telegraphone Co., 271
6-amino penicillanic acid (6-APA), 351, 353
Ampex, 207
Analogue computer, 342
Anderson, Prof. J., 61
Andersonian Institution, 61
Andrew, J. P., 313
Anglo-Iranian Oil Co., 236
Aniline dyes, 101, 155; mauve, 50, 63
Anschütz-Kaempfe, H., 254–5
Ansdale, R., 12, 356
Anti-Trust and Monopoly, Hearings before Sub-Committee U.S. Senate, *see* Hart Committee
Antz, H., 265
Appleton, Sir E., 79, 283
Arkel, van, 315
Arkwright, Sir R., 47
Armour Research Foundation, 271
Armstrong, E., 64, 77, 79, 81, 83, 84, 141, 144, 287–9
Armstrong, Sir W., 39, 154
Arnold, H. D., 287
Arvey Corporation, 332
Ashland Iron and Mining Co., 242
Atanasoff, J. V., 343
Atha, B., 239
A.B. Atlas Diesel, 251
Atom bomb, 213; development cost, 216
Atomic energy, 21, 67–9, 171, 200; expenditure on, 70, 196, 213
Atomic Energy Authority, 70
Auto-analyser, 207, 209
Autogiro, 84, 87, 258–9
Automatic guns, 100; machine, 97
Automatic Hook and Eye Co., 324
Automatic telephone dialling system, 97
Automatic transmissions, 66, 73, 84, 92, 161, 166, 167, 178; case history, 231–3
Automation, 35, 67, 145
Automobile overdrive, 84

Babbage, C., 38, 62, 202, 211, 342, 345
Babcock and Wilcox Tube Co., 241
Bachmann, W. S., 269
Bacon, F., 205
Bacon, Sir Francis, 65, 177
Badische Anilin- und Soda-Fabrik, 305

Baekeland, L. H., 28, 84, 233–4
Baird, J. L., 115, 145, 307, 309
Bakelite, 28, 66, 71, 73, 74, 84, 92, 161; Corporation, 234; case history, 233–4
Baker, J. R., 64
Ball-point pen, 25, 66, 67, 73, 84, 97, 166; case history, 234–5
Ballio, A., 354
Balzer, F., 348
Banting, Sir F. G., 84, 138, 260–2
Barbed wire, 30
Barber, J., 262
Bardeen, J., 83, 318–19
Barnard, D. P., 180
Barnes, Dr. T., 61
Barnes, W. B., 84
Barnett, M. F., 283
Batchelor, F. R., 352–4
Bathe, G. and D., 44
Battelle Development Corporation, 323
Battelle Memorial Institute, 182, 323
Bauer, K., 270
Baumhauer, A. G. von, 259
Baumhauer, H., 320
Beacham, T. E., 205
Beau de Rochas, A., 57
Bedford, Duke of, 47
Beecham Group, 123, 166, 208, 215, 351, 352, 353, 354
Beecham Research Laboratories, 352, 353
Begun, S. J., 270–1
Bell, A. G., 28, 49, 53, 62
Bell, H., 45
Bell Telephone Co., 288, 331
Bell Telephone Laboratories, 76, 111, 119, 215, 218, 271, 307, 317–19, 322, 342
Benz, C., 57
Berlin, Sir Isaiah, 27
Berlin, Free University of, 333
Berlin Institute of Technology, 339
Berlin Technical Academy, 59, 343
Berliner, E., 53–4, 58, 210
Bernal, J. D., 36–7, 96
Bernstein, J., 345
Berry, C. E., 343
Berry, H. M., 243
Bertelson, W. B., 330
Bertsch, H., 141, 305
Bessemer, Sir H., 50–1, 63, 96, 154, 339, 341
Bessemer converter, 51, 155
Bessemer process, 154, 338, 340
Best, C. H., 261
Betz., Prof. A., 205
Bevan, E., 56
Beveridge, W. I. B., 101
Bibby, C. 194
Bickerstaff, K. 337
Bicycloheptadiene, 333
Biles, Prof. Sir J., 60
Biplane, 59
Birdseye, C., 143
Biro, G., 234
Biro, L. J., 84, 234–5
Birtwhistle, W. K., 310
Black, J., 40 44, 62
Blackett, C., 45
Blackett, Prof. P., 203, 284
Blattnerphone, 270
Blenkinsop, J., 45
Bliss, D., 331
Blumlein, A. D., 309
BMW Engine Co., 88, 265–6, 355
Boehme, H. Th., Fettchemie, 305
Boer, de, 315
Boffey, H., 246
Böhler Gebrüder, 240
Bolton, E. K., 274, 276
Boot, H. A. H., 284
Booth, Prof., 49
Borchers, W., 300
Borg Warner Corporation, 142; Warner Gear Division, 232
Borsig, A. G., 355, 356
Boulton, M., 41, 115
Bowen, H. G., 285
Boyd, T. A., 313
Boye, Prof., 49
Boys, C. V., 108
Brabazon, Lord, 216
Brackett, Prof. C. F., 53, 62
'Brain Drain', 202
Bramah, J., 47
Bramo Engine Co., 265
Bramwell, Sir F., 51
Brandenberger, J. E., 84, 237–8
Branly, E., 286
Brassert Oxygen Technik A. G., 339, 341
Brattain, W. H., 318
Braun, Prof. F., 287, 307
Braun, W. von, 290–1
Braunmühl, H. J. von, 271
Bray, C. W., 242
Brearley, H., 300–1
Bréguet, A.-L., 293
Bréguet, L., 257–9
Breil, H., 280
Bridgman, P. W., 279
Bristol Myers Co., 353
British Advisory Council, *see* Scientific policy

British Admiralty, 330, 344
British Association for the Advancement of Science, 62, 199
British Broadcasting Corporation, 75, 270
British Hovercraft Corporation, 331
British Iron and Steel Research Association, 240
British Medical Journal, 349, 350
British Motor Corporation, 337
British Nylon Spinners, 277
British Oxygen Co., Ltd., 132
British Patent Office, 330
British Thomson-Houston Co. Ltd., 98, 126, 253, 263
Britten-Norman, 331
Broderick, T., 269
Brown., John, & Co., Ltd, 300
Brown, S. G., 84, 85, 255
Brown Bayley Ltd., 300
Brown Boveri Co., 343
Brunel, I. K., 39, 58
Brunner Mond Ltd., 257
Brush, C. F., 54
Brush Development Co., 271
Bryce, J. W., 342
Buckley, O. E., 111
Buick Dynaflow, 232
Bulban, E. J., 332
Bunsen, Prof. R. W. von, 62
Burbank, L., 99
Burck, G., 345
Burks, A. W., 344, 345
Burlington Co., 251
Burroughs Co., 344
Burton process, 212
Busemann, Prof. A., 205
Bush, V., 80, 347

Cadillac Co., 231
Calculating machines, 211
California, University of, 248–9
Callis, C. C., 313–14
Cambridge University, 344
Camm, Sir Sydney, 218, 219
Campbell, A., 84, 243–4
Campbell-Swinton, A. A., 307–8
Camras, M., 271–2
Canadian National Railways, 251
Caproni-Campini, 262
Carboloy Co., 320
Carbon black, 139
Carlson, C. F., 97, 98, 141, 206, 321–3
Carlson, W.L., 270
Carnot cycle, 58
Caro, H., 63
Carothers, W. H., 25, 28, 30, 77, 82, 84, 92, 141, 164, 274, 275–7, 297, 310
Carpenter, G. W., 270
Carrier, W. H., 84
Cartwright, E., 28, 47
Catalytic cracking of petroleum, 66, 67, 73, 74, 84, 98, 166, 167; development cost of Houdry process, 155, 214; case history, 235–7
Cayley, Sir George, 58, 59
Celanese, 216
'Celbenin', 353
'Cellophane', 66, 73, 84, 210, 214; case history, 237–8
'Cellophane' tape, 66, 75, 76, 126, 161
Chain, Prof. E. B., 12, 351, 352, 354
Chalmers, W., 84, 273–4
Chanute, O., 59
Charch, W. H., 238
Chase Brass and Copper Co., 240
Chemical industry, 117, 120, 122, 124; research expenditure, 199
Chemical Research Laboratory of D.S.I.R., 312
Chlordane, 11, 66, 75, 76, 207, 209, 217; case history, 332–4
Chromium plating, 66, 71, 73, 75, 84
Chrysler Co., 282
Church, A., 342
Churchill, Sir W., 171, 190
Cierva, J. de la, 84, 86–7, 257–60
Cierva Autogiro Co., 86, 258
Cinerama, 12, 66, 73
Citroën, 354
Clark University, 313
Clarke, Mrs., A. G., 12
Clarke, Mrs., C. A., 349
Clarke, Prof. C. A., 204, 348, 349, 351
Claude, G., 252
Coalmining industry, 133–4
Coats, A., 232
Cockerell, Sir C., 84, 204, 210, 329–32
Codrington, G., 251
Cold Metal Process Co., 240
Collins, A. M., 274
Collip, J. B., 261
Columbia Broadcasting System, 98, 268–9, 347
Columbia Records, 268–9
Columbia Steel Co., 242–3
Columbia University, 287
Commons, House of, Committee on Letters Patent (1871), 154
Competition: and research, 134–7; and development, 162–8; and innovation, 228

Comptoir de Textiles Artificiels, 238
Compton, A. H., 69, 253
Computers, 11, 67, 75, 76, 98, 206, 207, 208, 217, 218, 219. *see also* Electronic Digital computer
Conant, J. B., 35, 279
Connaught Laboratories, 261
Conrad, F., 288
Continuous casting of steel, 66, 76, 84, 85, 131; case history, 239–41
Continuous hot-strip rolling of steel, 66, 67, 75, 76, 126, 131; case history, 241–3
Continuous Metalcast Corporation, 240
Control Data, 207, 209, 217, 344
Controllable-pitch propeller, 97
Cooke, Sir W. F., 48
Coolidge, W. D., 112, 1[illegible]1, 249
Cooper, A. C., 225, 226
Cooper-Hewitt, P., 252
Corelli, M., 175
Corfam, 199, 213
Cornell University, 312
Corning Glass Works, 297–8
Cotton bleaching and printing industry, 133
Cotton gin, 28, 47
Cotton industry, 122, 133
Cotton machinery, 28, 46–7, 154
Cotton picker, 25, 66, 67, 73, 74, 84, 92, 166, 176; case history, 243–5
Cottrell, F. G., 84
Courtauld's Ltd., 155, 277
Cramer, S. W., 84
Crawford, J. W. C., 273
Cray, S., 207, 209, 344
Crease-resisting fabrics, 28, 66, 75, 122, 125, 161; case history, 245–8
Crerar Library, 12
Crocco, G. A., 257
Crompton, S., 46
Croning, J., 84, 295–6
Crooke's radiometer, 55
Crosfield, J., & Co., 256–7
Cross, C. F., 28, 55, 77, 310
Crown Casting Associates, 296
CRT machines, 346, 347
Crystal detector, 287
Curtiss-Wright Co., 330, 331
Cuscoleca, O., 341
Cuvier, Baron, 38
Cyclotron, 24, 28, 66, 67, 73, 74, 91, 171; case history, 248–9

Dacron, 105, 312; development costs, 214
Daimler, G., 57, 63
Daimler, K., 305
Daimler-Benz A.G., 265
Dalton, J., 44
Dantsizen, C., 300
Darwin, E., 41
Dassault Co., 218
Davis, E. W., 98, 167, 217
Davis, F. W., 84, 87, 281–2
Davy, Sir H., 43, 48, 52, 60, 62, 252
Dayton Engineering Laboratories, 312
DDT, 66, 75, 76, 332, 333; case history, 249–50
Dean, R. S., 316
De Dion-Bouton Ltd., 300
Deere & Co., 243
Defence, expenditure on, 196
Defoe, D., 181
Delrin, 216
Department of Scientific and Industrial Research, Chemical Research Laboratory, 312
Detergents, synthetic, 66, 75, 166; case history, 304–7
Deutsche Hydrierwerke, 305
Deutsche Wolle Co., 303
Development: distinction from invention, 28–32, 105; costs and expenditure on, 70, 155–61, 196–8, 212–17, 219–25; modern view of its importance, 152–3; definition of, 153, 194–5; in the nineteenth century, 154–5; monopoly, competition and, 162–168; pace of, 177–8; by large corporations, 212–17; by smaller firms, 217–19
Dewar, Prof. Sir J., 62, 204,
Dick, R., 65
Dickinson, H. W., 41
Dickson, J. T., 310–11
Dieldrin, *see* Chlordane
Diesel, R., 28, 57–8, 64, 141
Diesel-electric locomotives, development costs, 214
Diesel engine, 57–8, 64, 105
Digiset, 207, 347
Digital Equipment Corporation, 217
Dinga, E., 347
Discoveries, chance in, 63–4, 101–2
Dodds, Sir C., 352
Domestic gas refrigeration, 66, 73, 84, 97
Dominion Foundries & Steel Co. Ltd., 341
Doncaster, D., 269
Donkin, B., 58
Dornberger, W., 290–2
Dow Chemical Co., 298, 313
Dow-Corning Corporation, 298
Doyle, F. P., 352–4

Drink industry, 124
Dubos, R., 302
Duco lacquers, 63, 66, 75, 101
Duncan, C. J., 348
Dunlop, J. B., 97
Dunlop Rubber Co., 138–9
Dunwoody, H. H., 287
Du Pont de Nemours & Co., E. I., 30, 75, 105, 143, 161, 163–4, 165, 238, 273, 275–277, 305, 311–12, 313
Durfee, B. M., 342
Durrer, R., 339–41
Dwyer, H. P., 286
Dynamo, 52, 54

Earle, K. J. B., 337
Eastman, G. F., 95, 97
Eastman Kodak Co. Ltd., 98, 105, 111, 233, 267–8, 322
Eberhard Faber Co., 235
Eckert, J. P., 204, 343, 344, 345
Eckert-Mauchly Computer Corporation, 207, 217, 344
Economists and innovation, 19–20
Edison, T. A., 52–3, 62, 63, 79, 83, 94, 115, 141, 202, 210
Edison Illuminating Co., 144
Edlefsen, N. E., 249
Education, scientific, in the nineteenth century, 60–2
Edward, T., 65
Edwards, Sir R. S., 128
Einstein, A., 127
Eldred, B. E., 240
Electric generator, 64
Electric Lamp Manufacturers' Association, 138
Electric precipitation, 12, 66, 73, 74–5, 84, 92, 161
Electrical and Musical Industries Ltd., 155, 162, 214, 307, 309
Electrical engineering industry, 122
Electrical Equipment industry, 118, 196; research expenditure, 199, 222
Electricity, cost of, 70
Electrolytic detector, 287
Electromotive Co., 142, 251
Electron microscope, 66, 67, 71, 73, 166
Electronic devices, 68, 174
Electronic digital computer, 11, 66, 67; case history, 341–5
Electronic industries, 196, 200, 208, 219; development costs, 219
Ellehammer, J. C., 258
Elliott Bros. Research Laboratory, 344
Ellis, C., 84, 85, 204
Elster, J., 307
Empire Cotton Growing Corporation, 109
Engineer, The, 61
Engineering Research Associates, 344
Engineers, supply of, 70, 219, 222
English Electric Co., 344
English Steel Co., 232
Enos, J. L., 194, 199, 204, 207, 210, 212, 216, 225
Ensign, E. E., 295
'Espinasse, M., 76
Evans, Oliver, 43, 44
Eversharp Co., 235
Ewart, P., 38

Fahlbery, Dr., 53
Fairbairn, Sir W., 59
Fairey Rotodyne, 259
Falk, O. T., & Partners, 263
Faraday, M., 48, 52, 60, 299
Farnsworth, P., 28, 64, 77, 81, 83, 84, 85, 92, 97, 129, 141, 161, 307–10
Farnsworth Radio and Television Corporation, 309
Farquharson, J., 352
Fats, hardening of liquid, 66, 73, 84, 166; case history, 256–7
Feedback circuit, 84, 287
Ferguson, H., 84
Ferguson transmission, 232
Fermi, E., 127, 201
Fernseh, A. G., 309
Ferranti Co., 344
Fessenden, R., 84, 141, 286–7
Fibersilk Co., 238
Fink, C. G., 84
Finn, Dr. R., 349, 351
Firth, Thomas & Sons, Ltd., 300
Firth-Sterling Steel Co., 300, 320
Firth-Vickers Ltd., 263
Fischer, E., 297
Fischer, R., 266–7
Fitch, J., 45
Fleming, Sir Alexander, 28, 31–2, 83, 84, 111, 210, 278–9
Fleming, Sir Ambrose, 287
Flexible skirts, 330–1
Float glass, 11, 12, 66, 71, 73, 75, 76, 102, 123, 198, 208, 218; development costs, 214; case history, 334–7
Florey, Sir H., 32, 278, 349, 352
Fluid flywheel, 84, 231

Fluorescent lamp, 138, 252–4; development cost of hot cathode, 155
Fluorescent lighting, 66, 71, 75, 138, 166; case history, 252–4
Focke, H., 257–60
Focke-Wulf Ltd., 258
Food industry, 123, 124
Ford, H., 144
Ford, H. C., 255
Ford Motor Co., 145
Forest, L. de, 84, 87, 99, 141, 287
Forest, de, Wireless Co., 287
Fortune, 207
Föttinger, H., 98, 231–3
Foucault, L., 254, 256
Foulds, R. P., 246
Franklin, B., 45
Franklin, C. S., 287–8
Franklin Institute, 54, 61, 62
Franz, A., 265
Frawley Manufacturing Corporation, 235
Frear, D. E. H., 334
Freda, Dr. V. J., 349, 351
Freeman, C., 206, 208, 209, 216, 219, 225, 345
Freireich, E., 333
Freon refrigerants, 12, 63, 66, 75, 101, 143
Frequency modulation, 84, 144, 288–9
Fresnel, 50
Freund, H. R., 346
Frisch, O., 76
Froede, Dr. W., 356
Froude, J. A., 201
Fuel cell, 209
Fulton, R., 44, 45, 48
Furnas, C. C., 180

Gabor, D., 205
Galbraith, Prof. J. K., 35–6, 133, 227
Gale, Prof. L. D., 49, 62
Gamma globulin, 350
Gardinol Corporation, 305
Gas refrigeration, 66, 73, 84, 97
Gas turbine, 106–7, 155, 174
Gauss, K. F., 48
Geigy, J. R., & Co., 249–50
Geitel, H., 307
Gemmer Manufacturing Co., 282
General Dynamics Co., 331
General Electric Co., 142, 166, 251, 253–4, 271, 284, 286, 287, 298, 307, 320, 344; laboratories, 112, 315
General Foods Corporation, 143
General Machinery Corporation, 232
General Motors Corporation, 75, 76, 97, 119, 125, 142, 143, 161, 212, 214, 231, 233, 251, 282, 312–14
German Air Ministry, 88, 355
German Experimental Aeronautical Institute, 343
German State Broadcasting Service, 271
Germer, E., 253
Gibney, R. B., 318
Gibson, R. O., 279
Giesen, W., 300
Gilbert, D., 43, 62
Gilchrist, P., 51, 63
Gilchrist-Thomas, S., 28, 50, 51, 63
Gilfillan, S. C., 176
Gillette, K., 84, 97, 292–3
Gillette Safety Razor Co., 293
Glasgow University, 60
Glider, 58
Gloster Aircraft Co., 263
Goddard, R. H., 289–90
Godowsky, L., 84, 267–8
Goldmark, P., 98, 268–9
Goldmining, 211
Goldschmidt, R., 286
Goldstine, H. H., 343–4
Goodrich, B. F., Co., 139, 213, 325
Goodyear, C., 28, 49, 62, 63, 64, 77, 154
Goodyear Co., 139
Goodyear process, 337
Gorman, J. G., 12, 349, 351
Göttingen University, 264
Government-operated and -financed R and D, 70, 213, 226–8, 331
Graham, Prof. T., 50
Gramme, Z., 52
Graphic Arts Research Foundation, 347
Gray, E., 49
Greenaway, F., 69
Greenewalt, C. H., 175
Griffith, A. A., 262
Groves, L. R., 216
Grunow, H., 291
Guided missiles, 213, 222; development cost, 216
Guillaume, M., 262
Guillet, L., 299
Gulf Refining Co., 235
Gunther, —, 305
Gurney, Sir G., 59
Gustafson, E., 213
Gyro-compass, 66, 73, 84, 85, 145; case history, 254–6
Gyroscope, 254

Hackworth, T., 46
Hadfield, R., 299
Hagedorn, H. C., 138, 166, 261
Haglund, T., 339
Haldane, J. B. S., 174
Hall, C. M., 54, 62
Hallpike, A. W., 231
Haloid Corporation, 217, 323
Hamberg, D., 208, 226
Hamilton, Dr. Eugene, 350, 351
Hamilton, F. E., 342
Hamilton, H. L., 216, 251
Hamilton, W., 36
Hammond, J. H., Jr., 84
Hansen, W. W., 284
Hardening of liquid fats, *see* fats
Hardie, D. W. F., 50
Harding, C. P., 313
Hargreaves, J., 46
Harris, Dr., 262
Hart Committee, 194, 204, 205, 208, 209, 210, 211, 225, 226
Harvard University, 342
Harwood, J., 84, 129, 294–5
Harwood Self-winding Watch Co., 294
Hatfield, H. S., 52, 58, 71, 80, 82, 95, 189
Haultmann, H., 340
Hawker-Siddeley, 218
Haynes, E., 299–301
Hayward, L. H., 332
Heal, W. E., 336, 337
Healey, T. H., 194
Hedden, C., 205
Hedley, W., 45
Heilborn, J., 293
Heinkel, E., 264
Heinkel, E., A.G., 161, 264–6
Hele-Shaw, Prof., 205
Helicopter, 66, 73, 74, 84, 87; case history, 257–60
Hell, Dr. R., 207, 347
Hellbruegge, Dr. H., 340
Helmholtz, H. L. F. von, 52, 53, 54
Henery-Logan, K. R., 351
Hennessy, D., 12
Henry, Dr., 51
Henry, J., 48, 49, 52, 62
Henson, W. S., 59
Herbrand–Gödel–Kleene theory, 342
Heroult, P., 54, 95
Hertz, H., 171, 201, 283, 286, 307
Herzfeld, S. H., 332
Heterodyne circuit, 287
Hetzer, —, 305
Hexahalocyclopentadiene, 332, 333
Heylandt Co., 290
Hickman, C. N., 271
Higonnet, R. A., 204, 346
Hill, Prof. A. V., 284
Hill, J. W., 276
Hill, R., 273
Hinshelwood, Sir C., 201
Hitchcock, H. K., 306
H.M.S.O.: *Statistics of Science and Technology*, 198
Hobbs, H. F., 84, 232
Hochwald, C. A., 313
Hodgkinson, Prof. E., 60
Hofmann, A. W., 50, 62
Hollerith, H., 342
Holroyd, M., 195
Holtzer, Jacob, Co. Ltd., 240
Holzkamp, E., 280
Hook, C., 242
Hooke, R., 38, 76
Hopkins, —, 254
Hotpoint Electric Heating Co., 142
Houdry, E., 84, 98, 167, 204, 236–7
Houlton, R., 216, 218
House of Commons, 58
Houston, E., 63, 95
Hovercraft, 12, 79, 92, 204, 208, 210; development costs, 215. *See also* Air cushion vehicles
Hovercraft Developments Ltd., 330, 331
Hovermarine Co., 331
Howard, F., 313
Hull, A. W., 253–4, 284
Hülsmeyer, C., 283
Hunter, A., 37
Hunter, M. A., 315
Huxley, T. H., 194
Huygens, C., 38
Hyalsol Corporation, 305
Hyde, J. F., 298
Hydra-Matic transmission, 231–3
Hydraulic converter-coupling, 232–3
Hydrofoil boat, 88
Hyland, L. A., 285
Hyman, J., 332–4
Hyman, Julius, & Co., 333

IBM, 206, 207, 213, 217, 342, 344, 345
Iconoscope, 161, 308, 309
I.G. Farbenindustrie, 75, 209, 247, 267, 271, 277, 305–6
Illinois University, 332

Imperial Chemical Industries, 28, 76, 105, 116, 119, 126, 161, 273, 277, 279–80, 298, 306, 312; I.C.I. titanium process, 159, 165, 316–17
Incandescent lamp, 55
India Rubber, Gutta Percha & Telegraph Works, 139
Industrial research: effectiveness of, 107–116; survey of expenditure on, 117–19; factors leading firms to conduct it, 119–123; position of large firms in, 123–30; Incidence sporadic, 141–5; statistics of expenditure on, 146–51; invention in firms of different size, 205–9, 211, 212
Industrial Research Associations, 137–8, 184
Industrial Research Laboratories and Organisations, 63, 67, 102, 103–6, 140–1, 142–5
Industries, R and D performance by, 220
Innovation, advantage of attack from many angles, 211–12, 225–8
Ingersoll-Rand Co., 251
Inman, G., 253
Insecticides, 196
Insulin, 66, 67, 71, 73, 74, 84, 138, 157, 165–6, 171; case history, 260–2
Internal combustion engine, 57–8
International Harvester Co., 243–5
Intertype Fotosetter, 346
Iron and steel industry, 124, 131, 132, 133, 146–9, 208
Istituto Superiore di Sanità, Rome, 352
Ives, H., 307, 309

Jaspert, W. P., 207
Jenkins, C. F., 307, 309
Jessup, G. W., 282
Jet engine, 21, 24, 28, 29, 30, 64, 66, 67, 71, 73, 84, 91, 100, 105, 129, 144, 161, 178, 208, 217; case history, 262–6
Jewett, F. B., 80
Jewett, Prof. F. F., 54, 62
Jewkes, J., 199, 205
Josephson, M., 202
Joule, J. P., 62
Judson, W. L., 324
Judson Pneumatic Street Railway Co., 324
Junghans, Gebrüder, 239
Junghans, S., 84, 85–6, 239–41
Junkers Airplane Co., 264–6
Junkers Engine Co., 265–6
Jurgens Co., Ltd., 257

Kaario, T. J., 329
Kaempfe, Dr., 254
Kaempffert, W. B., 36
Kaiser, H. J., Co., 431
Kalachek, E. D., 208, 216
Kalling, Prof. B., 339
Kayser, E. C., 256
Kearns, Prof. C., 332, 334
Keicher, Herr, 255
Kelly, M. J., 317
Kelly Johnson, 218
Kelvin, Lord, 49, 60, 62, 174, 254
Kendall, M. G., 345
Kettering, C. F., 75, 80, 92, 110, 141, 142, 212, 251, 312–13
Keynes, J. M., 201
Kipling, R., 173
Kipping, Prof. F. S., 209, 297–8
Kleeberg, 233
Kluge, Prof. H., 232
Klystron, 284
Kodachrome, 66, 73, 74, 84, 92, 97, 105; case history, 266–8
Koppers Co., 165
Krafft, 305
Krais, 306
Kraus, C. A., 313–14
Krilium, 66
Kroll, W. J., 28, 77, 81, 84, 87–8, 115, 152, 315–17
Kroll titanium process, 145, 155, 164–5, 315–17
Krupps Ltd., 301, 320
Kuhml, A., 333
Kuhn, T. S., 200
Kuznets, Prof. S., 34

La Cellophane, Ltd., 238
Lachmann, G. V., 205
Laidlaw, Drew & Co., 263
Laise, C. A., 319
Lake, C. D., 342
Lanchester, F. W., 84, 87
Land, E. H., 84, 86, 141, 199, 226
Landsteiner, Dr., 348
Langer, E., 63
Langmuir, I., 82, 83, 112, 141, 287
Larkem, H. W., 12, 347
Latimer-Needham., C. H., 331
Lavoisier, A., 126
Lawrence, E. O., 28, 84, 91, 115, 248–9
Lazell, H. G. L., 352
Leach, Dr. E., 195
Learned, C., 204
Leblanc, N., 50
Leblanc process, 50
Lee, Sir, K., 122, 246, 248,

Lehane, Dr. D., 349
Leith, E., 205
Lenher, Prof. V., 313
Lenoir, E., 57
Leo Computers, 217, 344
Leonardo da Vinci, 82, 93, 156, 211
Leprince & Siveke, 256
Leupold, 43
Lever Bros. Ltd., 257
Levi, E., 13
Levine, Dr. P., 348, 349
Lewis, S., 114
Ley, W., 290
Leyland Co. Ltd., 232
Licensing of innovations, 166
Lidov, R. E., 333
Lieben, von, 287
Liebmann, A. J., 319
Lilienthal, Otto, 59
Lilly, Eli, & Co., 261
Lindner, F. A., 240
Linofilm, 347
Linotron, 347
Lister Institute, 349
Little, Arthur D., Inc., 182
Littlefield, B., 269
Liverpool Regional Blood Transfusion Service, 349
Livingston, R., 45
Ljungström Steam Turbine Co., 232
Lloyds, Messrs., 339
Lockheed Co., 218
Lockheed power steering, 282
Lodge, Sir O., 286
London General Omnibus Co., 231
London University, 60, 278, 344
Long-playing record, 66, 67, 76, 98; development costs, 155; case history, 268–9
Longsdon, R., 63
Lorenz, C., A.G., 270
Loud, J., 235
Lubbock, I., 263
Lucite, 76, 272–3
Ludlum Steel Co., 320
Ludovici, E. J., 27
Luft, A., 233
Lumitype, 346
Lunar Society, 41, 62
Lyons, J., & Co, 344
Lysenko, T., 183
Lysholm, A., 232

McAfee, A. M., 235
McBain, Prof. E. L, 305
McConnell, Dr. R. B., 349, 351
McCoy, C. B., 216
MacCurdy, E., 111
McGill University, 183, 273
McGregor, R. R., 298
Machinery and instruments, research expenditure on, 199
Mackinnon, D. W., 205
Mackintosh, R., 12, 204, 347
McLaurin, W. R., 36, 144, 288
McLeod, Prof. J. J. R., 261
McLouth Steel Corporation, 341
Maddala, G. S., 341
Magnetic recording, 66, 73, 84, 161, 166, 217–18; case history, 269–72
Magnetometer, 211
Magnetophon, 271
Magnetron, 284
Maillard, 356
Mallory, P. R., & Co., 322–3
Manchester Literary and Philosophical Society, 38, 61, 62
Manchester University, 344
Mannes, L., 84, 267–8
Mannesmann A. G., 240, 339
Mansfield, E., 225
Manville Bros., 324
Marconi, G., 84, 92, 115, 286, 288
Marconi Co., 145, 270, 287–9
Marschak, T., 218
Marsh, J. T., 246
Marshall, A. L., 298
Martin, D. H., 280
Martin, H. G., 234
Maser and laser, 209
Massachusetts Institute of Technology, 53, 207, 313; radiation laboratory of, 285, 318
Matschoss, C., 58
Mauch, H., 47
Mauchly, J., 204, 343–5
Maudslay, H., 47
Maurer, E., 299, 301
Maxim, Sir H., 55, 60
Maxwell, C., 286
Maybach, W., 57, 63
Mechanics' Institutes, 61
Mechanics' Magazine, 61
Mees, C. E. K., 111, 267–8
Mehring, von, 260
Meissner, 287
Mellon Institute, 182, 298
Mercedes, 57
Mercer, J., 77
Merck & Co., 302
Mergenthaler-Linotype, 207, 347
Messerschmitt A. G., 265

Methyl methacrylate polymers (perspex, etc.), 66, 75, 76; case history, 272–4
Metropolitan Vickers Co. Ltd., 126
Meyer, F., 253
Michels, A., 279
Midgley, T., 75, 77, 82, 97, 125, 141, 142, 312–14
Midland Silicones Ltd., 298
Milan Polytechnic Institute, 280
Miles, J., 339
Miller, A. A., 231
Miller, Prof. F. R., 260
Miller, P., 45
Miller, R., 194, 200, 204, 210, 216
Millikan, R. A., 171
Minkowski, 260
Minneapolis-Honeywell Co., 344
Minnesota Manufacturing & Mining Co., 217
Minnesota University, 217
Mirrophone, 271
Mississippi University Experimental Station, 244
Modern artificial lighting, 66, 75
Moissan, H., 319
Mond, L., 157
Monnartz, P., 300, 301
Monoplane, 59
'Monophoto', 346
Monopolies Commission, 137–8, 337
Monopoly, and research, 130–9; and development, 162–8; and innovation, 184–6, 208, 288
Monotype Corporation Ltd., 346
Monsanto Chemical Co., 165
Moore, D. McM., 252
Morgan, S. O., 317
Morrison, P. & E., 345
Morse, S., 48–9, 62
Moseley, M., 202
Mosley, R. F., Ltd., 300
Mote, S. C., 139
Motor-vehicle industry, 132, 199
Mottelay, P. F., 54
Moulton, A. E., 12, 84, 204, 209, 337–338
Moulton, Stephen, 337
Moulton bicycle, 11, 66, 67; case history, 337–8
Moulton Bicycles Ltd., 338
Mueller, M., 264–6
Mueller, W. F., 206, 210, 216
Mule-spinning, 46, 154
Müller, P., 141, 250
Munters, C., 84
Murdoch, W., 115
Murray, M., 45

Nasmyth, J., 51, 65
National Cash Register Co., 344
National Physical Laboratory, 284, 331, 344
National Research Associates Inc., 330
National Research and Development Corporation, 79, 207, 330, 331, 344, 347
National Science Foundation, 194, 197, 219, 221, 224
Natta, G., 209, 280
Naugle, H. M., 242–3
Naylor, J. H. C., 352–4
Nelson, R. R., 205, 208, 216, 225
Neoprene, 63, 66, 67, 75, 76, 102; case history, 274–5
Neuman, J. von, 97, 204, 343–5
Nevanlinna, H. R., 349
New York City Health Research Council, 350
Newcomen, 62
Nicholson's Journal, 58, 61
Nickerson, W., 84, 293
Nieuwland, J. A., 84, 274–5
Noble, E. G., 261
Noise, 196
Nordisk Insulinlaboratorium, 138, 166, 261
Normann, W., 84, 256–7, 305
Normanville, E. J. de., 84
Norris, W., 344
North-Eastern Railway, 251
Notre Dame University, 274
Nottingham University, 297
NSU Motorenwerke A.G., 88, 354, 355, 356
Nuclear physics, 68–70, 91, 196
Nuffield Foundation, 348
Nylon, 25, 28, 66, 71, 75, 100, 105, 119, 154, 158, 163–4, 171, 199, 310–11; development cost, 155, 214; case history, 275–278

Oberth, H., 289–90, 292
OECD: Reviews of National Science Policy, 195, 199; Overall Level and Structure of R and D Efforts, 196; Ministers Talk About Science, 225
Ohain, H. von, 84, 264–5
Oldknow, S., 38
Oligopoly: and technical progress, 130–7; and research in industry, 132–7
Oliphant, Prof. M. L., 284
'Orlon', development costs, 214

Ortho Pharmaceutical Co., 349
Ortho Research Foundation, 349, 350
Osram Gesellschaft, 253, 319–20
Österreichische-Alpine Montangesellschaft, 340–1
Otto, N. A., 57, 63
Oxygen steel-making, 11, 66, 67, 71, 75, 76, 132, 207, 208, 212, 217; case history, 338–41

Pacinnotti, A., 52
Page, R. M., 285
Painter, W., 292
Paramount Corporation, 309
Parking meter, 97
Parsons, Sir C., 41–2, 155
Paschke, H. D., 356
Patent Law, House of Lords Committee on (1851), 154
Patent Law, Royal Commission on (1865), 154
Patent lawyers, 191
Patent Office 33, 92
Patent statistics, 25, 88–91, 196, 198, 205, 209
Patents, 209, 225, 226
Patnode, W. I., 298
Patterson, R., 43
Paul, K. S., & Associates, 207, 215, 347
Paul, L., 46, 62
Paul–PM Machine, 347
Pechmann, von, 272
Peck, E. B., 313
Peck, M. J., 207, 208, 216
Pedersen, P. O., 270
Peel, Sir R., 46
Pénaud, A., 59
'Penbriten', 353
Penicillin, 28, 63, 66, 67, 71, 73, 75, 84, 101, 129, 144, 157, 171, 199; development of, 31–2, 155; development cost of, 155; case history, 278–9
Pennsylvania University, Moore School of Engineering, 343, 345
Penrose Journal, 347
Peoria Research Laboratory, 32
Perkin, W. H., 28, 50, 63, 64, 77
'Perlon', 75, 164, 277
Perrelet, A.-L., 293
Perrin, M. W., 279
Perspex, *see* Methyl methacrylate polymers
Pescara, Marquis R. de P., 84, 258
Pescara free-piston engine, 153, 161
Petro-Chemicals Ltd., 165
Petrol-electric traction, 250–1
Petroleum, *see* Catalytic cracking of
Petroleum industry, 200, 207–8
Petty, Sir W., 93
Pfleumer, F. 205, 271–2
Philco Corporation, 309
Phillips, H., 59
Phillips Metall-Glühlampenfabrik A.G., 315
Phillips Petroleum Co., 280
Photo-typesetting, 11, 66, 67, 73, 74, 207, 209, 217; development costs, 215, 347; case history, 345–8
Photon, 346; development cost, 215
Photon Inc., 346–7
Pickard, G. W., 287
Pierce-Arrow Motor Co., 281
Pilcher, P., 59, 60
Pilkington, L. A. B., 12, 218, 335, 337
Pilkington Bros., 123, 208, 214, 335
Pixii, H., 52
Planck, Max Institutes, 76, 165, 182, 280
Plastic glass, 84, 272–3
Plastic industry, 209
Platz, K., 305
Playfair, L., 62
Plexiglas, 76, 272–3
Pohl, R. W., 264
Poland, F. F., 240
Polanyi, Prof. M., 26; suggested reform of patent system, 188–9
Polariser, *see* Synthetic light
Polaroid Corporation, 217
Polaroid Land camera, 66, 73
Pollack, Dr. W., 349, 351
Polyethylene, 28, 63, 66, 75, 76, 84, 101, 105, 119, 145, 171; case hostory, 279-281
Pontalite, 272
Popular Science Monthly, 61
Porsche, F., 112
Porterfield, E., 269
Portevin, A. M., 299–300
Post, Emil, 342
Poulsen, V., 84, 269–72, 286
Pound, R., 203
Power Jets Ltd., 161, 263–4
Power loom, 47
Power steering, 66, 73, 84, 87, 92, 161, 166, 167; case history, 281–3
Price, T. H., 244
Price–Campbell Cotton Picker Corporation 243–5
Priestley, J., 41, 62
Princeton University, 207
Prindle, K. E., 238
Proctor & Gamble Ltd., 305

Propeller, variable-pitch, 59
Prudhomme, E. A., 236
Pullin, C. G., 259
Pupin, M., 287
Purdue University, 286, 318
Purdy, P., 12, 204, 347
Pye, Sir D., 263

Quick-freezing, 66, 67, 73, 161

Rabinow, J., 204, 232
Radar, 66, 77; case history, 283–6
Radio, 64, 66, 67, 71, 73, 84, 92, 171; case history, 286–9
Radio Corporation of America, 75, 144, 155, 161, 207, 214, 269, 271, 288, 307–9, 344, 347
Radioactive waste, 196
Radiometer, Crooke's, 55
Railroads, 200
Raistrick, Prof. H., 12, 32, 112, 278, 354
Raleigh Co., 338
Randall, J. T., 284
Rank Organisation Ltd., 323
Raper, K. B., 101
Rayon, viscose, 56, 97, 154; for tyre casings, 139; development costs, 155
Rayon and nylon industry, 122
Record, *see* Long-playing
Recording machines, 100
Recordon, L., 293
Rectron Co., 253
Refrigeration, *see* Domestic gas
Regenstein, J., 332, 333
Regent Street Polytechnic, 58
Remington Rand, 344, 345
Renard, C., 257
Rennie, G., 59
Republic Steel Corporation, 241
Research; expenditure on, 70, 196, 197, 199; definition of industrial, 105–6; difficulties of industrial, 107–11, 114–16; conflicting purposes of research and business, 111–114; survey of expenditure on industrial, 117–19; factors leading firms to conduct research, 119–23; position of large firms, 123–6; benefits of scale in, 125–7; teamwork in, 126–9; whether encouraged by monopoly or oligopoly, 130–7; statistics of industrial expenditure on, 146–51; institutionalisation of, 178–83; wide range of American institutes for, 181, 207; British unenterprising in organisations for, 181–2; desirability of variety of institutions for, 184–6; how individual inventors can be helped, 186–93; definition of basic, 194, definition of applied, 194–5; dispersion of R and D by industry and size of firm, 219–25
Retractable undercarriage, 59, 144
Reuter, E., 19
Reychler, A., 305
Reynolds, M., 235
Rhesus haemolytic disease treatment, 11, 12, 66, 67, 75, 204, 209; case history, 348–51
Riedel, K., 290
Riedel, W. J. H., 290
Riemschneider, R., 333
Rieseler, H., 232
Rinesch, R., 340
Risler, J., 253
Ritchie, G. G., 310
Ritchie, W., 48
Roberts, R., 154
Robertson, J. G., 351
Robinson, Prof. Sir R., 279
Robison, J., 40
Rochow, E. G., 298
Rockefeller Foundation, 13
Rocket engines, 208
Rockets, 11, 66, 77, 154; case history, 289–92
Roget, P. M., 59
Röhm, O., 272–3
Röhm and Haas, 76, 272–3
Rolex Watch Co., 294
Rolinson, G. N., 352–4
Roll, A. G. von, 339–40
Rolls-Royce Ltd., 105, 161, 264
Rosing, Prof. B., 308–9
Rosse, third Earl of, 41
Rossi, I., 239
Rossman, J., 80
Rossman, R., 84, 303
Rotary piston engine, 74, 88, 210, 354–6
Rotodyne, 259
Round, H. J., 287
Rover Ltd., 264
Rowe, A. P., 284
Royal Aircraft Establishment, 262
Royal College of Chemistry, 50
Royal Commission on Patent Law 1865, 154
Royal Institution, 43, 61
Royal Society, 43, 48, 59, 93; *Philosophical Transactions*, 61
Rubber: vulcanisation of, 49, 63–4, 171; tyres, 138–9; development cost, 154
Ruben, S., 204

Rumford, Count, 43
Rumsean Society, 45
Rumsey, J., 45
Rust, J. and M., 84, 92, 243–5
Rutgers University, 302
Rutherford, Lord, 171, 183, 202

Sabatier, P., 256
Sachs, H., 293
Safety razor, 25, 30, 66, 67, 73, 74, 97, 100; case history, 292–3
Sailing ship, 154
St. Petersburg Technological Institute, 307–8
Salerni, P. M., 89, 232
Sanden, Prof. von, 232
Sänger, E., 290
Sargent, G. J., 84
Saunders-Roe Co., 330, 331
Savery, T., 44
Sawers, D., 194, 200, 204, 210, 216
Schaffert, R. M., 323
Scheffer, C. E., 346
Schelp, H., 264–5
Scherer, F.M., 206
Schilling, 48
Schlack, P., 277
Schlaifer, R., 208, 211, 226
Schmidt, pulse-jet, 265
Schmookler, J., 194, 198, 200, 205, 209, 210, 225
Schneider, A., 232
Schneider, H., 232
Schneider torque converter, 142, 232
Schöller, C., 306
Schon, D. A., 12, 204, 205, 212, 217
Schrauth, W., 305
Schreyer, Dr., 343
Schroeter, K., 319–20
Schuler, M., 255
Schwarz, Prof., 339
Science: distinction between pure science and technology, 21, 26, 200; connection with invention in the nineteenth century, 37–8, 63–4; 'social purposes' of science, 169–71, 202; importance given to, 177–81; science, technology and economic growth, 225–7
Science-based industries, 200
Scientific policy, British Advisory Council on, 200, 203
Scientists: supply of, 23, 70, 219, 222; contacts with inventors in nineteenth century, 37–8
Scotch tape, 217
Scott, D.A., 261
Screw propellers, 59
Seech, F., 84, 235
Segel, A., 333
Select Committee on Science and Technology, 70
Self-winding wrist-watch, 66, 67, 73, 74, 84, 129, 144; case history, 293–5
Semi-synthetic penicillins, 11, 12, 66, 71, 73, 75, 76, 102, 123, 166, 199, 208; development costs, 215, 353; case history, 351–4
Senderens, J. B., 256
Sensaud de Lavaud, 356
Shaw, R. A., 330
Sheehan, J.C., 351
Sheets, H.F., 236
Shell Development Co., 333
Shell moulding, 66, 67, 77, 84, 161, 165; case history, 295–6
Shell Petroleum Co., 165, 263
Shepard, J., 288
Sheppard, Prof. P. M., 12, 204, 348, 349, 351
Shockley, W., 84, 141, 317–8
Shoenberg, I., 141, 162, 309
Siemens, Sir W., 50, 51, 155
Siemens, W. von, 48, 52
Siemens and Halske, 315
Siemens family, 63
Sikorsky, I., 258–60
Silicones, 66, 75, 76, 101, 178; case history, 296–9
Silliman, Prof. B, 49, 61, 62
Simon, F., 181
Simone, D. V. de, 204, 206
Sinclair, H., 84, 231
Skeggs, L. T., 205
Sloan, A.P., 143, 212
Small, W., 41
Smeaton, R., 38
Smiles, Samuel, 27, 46, 61, 64–5
Smith, A., 233
Smith, C. H., 351
Smith's transmission, 232
Smithsonian Institution, 61
Snepvangers, R., 269
Social scientists, 22, 34
Société Bertin, 331
Société Française Radioélectrique, 283
Society of Arts, *Transactions*, 61
Socony-Vacuum Oil Co., 236
Soloway, S., 333
Solvay process, 50, 155
Soundmirror, 271

Space travel, 196
Spanner, H., 253
Spannhake, E. W., 232
Spannhake, Prof. W., 232
Spencer, H., 63, 93–4
Sperry, E. A., 255
Spinning box, 56
Sprengel mercury pump, 55
Stafford, C. D. W., 12
Stainless steels, 66, 67, 71, 76; case history, 299–301
Standard Oil Co. of Indiana, 212
Standard Oil Co. of New Jersey, 235, 236, 313
Stanford University, 207, 284
Stanton-Jones, R., 12, 331–2
Staphylococcus aureus, 351
Staratt, F. W., 341
Starr, J. W., 55
Staudinger, Prof. H., 297–8
Steam-engine, 21, 29, 40–1, 59; the high-pressure, 43–4, 64, 115
Steam locomotive, 45–6, 115
Steam turbine, 37, 41–2; development cost, 155
Steamship, 45
Stearn, C. H., 55, 63
Steckel, A., 84
Steel: invention of cheap production methods, 50–1, 64. *See also* Continuous casting of; Continuous hot-strip rolling of; *and* Oxygen steel-making
Steel Company of Wales, 240
Steinmetz, C. P., 112, 141
Stephenson, G., 45–6, 93
Stephenson, R., 46
Stevens, Col. J. C., 45
Stibitz, R., 342, 345
Stille, K., 270
Stine, C. M. A., 275
Stokes, Sir G., 252
Stone, 286
Stoney, G. G., 41
Story, H., 233
Strachey, Lytton, 195
Strauss, B., 299, 301
Strauss, E., 111
Strauss, F., 332
Streptomycin, 63, 66, 73, 74, 84; case history, 301–2
Stringfellow, J., 59
Stromberg Co., 166
Sturgeon, W., 60
Suess, T. E., 340, 341
Sullivan, E. C., 297
Sulzer Bros., 105, 251, 303–4
Sulzer loom, 66, 73, 84, 144; case history, 303–4
Sun Oil Co., 236
Sundback, G., 84, 324–5
Superchargers, 166
Superheterodyne circuit, 84, 288
Supersonic aircraft, 200, 213
Supply, Ministry of, 312
Svennelson, I., 225
Swallow, J. C., 279
Swan, Sir J. W., 55–6, 63
Swedish General Electric Co., 251
Swinburne, Sir J., 56
Sylvania Industrial Corporation, 238
Symington, W., 45
Synchro-mesh, 99, 231–2
Synthetic alizarin, 63
Synthetic detergents, 196
Synthetic light polariser, 12

Taconite, 98, 167, 178, 217
Tankard, J., 246
Taxation, high, effects on research and invention, 191–3
Taylor, A. H., 283, 285
Taylor, W. C., 297
Technical advance and economic wealth, 198–203
Technichon Instrument Corporation, 207
Technicolor Motion Pictures Ltd., 267
Technologists, supply of, 23
Technology: Connection with science, 21, 26; division into invention and development 28–32; importance given to, 177–178
'Teflon', 101
Telecommunications industry, 118
Telecommunication Research Establishment, 285
Telefunken Co., 287, 288, 289
Telegraph: electric, 48–9, 64; harmonic, 53
Telegraphie-Patent-Syndikat, 270
Telegraphone, 270
Telephone, 48, 53
Television, 64, 66, 67, 75, 84, 85, 92, 97, 129, 145, 161–2, 166, 173, 178; development costs, 155, 214; case history, 307–10
Temple, F. du, 59
Temporary National Economic Committee, 80
'Terylene', 28, 66, 75, 76, 105, 126, 129, 143, 155, 158, 161; development costs, 155, 214; case history, 310–12
Tesla, N., 286

Tetraethyl lead, 63, 66, 75, 143; case history, 312–14
Texas Instruments, 207, 217
Textil-Finanz A.G., 303
Textile industry, 91, 124
Thermodynamics, 64
Thomas, S., 345
Thomas process, 338, 340
Thompson, E. A., 84, 99, 231
Thomson, E., 54, 62, 63
Three-dimensional films, 166
Till, I., 36
Times, The, 174
Titanium, 66, 71, 73, 74, 84, 87–8, 92, 164, 167; I.C.I. process, 159, 165; case history, 314–17
Tizard, Sir H., 284
Tootal Broadhurst Lee Co., 28, 122, 245–7
Topham, F., 55–6
Toronto University, 260–1
Torque converter, 142, 232–3
Townsend, A. J., 242
Toyo Kogyo, 354
Tracked Hovercraft Co. 331
Traffic congestion, 196
Tranquilliser drugs, 214
Transistor, 11, 29, 66, 71, 75, 76, 91, 97, 100, 101, 119, 171, 203, 207, 217, 344; development costs, 215; case history, 317–19
Transo Envelope Co., 332
Trenkler, H., 340
Trevithick, R., 43, 45, 63
Trilok Research Society, 232
Triode tube, and valve, 87, 92, 287, 318
Troland, L. T., 267
Tungsten carbide tools, 66, 67, 76; case history, 319–21
Tunnel diode, 209
Turbo-jet engine, 64, 86
Turing, A. M., 342
Turnbull, W. R., 205
Tuve, M. A., 283
Twitchell, E., 305
Tyndall, Prof. J., 38
Tyre Manufacturers' Conference, 138
Tyres, 137, 138–9
Tytus, J. B., 242–3

Unilever Ltd., 206, 305
Union Carbide Co., 165, 316
Union Pacific Co., 251
United Aircraft Corporation, 259
United Kingdom, expenditure on R and D in, 198
United Shoe Machinery Corporation, 132
United States Federal Government: and penicillin, 32; Commissioner of Patents, 49, 154; expenditure on R and D, 196–8; Naval laboratory, 285; Navy, 285, 288, 330, 342; Office of Scientific Research and Development, 285; Signal Corps, 323; Bureau of Mines, 316; Rubber Reserve, 332; Department of Agriculture, 333; Patent Office, 336; Army, 343, 345; Government Printing Office, 346, 347
United States Steel Corporation, 132, 212
United States Weather Bureau, 286
Universal Fastener Co., 324
University autonomy, 180–1
University Grants Committee, 203
University research, 75, 204
Unsplinterable glass, 25
Upatnieks, J., 205
Utah Radio Products, Co., 271

Vacuum Oil Co., 236
Vacuum tube, 87, 92, 287–8, 317
Vainio, T., 349
Valier, M., 290
Van den Bergh Ltd., 257
Varian, R. H., 284
Varian, S. F., 284
Velox photographic paper, 233
Velsicol, 207, 217, 332, 333
Vereinigte Oesterreichische Eisen- und Stahlwerke Aktiengesellschaft (VOEST), 207, 340–1
Vickers, H., 281–2
Vickers Inc., 281–2
Vickers-Armstrong Ltd., 232, 331
Video tape recording, 207, 214
Vitruvius, 156
Vivian, A., 63
Voigtländer und Lohmann Metallfabrikation G. M. B. H., 319
Volta Laboratories, 53
Vortex theory of flight, 87
Vosper-Thorneycroft Co., 331
Vulkan Shipyard, 231

Wagner, H., 264–5
Waksman, S. A., 63, 84, 302
Walker, L., 324
Walker Dishwater Corporation, 142
Walker's *Annals of Philosophy*, 61
Wallis, B. N., 97, 141
Walter, H., 290
Walther ram jet, 265

Walton, F., 132
Walton, R., 207
Wankel engine, 11, 12, 66, 73, 88, 208; case history, 354–6
Wankel, F., 12, 84, 88, 204, 210, 354–6
Warner & Swasey Ltd., 304
Warren Telecron Co. 142
Wartman, F. S., 316
Watson, T. J., 213
Watt, James 26, 28, 38, 41–2, 43, 62, 115
Watt, Sir R. Watson-, 284–5
Weber, W., 271
Weber, W. E., 48
Wedgwood, J., 65
Wedgwood, T., 41, 63
Weiland, C., 330
Weinrich, H., 265
Weir, G. & J., 87, 258–9
Weir, Lord, 86, 258
Welfare services, expenditure on, 196
Wells, H. G., 170, 171
Wells, R. D., 313
Wenham, F.H., 59
Western Electric Co., 272; Laboratories, 218
Westinghouse Co., 288, 308–9, 324
Westland Co., 331
Wheatsone, Sir C., 48, 52, 60
Whinfield, J. R., 28, 77, 129, 209, 310–12
Whitby, G. S., 273
Whitehead, A. N., 52, 182
Whitney, Eli, 28, 48, 93
Whitney, W.R., 300
Whittle, Sir F., 28, 40, 42, 77, 81, 84, 86, 94, 98, 102, 129, 141, 187, 210, 262–6
Wideroe, R., 248
Wieland, Herr, 85
Wieland-Werke A.G., 85–6, 239
Wiener, Dr., 348
Wiener, N., 68
Wile, F.W., 58
Wilkes, M. V., 344
Wilkins, A. F., 284
Willans, P. W., 309
Williams, E. R., 241
Williams, Prof. F. C., 344
Williams, S. B., 342
Willows, R. S., 246
Wilson, C., 206
Wilson, R. E., 313
Wilson, W. G., 60, 231
Wilson epicyclic gearbox, 60, 231
Wimperis, H. E., 284
Wind tunnels, 59, 200
Windscreen wiper, 24, 102
Winton, A., 251
Winton Gas Engine & Manufacturing Co., 142, 251
Wireless telegraphy, 144
Wisconsin University, 313
Wise, T. A., 207
Wittwer, M., 306
Wolfenbüttel Gynasium, 307
Wright, Sir A., 278
Wright brothers, 59
Wyatt, J., 46, 62

Xerography, 11, 28, 66, 73, 84, 92, 97, 98, 167, 217; case history, 321–3
Xerox Co., 217

Yankee Network, 289
Young, L., 283, 285

Zeidler, O, 250
Ziegler, Prof. K., 84, 109, 165, 209, 279–81
Ziolkovsky, K. E., 289
Zip fastener, 24, 30, 66, 67, 73, 74, 84, 100, 171; case history, 324–5
Zirconium, 88, 316
Zuckermann, Sir S., 227
Zuse, F., 204, 343, 345
Zworykin, V., 85, 161, 307–9